Shock Wave Reflection Phenomena

Gabi Ben-Dor

Shock Wave Reflection Phenomena

With 150 Figures

Springer-Verlag
New York Berlin Heidelberg London Paris
Tokyo Hong Kong Barcelona Budapest

Gabi Ben-Dor
Department of Mechanical Engineering
Ben-Gurion University of the Negev
Beer-Sheva, Israel

Library of Congress Cataloging-in-Publication Data
Ben-Dor, Gabi, 1950–
Shock wave reflection phenomena / Gabi Ben-Dor.
p. cm.
Includes bibliographical references and index.
ISBN 0-387-97707-4
1. Shock waves. 2. Fluid dynamics. I. Title.
TL574.S4B46 1991
629.132′3—dc20 91-30180

Printed on acid-free paper.

Production managed by Hal Henglein; manufacturing supervised by Jacqui Ashri.
Camera-ready copy prepared by the author.
Printed and bound by Braun-Brumfield, Inc., Ann Arbor, MI.
Printed in the United States of America.

9 8 7 6 5 4 3 2 1

ISBN 0-387-97707-4 Springer-Verlag New York Berlin Heidelberg
ISBN 3-540-97707-4 Springer-Verlag Berlin Heidelberg New York

to **Professor Ozer Igra**

who introduced me to the world of shock tubes,

to **Professor Irvine Israel Glass**

who led me into the world of shock wave reflection phenomena, and

to **Ms. Edna Magen**

my dear wife who gave me all I ever dreamed of.

Preface

The phenomenon of shock wave reflection was first reported by the distinguished philosopher Ernst Mach in 1878. Its study was then abandoned for a period of about 60 years until its investigation was initiated in the early 1940s by Professor John von Neumann and Professor Bleakney. Under their supervision, 15 years of intensive research related to various aspects of the reflection of shock waves in pseudo-steady flows were carried out. It was during this period that the four basic shock wave reflection configurations were discovered. Then, for a period of about 10 years from the mid 1950s until the mid 1960s, investigation of the reflection phenomenon of shock waves was kept on a low flame all over the world (e.g. Australia, Japan, Canada, U.S.A., U.S.S.R., etc.) until Professor Bazhenova from the U.S.S.R., Professor Irvine Glass from Canada, and Professor Roy Henderson from Australia re-initiated the study of this and related phenomena. Under their scientific supervision and leadership, numerous findings related to this phenomenon were reported. Probably the most productive research group in the mid 1970s was that led by Professor Irvine Glass in the Institute of Aerospace Studies of the University of Toronto. In 1978, exactly 100 years after Ernst Mach first reported his discovery of the reflection phenomenon, I published my Ph.D. thesis in which, for the first time, analytical transition criteria between the various shock wave reflection configurations were established.

For reasons which for me are yet unknown, the publication of my Ph.D. findings triggered intensive experimental and analytical studies of the shock wave reflection phenomenon over a variety of geometries and properties of the reflecting surface and in a variety of gases. The center of the experimental investigation was shifted from Canada to Japan, in general, and to Professor Kazuyoshi Takayama's research laboratory, in particular. Under his supervision flow visualization techniques reached such a stage that the phrase "*cannot be resolved experimentally* " almost ceased to exist in the scientific dictionary.

In the same year that I published my Ph.D. thesis, I published my first journal paper related to the shock wave reflection phenomenon. This paper, entitled "Nonstationary Oblique Shock Wave Reflections: Actual Isopycnics and Numerical Experiments" was co-authored by myself and my Ph.D. supervisor, Professor Irvine Glass. In the conclusion to this paper we wrote "*Undoubtedly, numerical codes will evolve in the future which will reliably predict not only RR and SMR but also CMR and DMR in real gases* ." I wish my lottery predictions were as successful as this prediction, since probably the most remarkable progress in the study of the shock wave reflection phenomenon in the past decade was made by American computational fluid dynamicists, who demonstrated that almost nothing is beyond their simulation capability. At one time it was feared that the computational fluid dynamicists would put the experimentalists out of business. Fortunately, this did not occur, and today experimentalists, computational fluid dynamicists, and theoreticians work together in harmony under the orchestration of Professor John Dewey, who realized, exactly 10 years ago, that scientists interested in the reflection phenomenon of shock waves will benefit the most if they meet once every year or two and exchange views and ideas. In 1981 he initiated the International Mach Reflection Symposium, which became the framework for excellent cooperation between scientists from all over the world who are interested in better understanding the shock wave reflection phenomenon.

As a final remark I would like to point out that this book comes as close as possible to summarizing almost all that I know about shock wave reflection phenomena from a phenomenological point of view. Fifteen years ago, when I first met Professor Irvine Glass, I knew nothing about the reflection of shock waves. When he assigned me the investigation of this phenomenon, I thought that it would take a lifetime to understand and explain it. Now I can state wholeheartedly that I was lucky to have been assigned to investigate this fascinating phenomenon, to have met and worked under the supervision of Professor Irvine Glass, and to have become a part of a wonderful group of scientists from all over the world with whom I have been collaborating throughout the past years, and with whom I hope to continue collaborating in the future.

Acknowledgment

I would like to thank Professor Kalman Schulgasser from the Department of Mechanical Engineering of the Ben-Gurion University of the Negev for reading the text of this book and for his editing suggestions.

I would also like to thank Ms. Tamar Bar-On for typing this book, Ms. Orna Nirenberg for drawing the figures, and Ms. Sima Cohen, my personal secretary, for bearing with me throughout the course of this project.

Contents

List of Abbreviations

D_iMR	-	Direct-Mach Reflection
DMR	-	Double-Mach Reflection
DMR^+	-	Positive Double-Mach Reflection
DMR^-	-	Negative Double-Mach Reflection
I_nMR	-	Inverse-Mach Reflection
IR	-	Irregular Reflection
MR	-	Mach Reflection
NR	-	No Reflection
R	-	Reflection Point
RR	-	Regular Reflection
S_tMR	-	Stationary-Mach Reflection
SMR	-	Single-Mach Reflection
TDMR	-	Terminal Double-Mach Reflection
TMR	-	Transitional-Mach Reflection
T	-	Triple Point
TRR	-	Transitioned Regular Reflection
vNR	-	von Neumann Reflection
i	-	Incident Shock Wave
r	-	Reflected Shock Wave
m	-	Mach Stem
s	-	Slipstream

1

General Introduction

List of Symbols

Latin Letters

h_i - enthalpy in state (i)
M_i - flow Mach number in state (i)
M_s - incident shock wave Mach number
p_i - static pressure in state (i)
T_i - static temperature in state (i)
u_i - flow velocity in state (i) with respect to R in RR or T in MR.
V_i - flow velocity in state (i) in a laboratory frame of reference
V_s - incident shock wave velocity in a laboratory frame of reference

Greek Letters

γ - specific heat capacities ratio
$\delta_{max}(M)$ - maximum flow deflection angle for a flow having Mach number M through an oblique shock wave
θ_i - deflection angle of the flow while passing across an oblique shock wave into state (i)
θ_w - reflecting wedge angle
ρ_i - flow density in state (i)

ϕ_i - angle of incidence between the flow and the oblique shock wave across which the flow enters into state (i)

χ, χ' - first and second triple point trajectory angles, respectively.

Subscripts

0 - flow state ahead of the incident shock wave, i.
1 - flow state behind the incident shock wave, i.
2 - flow state behind the reflected shock wave, r.
3 - flow state behind the Mach stem, m.
m - maximum deflection
s - sonic

Superscripts

R - with respect to the reflection point R
S - strong solution
T - with respect to the triple point T
W - weak solution

When a shock wave propagating in a medium having a given acoustic impedance obliquely encounters another medium, having a different acoustic impedance, it experiences a reflection which is known in the literature as oblique shock wave reflection.

1.1 Introduction and Historical Background

Probably the first scientist to notice and record the reflection phenomenon of shock waves was the distinguished philosopher Ernst Mach, who reported his discovery as early as 1878. In his ingenious experimental study, which was recently surveyed by Reichenbach (1983) and Krehl & van der Geest (1991), he recorded two different shock wave reflection configurations. The first, a two shock wave configuration is known today as regular reflection, and the second, a three shock wave configuration, was named after him, and is known today as Mach reflection.

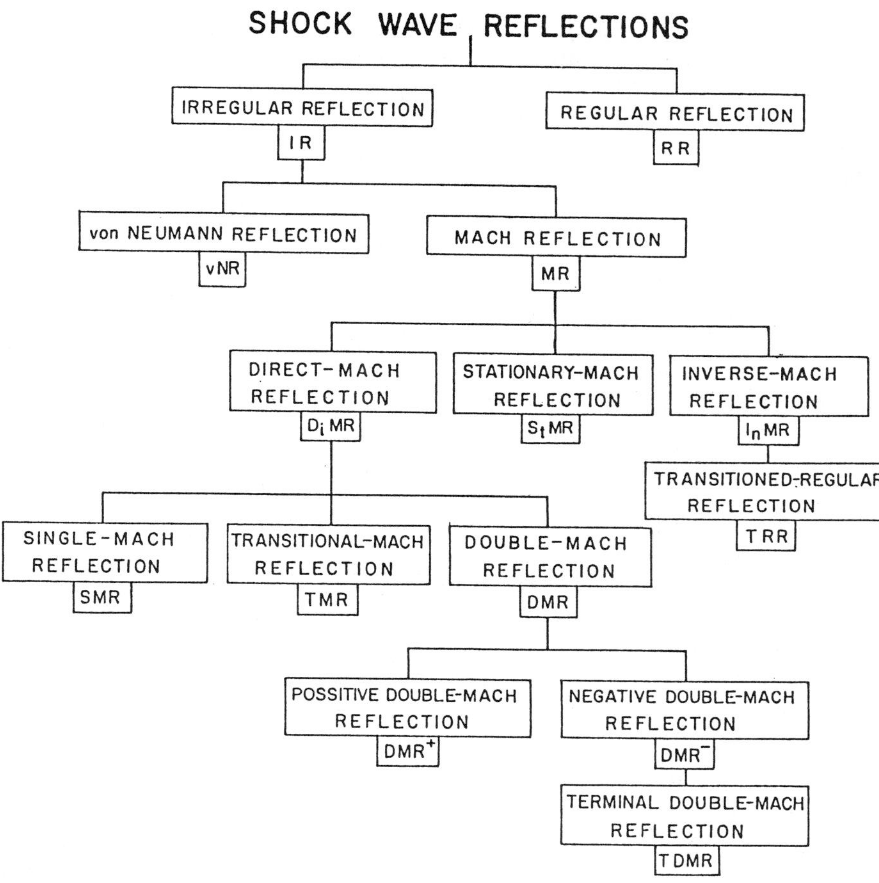

Figure 1.1 The various shock wave reflections

Intensive research of the reflection phenomena of shock waves was re-initiated in the early 1940's by von Neumann. Since then it has been realized that the Mach reflection wave configuration can be further divided into more specific wave structures.

A general illustration of the various shock wave reflections is given in figure 1.1. In general, the reflection of shock waves can be divided into regular reflection (RR) or irregular reflections (IR). The

RR wave configuration consists of two shock waves: the incident shock wave - i, and the reflected shock wave - r. These two shock waves intersect at the reflection point - R, which is located on the reflecting surface. A schematic drawing of the wave configuration of an RR is shown in figure 1.6. All the other wave configurations which are obtained when an incident shock wave reflects over a surface are termed irregular reflections, IR. The IR can be divided, in general, into two categories: von Neumann reflection - vNR, and Mach reflection - MR. The MR wave configuration consists of three shock waves, namely; the incident shock wave - i, the reflected shock wave - r, the Mach stem - m, and one slipstream - s. These four discontinuities intersect at a single point called the triple point - T, which is located above the reflecting surface. A schematic drawing of the wave configuration of an MR is shown in figure 1.7. The reflection point - R is at the foot of the Mach stem where it touches the reflecting surface. Colella & Henderson (1990) recently found that there are cases in which the reflected shock wave - r degenerates to a compression wave near the triple point. In such cases the reflection is not an MR. They termed it von Neumann reflection - vNR.

Following the re-initiation of the investigation of the shock wave reflection phenomena in the early 1940's, Courant & Friedrichs (1948) indicated that, theoretically, three different types of MR are possible, depending on the direction of propagation of the triple point. If the triple point moves away from the reflecting surface, then the MR is called direct, D_iMR; if it moves parallel to the reflecting surface, then it is called stationary, S_tMR; and if it moves towards the reflecting surface, then it is called inverse, I_nMR. (Courant & Friedrichs originally termed it inverted-Mach reflection.) The existence of these three types of MR was later validated experimentally by Ben-Dor & Takayama (1986/7). Schematic drawings of a D_iMR, a S_tMR, and an I_nMR are shown in figure 4.7. Since the I_nMR is an MR in which the triple point moves towards the reflecting surface, it terminates as soon as its triple point interacts with the reflecting surface. The termination of the I_nMR leads to the formation of a new wave configuration, first mentioned by Ben-Dor & Takayama (1986/7). The wave configuration of this reflection consists basically of an RR followed by an MR. A schematic drawing of this wave configuration is shown in figure 4.8c. Since it is formed following the termination of an I_nMR, and since it has the basic structure of an RR, it is called transitioned regular reflection - TRR.

While experimentally investigating the reflection phenomena, Smith (1945) noted that in some cases he observed a kink in the reflected shock wave of an MR. However, only after White (1951) discovered a new type of reflection which he called double-Mach reflection, DMR, was the wave configuration observed by Smith (1945), i.e., an MR with a kink in the reflected shock wave, recognized as yet another type of reflection. Throughout the past 50 years it has been referred to as a complex-Mach reflection, as opposed to the simple-Mach reflection with a reflected shock wave without a kink. However, since the so-called simple-Mach reflection is not simple at all, it was later renamed and is known today as single-Mach reflection - SMR. Similarly, since the so-called complex-Mach reflection is less complex than some of the other reflection configurations and since, as will be shown subsequently, its wave configuration can be viewed as an intermediate wave configuration between the SMR and the DMR, it is called transitional-Mach reflection* - TMR. Schematic drawings of an SMR, a TMR and a DMR are shown in figures 2.5, 2.8 and 2.12, respectively.

The reflection structure discovered by White (1951) was termed double-Mach reflection - DMR, because its structure (see schematic drawing in figure 2.12) consists of two triple points. Ben-Dor (1981) showed that the trajectory angle of the second triple point - χ' could be either larger or smaller than the trajectory angle of the first triple point - χ, depending on the initial conditions. Lee & Glass (1984) termed the DMR when $\chi' > \chi$ as DMR^+ and the DMR when $\chi' < \chi$ as DMR^-. Schematical drawings of a DMR^+ and a DMR^- are shown in figures 2.17a and 2.17b, respectively. Lee & Glass (1983) also argued that there are conditions for which the second triple point, T', could be located on the reflecting surface, i.e., $\chi' = 0$. Such a reflection configuration was termed by them as a terminal double-Mach reflection - TDMR. A schematic drawing of a TDMR is shown in figure 2.19.

In summary there are ten different wave configurations which are associated with the reflection of a shock wave over an oblique surface, namely: RR, vNR, S_tMR, I_nMR, TRR, SMR, TMR, DMR^+, DMR^- and TDMR.

As will be shown subsequently, in steady flows only RR and SMR are possible. Pseudo-steady flows give rise, in addition to RR and SMR, to vNR, TMR, DMR^+, DMR^- and TDMR. S_tMR, I_nMR and TRR are possible only in unsteady flows. As a result of this fact the presentation of the shock wave reflection phenomena will be divided in this book into three parts: the reflection in pseudo-steady flows, in steady flows and in unsteady flows.

---

* This name was originally suggested by Professor I. I. Glass

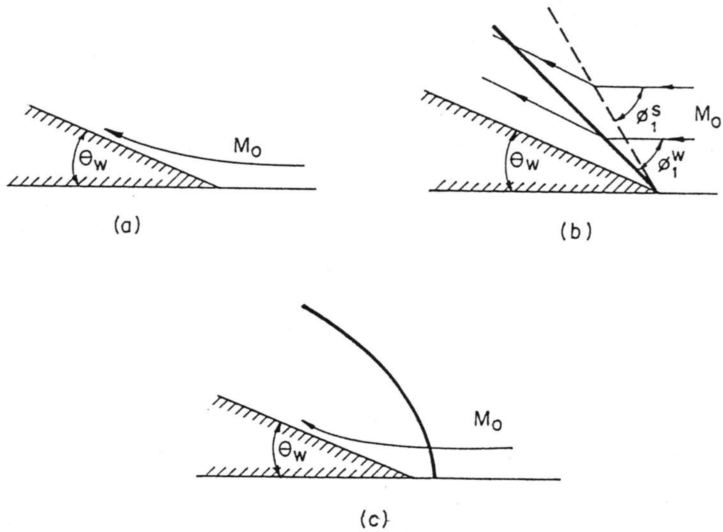

Figure 1.2 Flow over a wedge in a steady flow; a) $M_0 < 1$, subsonic turning; b) $M_0 > 1$ & $\theta_w < \delta_{max}(M_0)$, attached shock wave; c) $M_0 > 1$ & $\theta_w > \delta_{max}(M_0)$, detached shock wave.

1.2 Reasons for the Reflection

Now that the shock wave reflection phenomenon has been briefly introduced, it is appropriate to explain the physical reasons for its occurrence.

The major reason for the occurrence of the reflection phenomenon is because of the following gas dynamic phenomenon. Consider figure 1.2 where three different cases are illustrated in which a flow with Mach number M_0 propagates towards a wedge having an angle θ_w to the oncoming flow.

In the first case, figure 1.2a, the flow is subsonic, i.e., $M_0 < 1$. Therefore it "knows" well in advance that an obstacle is awaiting it, and hence the streamlines are adjusted well before the flow particle reaches the wedge to negotiate it through a continuous subsonic turning. However, if the flow is supersonic, it is unaware of the obstacle ahead of it, and the only way by which it can negotiate the obstacle which suddenly disturbs its propagation, is with the aid of a shock wave (as shown in figure 1.2b) which deflects the flow particle instantaneously to become parallel to the wedge surface. It is well known [Liepmann & Roshko (1957)] that for a given combination of M_0 and θ_w, there are two possible attached shock waves by which the required flow deflection can be obtained, provided $\theta_w < \delta_{max}(M_0)$, where $\delta_{max}(M_0)$ is the maximum possible deflection of a flow having a Mach number M_0 by a shock wave. The two shock waves are illustrated in figure 1.2b. The first shock, known as the weak shock solution, is shown by the solid line. The angle of incidence between it and the oncoming flow is ϕ_1^w. The second shock, known as the strong shock solution, is shown by a dashed line. The angle of incidence between it and the oncoming flow is ϕ_1^s which is always greater than ϕ_1^w. In the case corresponding to the weak shock solution, the flow behind the shock wave is supersonic, while it is subsonic in the case corresponding to the strong shock solution. It is an experimental fact, that unless special measures are taken, the shock wave corresponding to the weak shock solution is usually obtained.

Figure 1.2c illustrates the case when $\theta_w > \delta_{max}(M_0)$. For this case an attached shock wave is impossible. Instead, a curved detached shock wave is obtained. The flow passing through the foot of the shock wave becomes subsonic and, as such, it negotiates the wedge surface through a continuous subsonic turning.

1.2.1 Reason for the Reflection in Steady Flows

Figure 1.3 illustrates a wedge with an angle of $\theta_w < \delta_{max}(M_0)$ mounted in a steady supersonic flow, i.e., $M_0 > 1$. This situation is analogous to the case shown in figure 1.2b, and hence, upon encountering the wedge, an attached oblique shock wave is generated at the leading edge of the wedge. The flow passing through this shock wave is deflected by an angle of $\theta_1 = \theta_w$ to become parallel to the reflecting wedge surface. Since, as mentioned earlier, this oblique shock wave arises from the weak shock solution, the flow behind it, in state (1), is supersonic, i.e., $M_1 > 1$. Thus, we have in state (1) a situation in which a supersonic flow is directed towards the bottom wall. This situation is again analogous to either one of the two cases shown in figures 1.2b and 1.2c, i.e., a supersonic flow having a Mach number M_1 which must negotiate an imaginary wedge with an angle θ_w. If $\theta_w < \delta_{max}(M_1)$, where $\delta_{max}(M_1)$ is the maximum possible deflection of a flow having a Mach number M_1, the flow will be deflected away from the bottom wall and become parallel to it by an "attached" shock wave. This shock wave which should emanate from the reflection point, point R, results in an RR configuration. However, if $\theta_w > \delta_{max}(M_1)$, the

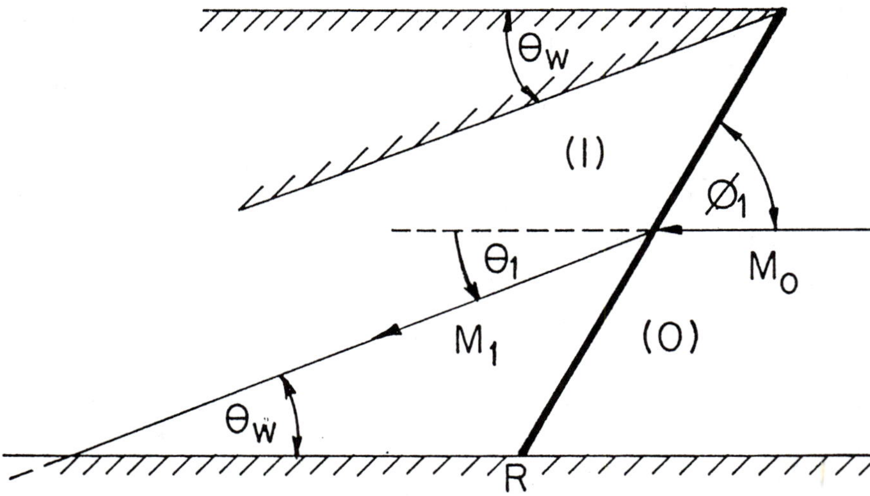

Figure 1.3 Schematic illustration of the reason for shock wave reflection in steady flows.

flow will be deflected away from the bottom wall by a "detached" shock wave which, in turn, results in an MR configuration.

1.2.2 Reasons for the Reflection in Unsteady Flows

Consider a planar incident shock wave, i, having a Mach number M_S, colliding with a sharp compressive wedge having an angle θ_w, as shown in figure 1.4a. Denote the flow states ahead and behind it by (0) and (1), respectively, and attach a frame of reference to the reflection point R where the incident shock wave touches the reflecting wedge surface. In this coordinate system the flow in state (0) moves parallel to the reflecting wedge surface and approaches the incident shock wave with a supersonic velocity $u_0 = V_S \operatorname{cosec} \phi_1$, or a flow Mach number $M_0 = M_S \operatorname{cosec} \phi_1$, where $\phi_1 = 90^\circ - \theta_w$ is the angle of incidence. While passing through the incident shock wave the flow is deflected towards the wedge surface by an angle of θ_1 from its original direction and its dynamic and thermodynamic properties are changed. The deflection results in a situation in which the flow in state (1) approaches the wedge surface obliquely at an angle θ_1. Unlike the previous case of a steady flow, where ϕ_1 is always the one appropriate to the weak shock solution, and hence the flow in state (1) is always supersonic,

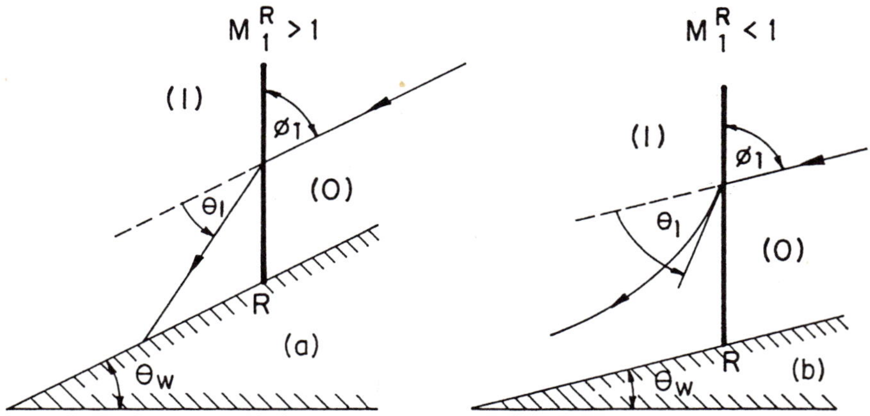

Figure 1.4 Schematic illustrations of the reason for shock wave reflection in pseudo-steady flows: a) $M_1^R > 1$; b) $M_1^R < 1$.

i.e., $M_1 > 1$, here ϕ_1 is controlled by θ_w since $\phi_1 = 90^\circ - \theta_w$. Thus, if θ_w is small enough, ϕ_1 can be large enough to fall into the strong shock solution domain. Hence, in a frame of reference attached to the reflection point R, M_1^R can be either supersonic or subsonic, depending upon the value of θ_w.

Let us first consider the situation when $M_1^R > 1$ which is analogous to either one of the two steady cases shown in figures 1.2b and 1.2c. If $\theta_1 < \delta_{max}$ (M_1^R), then an attached oblique shock wave emanating from reflection point R will deflect the flow away from the wedge surface, while forming an RR wave configuration, and if $\theta_1 > \delta_{max}$ (M_1^R), the flow deflection will be achieved by a detached shock wave, which will form an MR wave configuration.

If, however, $M_1^R < 1$, the analogy to figure 1.2a suggests that the flow should continuously turn to negotiate the wedge surface, as shown schematically in figure 1.4b, without any reflection structure. In reality, however, this is not the case. For all the combinations of M_0 and θ_w which result in $M_1^R < 1$, an MR structure is obtained. The exact reason for this lies probably in the following explanation.

Consider figure 1.4b, where the subsonic flow obtained behind the incident shock wave is seen to continuously turn in order to

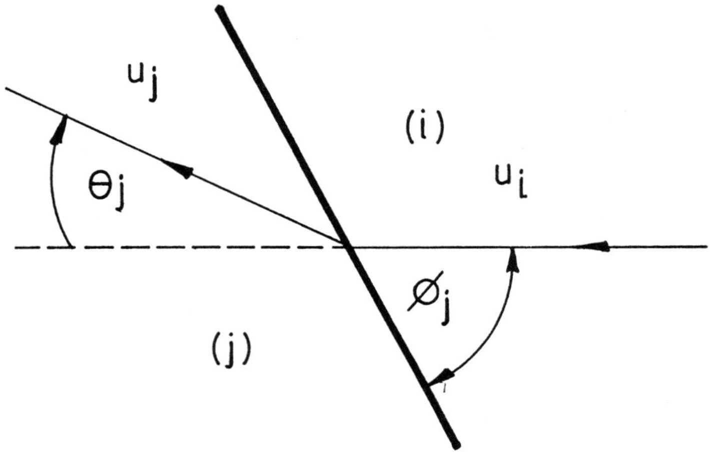

Figure 1.5 Definition of parameters across an oblique shock wave.

negotiate the wedge. Although it was noted earlier that this situation is analogous to the one shown in figure 1.2a, there is one important difference. While in figure 1.2a the flow particle "knows" about the obstacle awaiting it when it is far away from the wedge, and hence can start adjusting its streamline to negotiate the obstacle long before actually encountering the wedge, in the situation shown in figure 1.4b the flow particle near the wedge surface does not "know" about the obstacle until it passes through the incident shock, i, and hence must negotiate a sudden new boundary condition. Perhaps this sudden change in the boundary condition is the reason for obtaining a reflection for the situation where the flow Mach number behind the incident shock wave is subsonic with respect to the reflection point R.

1.3 Analytical Approaches for Describing Regular and Mach Reflections

The analytical approaches for describing the RR and the MR wave configurations were both initiated by von Neumann (1943a & b). The one describing the RR is known as "the two-shock theory" while the one describing the MR is known as "the three-shock theory". Both theories make use of the inviscid conservation equations across an oblique shock wave, together with appropriate boundary conditions.

Consider figure 1.5 where an oblique shock wave is seen. The flow states ahead and behind the oblique shock wave are (i) and (j), respectively. The angle of incidence between the oncoming flow and the oblique shock wave is - ϕ_j. While passing through the oblique shock wave the flow is deflected by an angle of - θ_j. The conservation equations, relating states (i) and (j) for a steady inviscid flow, are:

-conservation of mass:

$$\rho_i u_i \sin \phi_j = \rho_j u_j \sin (\phi_j - \theta_j) \tag{1.1}$$

-conservation of normal momentum:

$$p_i + \rho_i u_i^2 \sin^2 \phi_j = p_j + \rho_j u_j^2 \sin^2 (\phi_j - \theta_j) \tag{1.2}$$

-conservation of tangential momentum:

$$\rho_i \tan \phi_j = \rho_j \tan (\phi_j - \theta_j) \tag{1.3}$$

-conservation of energy:

$$h_i + \frac{1}{2} u_i^2 \sin^2 \phi_j = h_j + \frac{1}{2} u_j^2 \sin^2 (\phi_j - \theta_j) \tag{1.4}$$

where u is the flow velocity in a frame of reference attached to the oblique shock wave and ρ, p, and h are the flow density, static pressure and enthalpy, respectively.

If thermodynamic equilibrium is assumed on both sides of the shock wave, then two thermodynamic properties are sufficient to define a thermodynamic state, e.g., $\rho = \rho(p, T)$ and $h = h(p, T)$, where T is the flow temperature. Consequently, the above set of four equations contains eight parameters, namely, p_i, p_j, T_i, T_j, u_i, u_j, ϕ_j and θ_j. Thus, if four of these parameters are known, the set of equations is solvable in principle.

1.3.1 Two-Shock Theory for an Inviscid Flow

The two-shock theory is, in fact, the analytical model for describing the flow field in the vicinity of the reflection point of an RR. The wave configuration of an RR is shown schematically in

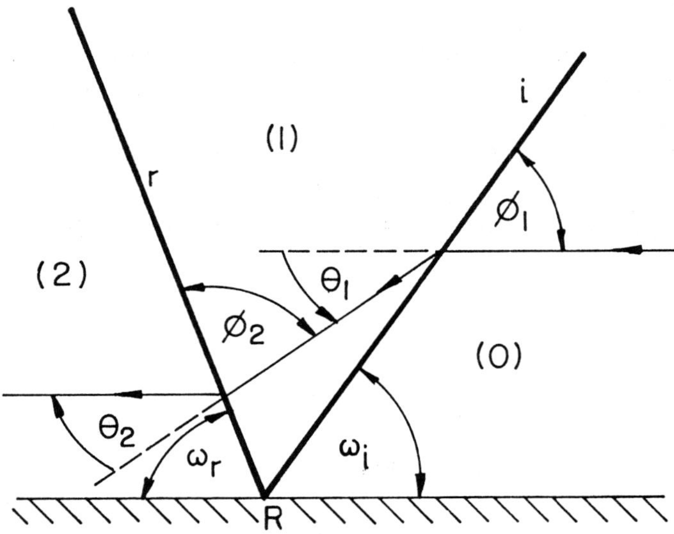

Figure 1.6 The wave configuration of a regular reflection - RR.

figure 1.6. It consists of two discontinuities: the incident shock wave, i, and the reflected shock wave, r. These two shock waves intersect at the reflection point, R, which is located on the reflecting surface. The RR is, in general, not linear, i.e., $\omega_i \neq \omega_r$.

Applying the oblique shock wave equations, given in section 1.3, to the wave configuration of an RR, results in the following set of governing equations for an RR in an inviscid flow.

Across the incident shock wave, i:

$$\rho_0 u_0 \sin\phi_1 = \rho_1 u_1 \sin(\phi_1 - \theta_1) \tag{1.5}$$

$$p_0 + \rho_0 u_0^2 \sin^2\phi_1 = p_1 + \rho_1 u_1^2 \sin^2(\phi_1 - \theta_1) \tag{1.6}$$

$$\rho_0 \tan\phi_1 = \rho_1 \tan(\phi_1 - \theta_1) \tag{1.7}$$

$$h_0 + \frac{1}{2} u_0^2 \sin^2\phi_1 = h_1 + \frac{1}{2} u_1^2 \sin^2(\phi_1 - \theta_1) \tag{1.8}$$

Across the reflected shock wave, r:

$$\rho_1 u_1 \sin\phi_2 = \rho_2 u_2 \sin(\phi_2 - \theta_2) \tag{1.9}$$

$$p_1 + \rho_1 u_1^2 \sin^2 \phi_2 = p_2 + \rho_2 u_2^2 \sin^2 (\phi_2 - \theta_2) \qquad (1.10)$$

$$\rho_1 \tan \phi_2 = \rho_2 \tan (\phi_2 - \theta_2) \qquad (1.11)$$

$$h_1 + \frac{1}{2} u_1^2 \sin^2 \phi_2 = h_2 + \frac{1}{2} u_2^2 \sin^2 (\phi_2 - \theta_2) \qquad (1.12)$$

In addition to these eight equations, there is also the condition that the flow behind the reflected shock wave, in state (2), must be parallel to the reflecting wedge surface. Hence, if the flow is assumed to be inviscid, then

$$\theta_1 - \theta_2 = 0 \qquad (1.13)$$

Thus, the two-shock theory which describes the flow field in the vicinity of the reflection point of an RR consists of a set of nine equations. If thermodynamic equilibrium is assumed in all the flow states, then the above set of nine equations contains 13 parameters, namely: p_0, p_1, p_2, T_0, T_1, T_2, u_0, u_1, u_2, ϕ_1, ϕ_2, θ_1 and θ_2. Consequently, four of these parameters must be known in order to have a closed set which, in principle, can be solved.

Henderson (1982) showed, that if the gas is assumed to obey the equation of state of a perfect gas, i.e., $p = \rho RT$, then equations (1.5) to (1.13) can be reduced to a single polynomial of the order six. Although a polynomial of order six yields six roots, Henderson (1982) showed that from simple physical considerations four of the six roots can be discarded. This implies that equations (1.5) to (1.13) do not result in a unique solution for a given set of initial conditions. This will be further illustrated and discussed in section 1.4.1.

1.3.2 Three-Shock Theory for an Inviscid Flow

The three-shock theory is, in fact, the analytical model for describing the flow field in the vicinity of the triple point of an MR. The wave configuration of an MR is shown schematically in figure 1.7. It consists of four discontinuities; three shock waves; (the incident shock wave, i, the reflected shock wave, r, and the Mach stem, m) and one slipstream, s. These four discontinuities intersect at a single point, known as the triple point, T, which is located above the reflecting surface. The Mach stem is usually curved (depending upon the initial conditions it can be either concave or

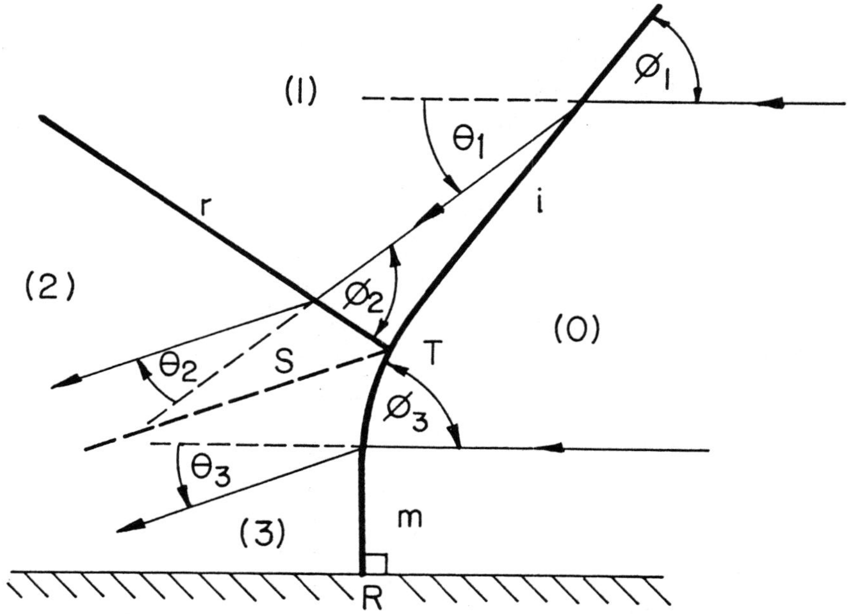

Figure 1.7 The wave configuration of a Mach reflection - MR.

convex). Its foot is perpendicular to the reflecting surface at the reflection point, R.

Applying the oblique shock wave equations, given in section 1.3, to the wave configuration of an MR results in the following set of governing equations for an MR in an inviscid flow.

Across the incident shock wave, i:

$$\rho_0 u_0 \sin \phi_1 = \rho_1 u_1 \sin (\phi_1 - \theta_1) \qquad (1.14)$$

$$p_0 + \rho_0 u_0^2 \sin^2 \phi_1 = p_1 + \rho_1 u_1^2 \sin^2 (\phi_1 - \theta_1) \qquad (1.15)$$

$$\rho_0 \tan \phi_1 = \rho_1 \tan (\phi_1 - \theta_1) \qquad (1.16)$$

$$h_0 + \frac{1}{2} u_0^2 \sin^2 \phi_1 = h_1 + \frac{1}{2} u_1^2 \sin^2 (\phi_1 - \theta_1) \qquad (1.17)$$

Across the reflected shock wave, r:

$$\rho_1 u_1 \sin \phi_2 = \rho_2 u_2 \sin (\phi_2 - \theta_2) \qquad (1.18)$$

$$p_1 + \rho_1 u_1^2 \sin^2 \phi_2 = p_2 + \rho_2 u_2^2 \sin^2 (\phi_2 - \theta_2) \tag{1.19}$$

$$\rho_1 \tan \phi_2 = \rho_2 \tan (\phi_2 - \theta_2) \tag{1.20}$$

$$h_1 + \frac{1}{2} u_1^2 \sin^2 \phi_2 = h_2 + \frac{1}{2} u_2^2 \sin^2 (\phi_2 - \theta_2) \tag{1.21}$$

Across the Mach stem, m:

$$\rho_0 u_0 \sin \phi_3 = \rho_3 u_3 \sin (\phi_3 - \theta_3) \tag{1.22}$$

$$p_0 + \rho_0 u_0^2 \sin^2 \phi_3 = p_3 + \rho_3 u_3^2 \sin^2 (\phi_3 - \theta_3) \tag{1.23}$$

$$\rho_0 \tan \phi_3 = \rho_3 \tan (\phi_3 - \theta_3) \tag{1.24}$$

$$h_0 + \frac{1}{2} u_0^2 \sin^2 \phi_3 = h_3 + \frac{1}{2} u_3^2 \sin^2 (\phi_3 - \theta_3) \tag{1.25}$$

In addition to these equations, there are also two boundary conditions which arise from the fact that states (2) and (3) are separated by a contact surface across which the pressure remains constant, i.e.,

$$p_2 = p_3 \tag{1.26}$$

Furthermore, if the flow is assumed to be inviscid and if the contact surface is assumed to be infinitely thin, i.e., a slipstream, then the flows on both sides of the slipstream are parallel, i.e.,

$$\theta_1 - \theta_2 = \theta_3 \tag{1.27}$$

Thus, the three-shock theory, which describes the flow field in the vicinity of the triple point of an MR, consists of 14 equations. Again, if thermodynamic equilibrium is assumed in states (0), (1), (2) and (3), then the above set of 14 equations contains 18 parameters, namely: p_0, p_1, p_2, p_3, T_0, T_1, T_2, T_3, u_0, u_1, u_2, u_3, ϕ_1, ϕ_2, ϕ_3, θ_1, θ_2, and θ_3. Consequently, four of these parameters must be known in order to have a set of equations which, in principle, can be solved.

Henderson (1982) showed that if the gas is assumed to behave as a perfect gas, then equations (1.14) to (1.27) can be reduced to a single polynomial of order ten, with the pressure ratio p_3/p_0 as the polynomial variable. The polynomial coefficients were taken to be a function of the specific heat capacities ratio, $\gamma = C_p/C_v$, the flow Mach number in state (0), M_0, and the pressure ratio across the incident shock wave, p_1/p_0. Although a polynomial of degree ten yields ten roots, Henderson (1982) showed that from simple physical considerations and the possibility of double roots, seven out of the ten roots can be discarded. This implies that equations (1.14) to (1.27) do not yield a unique solution for a given set of initial conditions. This is further illustrated and discussed in section 1.4.2.

1.4 Shock Polars

Kawamura & Saito (1956) were the first to suggest that owing to the fact that the boundary conditions of an RR and an MR are expressed in terms of the flow deflection angles, θ, and the flow static pressures, p, the use of (p, θ)-shock polars might be of great advantage in better understanding the shock wave reflection phenomenon.

The graphical presentation of the relationship between the pressure, p, obtained behind an oblique shock wave and the angle, θ, by which the flow is deflected while passing through an oblique shock wave for a fixed value of the flow Mach number, M_0, is called a pressure-deflection shock polar. A typical pressure-deflection shock polar is shown in figure 1.8. Four special points are indicated in the shock polar shown in figure 1.8:

- Point "a" illustrates a situation in which the flow state behind the oblique shock wave is identical in all its properties to the flow state ahead of it. This situation is obtained when the angle of incidence between the oblique shock wave and the oncoming flow, ϕ_1, is identical to the Mach angle $\mu = \sin^{-1}(1/M_0)$ In this case the pressure jump across the oblique shock wave is minimal, i.e., $p_1/p_0 = 1$, and the flow deflection, θ_1, is zero.

- Point, "b" indicates the flow state which is obtained from state (0) by passing through the strongest oblique shock wave, a normal shock wave, i.e., $\phi_1 = 90°$. In this case, the pressure jump across the normal shock wave, p_1/p_0, is maximal and the flow deflection,

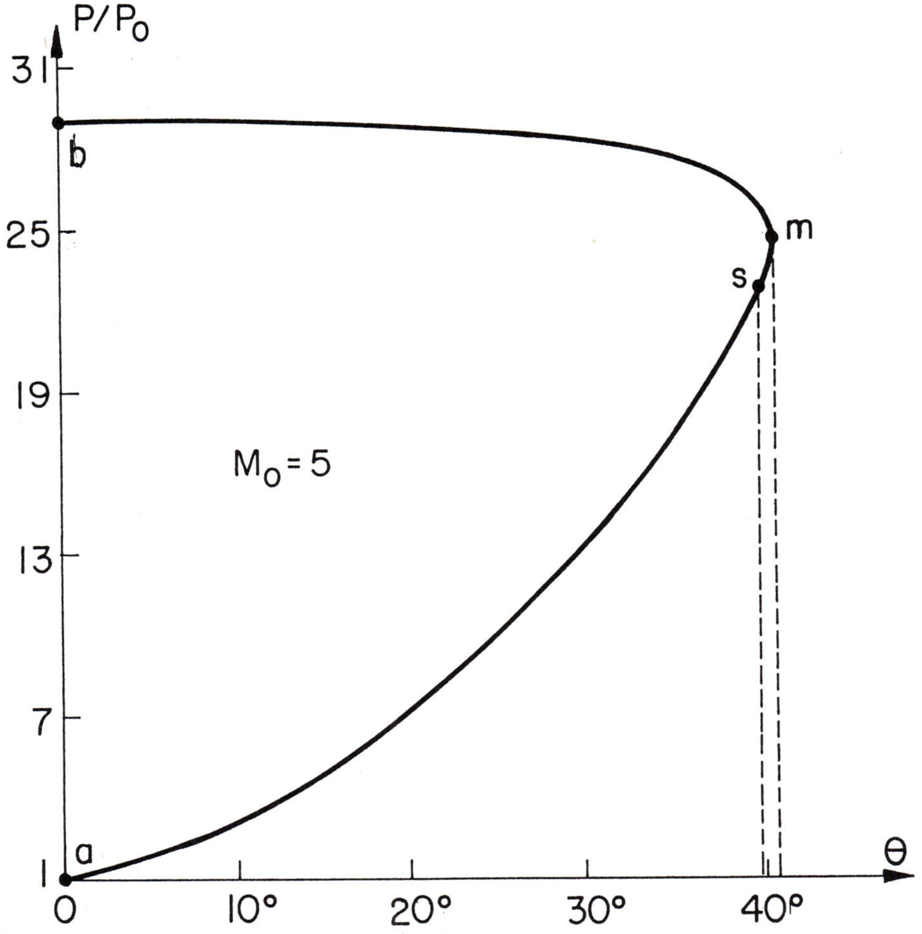

Figure 1.8 A typical (p,θ)-shock polar for $M_0 = 5$ and $\gamma = 1.4$.

θ_1, is again zero.

- Point, "s" divides the shock polar into two sections. Along the section between points "a" and "s" the flow Mach number behind the oblique shock wave, M_1, is supersonic, i.e., $M_1 > 1$, and along the section between points "s" and "b" the flow Mach number behind the oblique shock wave is subsonic, i.e., $M_1 < 1$. Thus, point "s" indicates the situation when the flow behind the oblique shock wave is exactly sonic, i.e., $M_1 = 1$. The "a" - "s" and "s" - "b" portions of the shock polar are known as the weak and strong shock polar portions, respectively.

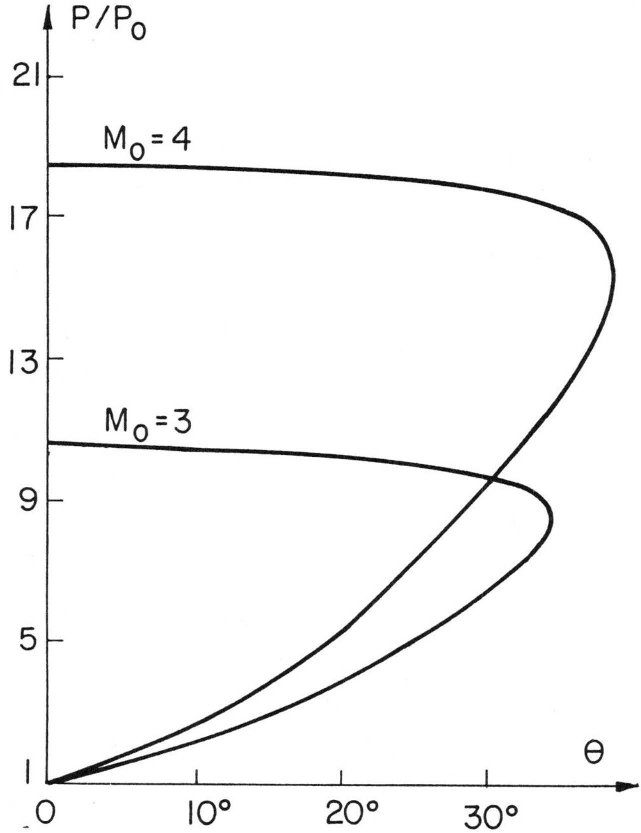

Figure 1.9 The dependence of the (p,θ)-shock polar on the incident flow Mach number for $\gamma = 1.4$.

- Point "m" is known as the point of maximum deflection or the detachment point. It indicates the maximum flow deflection by which a given supersonic flow can be deflected by an oblique shock wave. As can be seen from figure 1.8 the difference between θ_s and θ_m is very small. It is equal to zero for $M_0 = 1$ and approaches zero as M_0 approaches infinity, i.e., $M_0 \rightarrow \infty$. Depending on the specific heat capacities ratio, γ, it assumes maximum values of a few degrees at moderate values of M_0. For this reason it is common to treat both of these points as practically a single point.

It is well known that the higher the flow Mach number, M_0, the larger is the corresponding shock polar. This is illustrated in figure 1.9 where the shock polars for a perfect diatomic gas and for $M_0 = 3$ and $M_0 = 4$ are accurately drawn to scale.

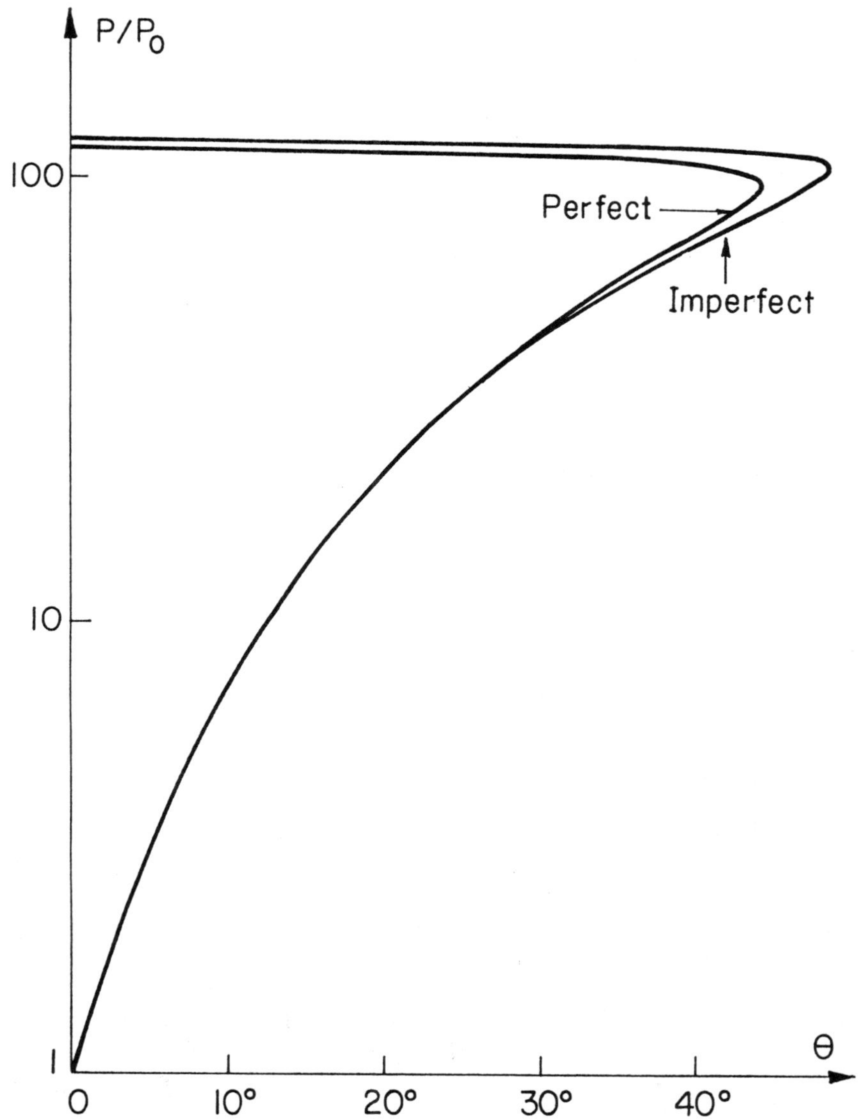

Figure 1.10 The dependence of the (p,θ)-shock polar on real gas effects for nitrogen at $M_0 = 10$, $p_0 = 15$ torr and $T_0 = 300K$.

Accounting for real gas effects also changes the shock polar as can be seen in figure 1.10. Both the maximum deflection angle and the pressure jump across the oblique shock wave increase as a result of the relaxation of the internal degrees of freedom of the gas. For the particular case illustrated in figure 1.10, i.e., $M_0 = 10$,

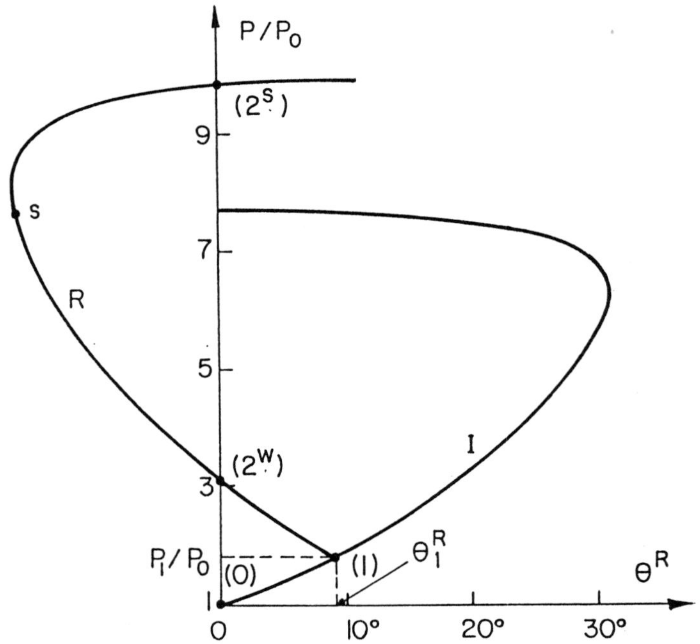

Figure 1.11 Shock polar solution of a regular reflection.

T_0 = 300K and p_0 = 15 torr, θ_m = 42.7° for a perfect nitrogen and θ_m = 49.0° for a nitrogen in dissociational equilibrium.

As mentioned earlier, in many phenomena involving shock wave interactions there is a possibility of more than one theoretical solution. In such cases shock polars can be useful in discarding unrealistic solutions and indicating the real solution.

1.4.1 Shock Polar Presentation of the Flow Field in the Vicinity of the Reflection Point of an RR

Figure 1.11 presents the (p, θ)-polar solution of the flow field in the vicinity of the reflection point of an RR. The flow deflection angles, θ, are measured with respect to the trajectory of the reflection point, R. State (0) is at the origin, where $p = p_0$ and $\theta^R = \theta_0^R = 0$. The locus of all the flow states which can be obtained from state (0) by passing through any oblique shock wave is represented by the I-polar. Consequently, state (1) which is obtained from state (0) by passing through the incident shock wave, i, is on the I-polar at the point $p = p_1$ and $\theta^R = \theta_1^R$. The locus of all the flow states which can be obtained from state (1) by passing through any oblique shock wave is represented by the R-polar. Note that since the flow deflection through the reflected shock wave is in a direction opposite to that of the flow deflection through the incident shock wave, the R-polar is opposite to the I-polar. Consequently, state (2), which is obtained from state (1) by passing through the reflected shock wave is on the R-polar. The boundary condition for an RR implies that $\theta^R = \theta_2^R = 0$, therefore, state (2) is obtained at the point where the R-polar intersects the p-axis (i.e., $\theta^R = 0$).

As can be seen in figure 1.11 two such points are obtained, namely, (2^w) and (2^s). Both of these points indicate possible solutions of equations (1.5) to (1.13). Point (2^w) is known as the "weak shock solution" and point (2^s) is known as the "strong shock solution". None of these solutions can be discarded on theoretical grounds. However, it is an experimental fact that the weak shock solution is the one which usually occurs. Consequently, the flow state behind the reflected shock wave is represented by point (2^w) of figure 1.11. In the following this state will be labeled as (2) only.

1.4.2 Shock Polar Presentation of the Flow Field in the Vicinity of the Triple Point of an MR

Figure 1.12 presents the (p, θ)-polar solution of the flow field in the vicinity of the triple point of an MR. The flow deflection angles are measured with respect to the trajectory of the triple point, T. The locus of all the flow states which can be obtained from state (0) by passing through any oblique shock wave is represented by the I-polar. Thus, state (0) is at the origin, and state (1) behind the incident shock wave, i, lies on the I-polar. State (1) is the origin of the R-polar which is drawn in a direction opposite to the I-polar. State (2), behind the reflected shock wave, lies on the R-polar, and state (3), behind the Mach stem, lies on the I-polar since it is also obtained from state (0) by passing through the Mach stem. Since the pressures in states (2) and (3) are equal, i.e., $p_2 = p_3$ [equation (1.26)], and the flows in these states are parallel, i.e., $\theta_2^T = \theta_3^T$ [Equation (1.27)], states (2) and (3) are obtained at the point where the I- and R-polars intersect.

Figure 1.13 presents three cases of different possible solutions of the flow field in the vicinity of the triple point of an MR. Unlike figure 1.12, where only the right branch of the I-polar is drawn, here the left branch is also shown. The three different possible MR solutions are indicated by points "a", "b" and "c". In the MR at point "a" the net deflection of the flows in states (2) and (3) is positive,

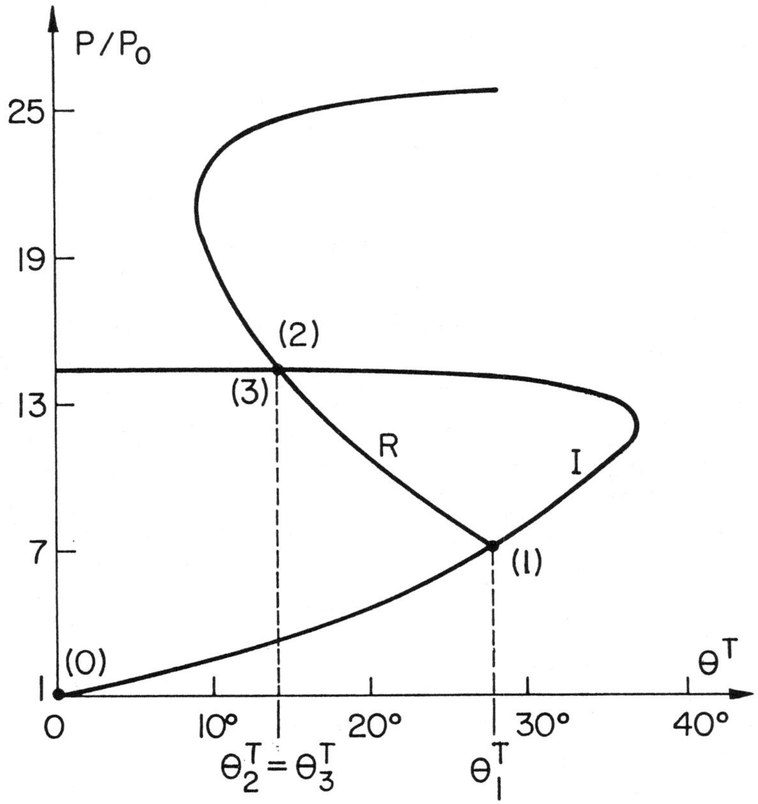

Figure 1.12 Shock polar solution of a Mach reflection.

(i.e., +7°) with respect to their original direction in state (0). In the MR at point "b" the net deflection of the flows in states (2) and (3) is zero, and in the MR at point "c" the net deflection of the flows in states (2) and (3) is negative (i.e., -8°). Courant & Friedrichs (1948) termed these three different types of MR as:

direct - Mach reflection (D_iMR) at point "a",
stationary - Mach reflection (S_tMR) at point "b", and
inverse - Mach reflection (I_nMR) at point "c".

Note that the I-R_{II} shock polar combination (in figure 1.13) indicates, in addition to the S_tMR at point "b", also a possible RR solution at point "b" since the R_{II}-polar intersects the p-axis at that point. Similarly, the I-R_{III} shock polar combination indicates, in addition to the I_nMR at point "c", a possible RR solution at point "d" where the R_{III}-polar intersects the p-axis. Thus, it is again evident that different reflection configurations can be theoretically obtained for the same initial conditions.

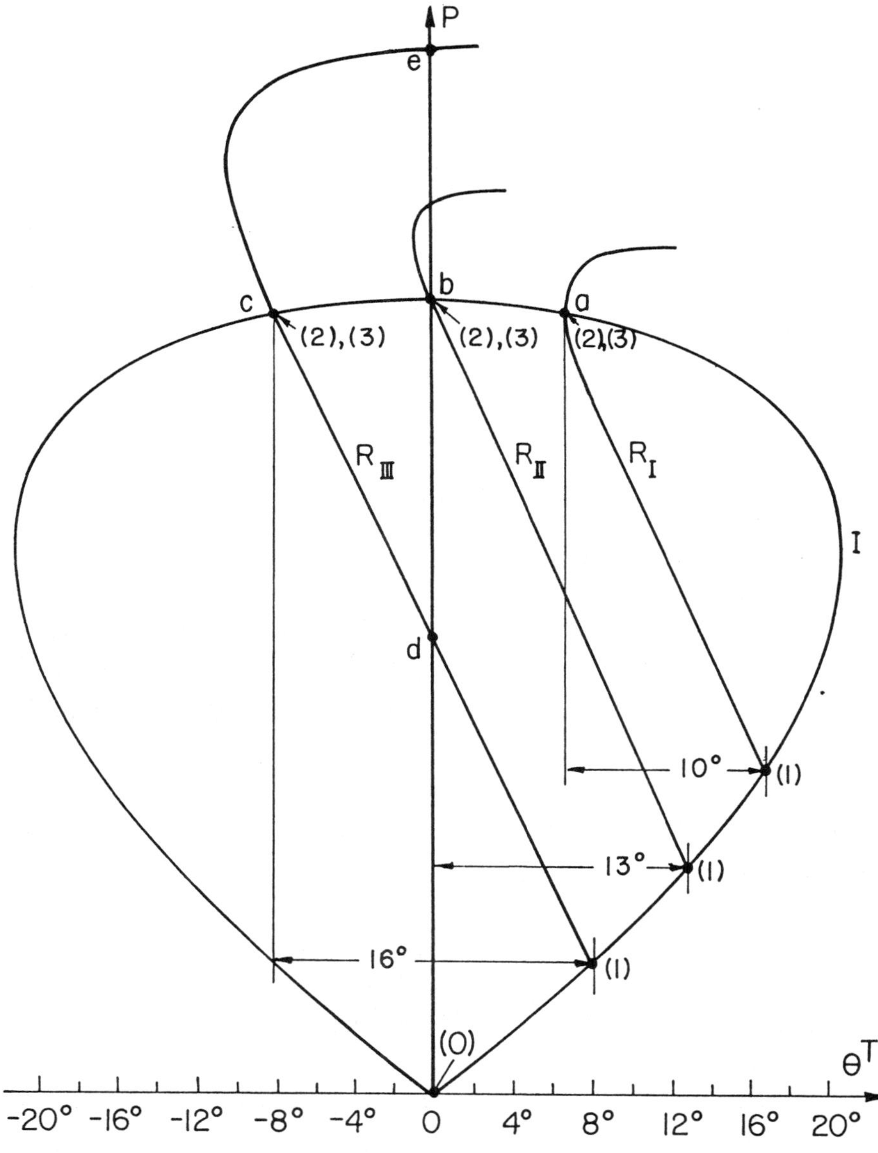

Figure 1.13 Three different possible MR solutions resulting in a direct-Mach reflection at point "a", a stationary-Mach reflection at point "b" and an inverse-Mach reflection at point "c".

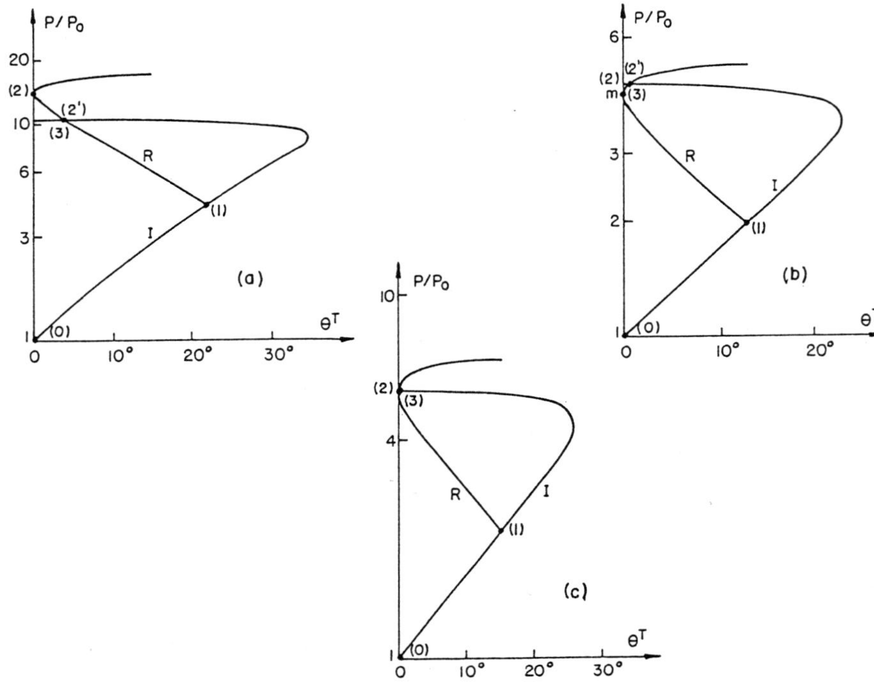

Figure 1.14 Three different I-R polar combinations: a) the R-polar is tangent to the p-axis outside the I-polar (imperfect nitrogen, $M_0 = 3$, $p_0 = 15$ torr, $T_0 = 300$K, $\theta_w = 50.01^o$ and $M_s = 1.93$); b) the R-polar is tangent to the p-axis inside the I-polar (imperfect nitrogen, $M_0 = 2$, $p_0 = 15$ torr, $T_0 = 300$K, $\theta_w = 47.32^o$ and $M_s = 1.36$); c) the R-polar is tangent to the p-axis at the normal shock point of the I-polar (imperfect nitrogen, $M_0 = 2.185$, $p_0 = 15$ torr, $T_0 = 300$K, $\theta_w = 48.46^o$ and $M_s = 1.449$).

1.5 Suggested RR⇔IR Transition Criteria

The search for the transition criterion from RR to IR and from IR to RR has been the objective of numerous analytical, numerical and experimental studies since von Neumann re-initiated the investigation of the oblique shock wave reflection phenomena in the early 1940's. The reason for the continuous search for the correct

RR⇔IR transition criterion was due to the fact that the agreement between the various proposed criteria and the experimental results was never satisfactory enough in the entire range of incident shock wave Mach numbers and reflecting wedge angles.

Most of the RR⇔IR transition criteria, which are discussed subsequently, were initiated by von Neumann in the early 1940's. In the following, the various RR⇔IR transition criteria are discussed in detail using shock polars for explanatory purposes.

1.5.1 Detachment Criterion

Consider figures 1.12 and 1.13 and note that while in the case of an RR the R-polar intersects the p-axis at two points, in the case of an MR the R-polar does not intersect the p-axis at all. The limiting case, i.e., the case in which the R-polar is tangent to the p-axis and hence intersects it at a single point, is known as the detachment criterion. Two I-R shock polar combinations for this situation are shown in figures 1.14a and 1.14b. A slight shift of the R-polar to the right will result in a situation in which the R-polar will not intersect the p-axis and hence a solution of an RR will be impossible.

Since at the detachment criterion, the flow deflection by the reflected shock wave is maximal, i.e., $\theta_2 = \theta_{2m}$, the transition line arising from the detachment criterion can be calculated using the two-shock theory, i.e., equations (1.5) to (1.13), and replacing θ_2 by θ_{2m}.

Depending on the initial conditions, the point of tangency between the R-polar and the p-axis can be either outside or inside the I-polar as shown in figures 1.14a and 1.14b, respectively. The limiting situation between these two cases is shown in figure 1.14c where the R-polar is tangent to the p-axis exactly at the normal shock wave point of the I-polar.

Incident shock waves, which result, at detachment, in a situation in which the point of tangency of the R-polar to the p-axis is outside the I-polar (such as shown in figure 1.14a) are usually referred to as "strong" shock waves, while those resulting in a situation in which the point of tangency of the R-polar to the p-axis are inside the I-polar (such as shown in figure 1.14b) are known as "weak" shock waves. This categorization was originally suggested by Henderson (1982) for reasons which will be given subsequently.

Three interesting I-R shock polar combinations are possible in the "weak" shock wave domain. They are shown in figures 1.15a to 1.15c. Figure 1.15a presents an I-R shock polar combination of an MR for which the net deflection of the flows in state (2), with respect to the triple point, is smaller than that in state (1). Hence the

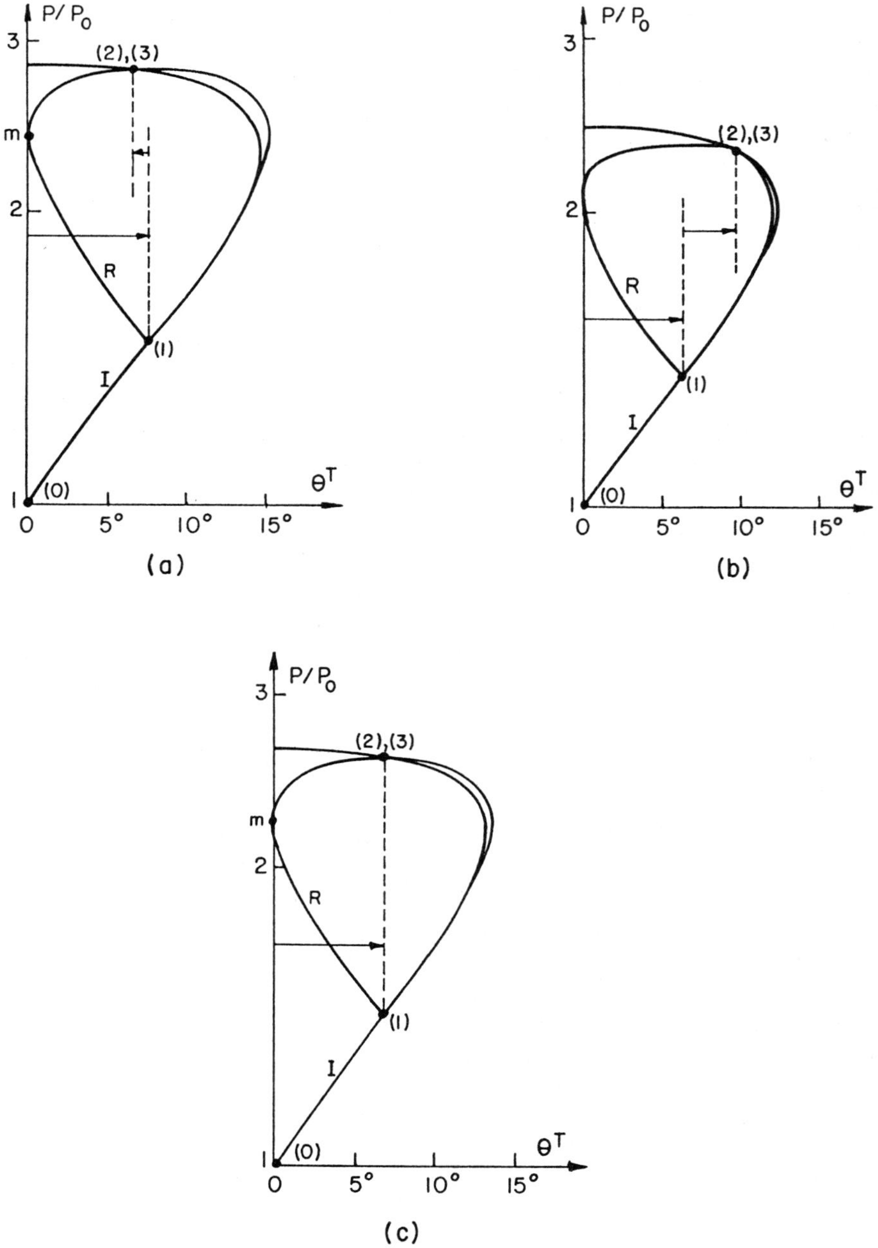

Figure 1.15 (p,θ)-shock polar solutions of a Mach reflection for: a) $\theta_2^T < \theta_1^T$ ($M_0 = 1.6$, $\theta_w^C = 42.12^\circ$ and $\gamma = 1.4$); b) $\theta_2^T > \theta_1^T$ ($M_0 = 1.5$, $\theta_w^C = 40.33^\circ$ and $\gamma = 1.4$); c) $\theta_2^T = \theta_1^T$ ($M_0 = 1.55$, $\theta_w^C = 48.50^\circ$ and $\gamma = 1.4$).

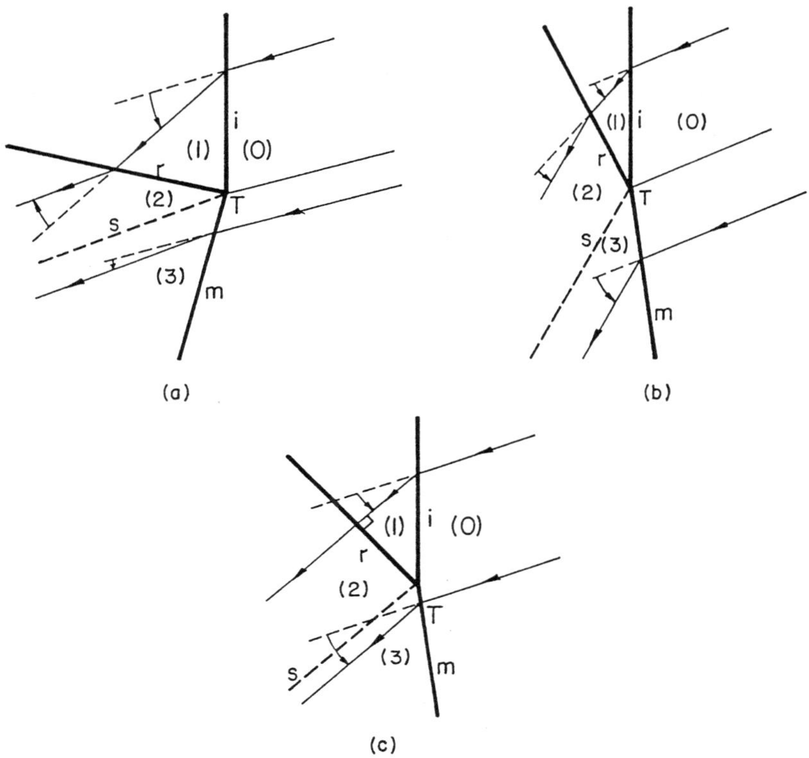

Figure 1.16 The wave configurations of the three Mach reflections shown in figures 1.15a to 1.15c, respectively.

flow originating from state (0) at a point above the triple point trajectory is first deflected towards the wedge surface by the incident shock wave, and then it is deflected away from the wedge surface by the reflected shock wave to result in a situation in which $\theta_2^T < \theta_1^T$. The I-R polar combination shown in figure 1.15b illustrates a different MR case. It is seen that the flow which is deflected towards the wedge surface while passing through the incident shock wave is further deflected towards the wedge when it passes through the reflected shock wave to result in a situation in which $\theta_2^T > \theta_1^T$. The limiting I-R polar combination between these two cases is shown in figure 1.15c which indicates that the flow passing through the reflected shock wave is not deflected at all, and hence $\theta_2^T = \theta_1^T$. This can occur only if the reflected shock wave is normal to the flow in state (1) ahead of it. The three MR shock wave configurations appropriate to these three shock polar combinations are schematically illustrated in figures 1.16a to 1.16c, respectively.

It should be pointed out here that if one wishes to use the three-shock theory in order to calculate the flow properties in the vicinity of the triple point of the three different MR wave configurations shown in figures 1.16a to 1.16c, then equation (1.27) should be modified. For the MR shown in figure 1.16a, equation (1.27) gives the appropriate condition across the slipstream. However, for calculating the MR shown in figure 1.16b, equation (1.27) should be replaced by

$$\theta_1 + \theta_2 = \theta_3 \tag{1.28}$$

since the deflection angles of the flow passing through the incident and reflected shock waves are in the same direction. For the limiting MR shown in figure 1.16c, either equation (1.27) or equation (1.28) can be used since for this case, the reflected shock wave is normal to the oncoming flow, and hence $\theta_2 = 0$. Thus, equations (1.27) and (1.28) degenerate to

$$\theta_1 = \theta_3 \tag{1.29}$$

In addition, for this case the three-shock theory can be further reduced, since one of the dependent variables, namely, the angle of incidence between the flow in state (1) and the reflected shock wave, ϕ_2, is known, i.e., $\phi_2 = 90°$. Thus, the three-shock theory can be significantly simplified for this case.

As will be discussed subsequently, the MR wave configuration shown in figure 1.16b, which based on the foregoing theoretical arguments is possible, does not materialize in reality. Instead, another type of reflection, the von Neumann reflection, vNR, is obtained. The major difference between an MR and a vNR is that the reflected shock wave, r, of an MR is replaced by a compression wave in a vNR. This fact was first hypothesized by Colella & Henderson (1990).

1.5.2 Mechanical Equilibrium Criterion

This criterion, originally suggested by von Neumann (1943), was re-initiated under this name by Henderson & Lozzi (1975) for the following reason. Consider the I-R polar combinations shown in figures 1.14a and 1.14b, both of which correspond to the detachment criterion. These I-R polar combinations suggest that if the RR⇔IR transition occurs at detachment, then the termination of

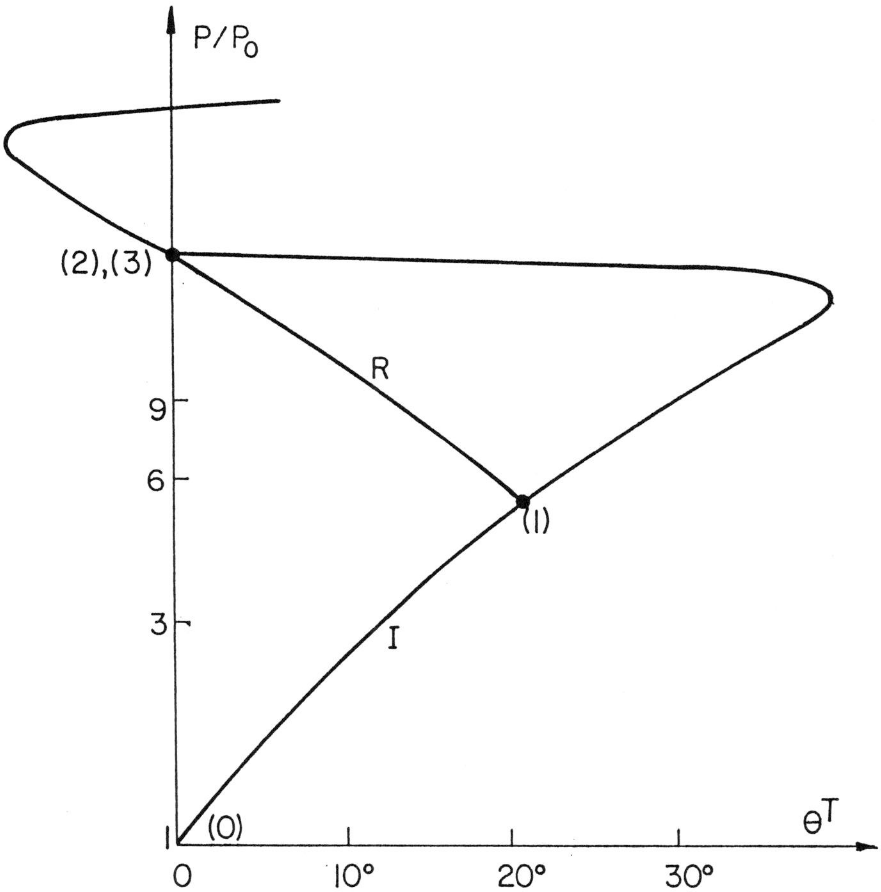

Figure 1.17 I-R polar combination illustrating the mechanical equilibrium criterion.

an RR and the formation of an IR must be associated with a sudden pressure change, from p_2 to p_2'. Henderson & Lozzi (1975) argued that such a sudden pressure change must be supported by either a compression wave (or a shock wave) or an expansion wave, depending upon whether p_2 is greater than (as shown in figure 1.14a) or smaller than (as shown in figure 1.14b) p_2'. Since neither of these additional waves has ever been observed experimentally, Henderson & Lozzi (1975) concluded that the detachment criterion is aphysical. Alternatively, they suggested a transition which corresponds to the shock polar combination shown in figure 1.17.

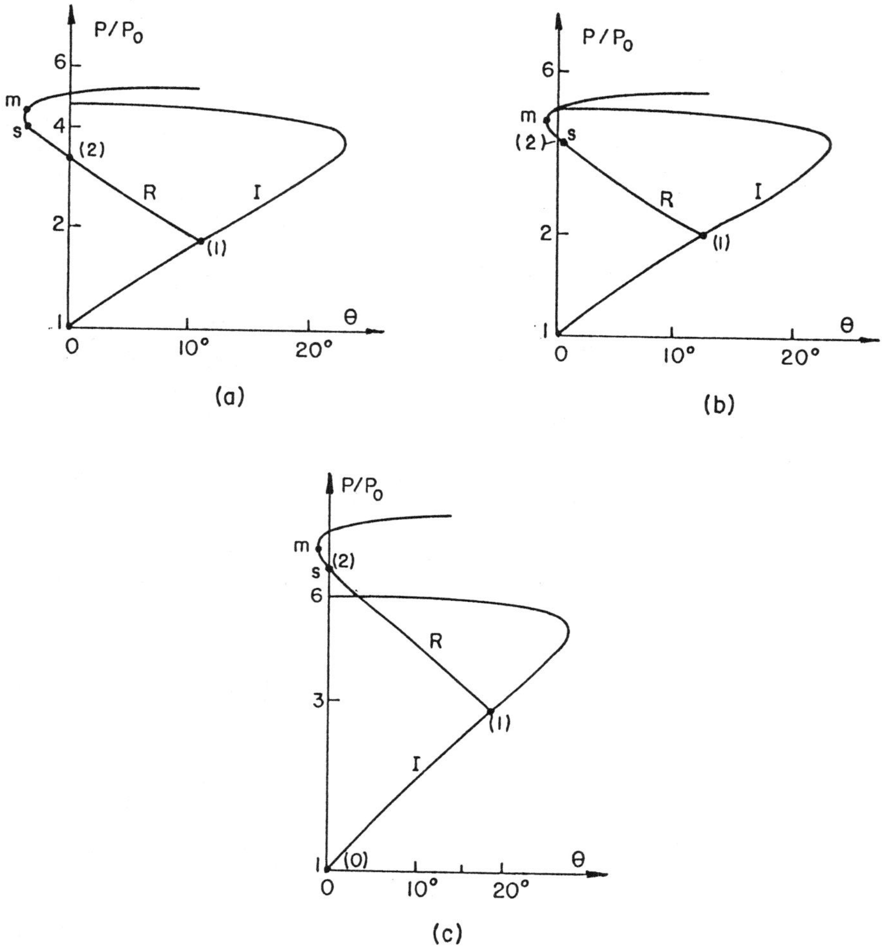

Figure 1.18 (p,θ)-shock polar solutions of three different regular reflections: a) supersonic flow behind the reflected shock wave ($M_0 = 2$, $p_0 = 15$ torr, $T_0 = 300$K, $\theta_w = 49.59^o$ and $M_s = 1.3$); b) subsonic flow behind the reflected shock wave ($M_0 = 2$, $p_0 = 15$ torr, $T_0 = 300$K, $\theta_w = 47.46^o$ and $M_s = 1.35$); c) sonic flow behind the reflected shock wave.

In this polar combination the R-polar intersects the p-axis exactly at the normal shock point of the I-polar. Consequently, both an RR and an MR are theoretically possible at that point. If this is the RR⇔IR transition point, then from the pressure point of view, the transition would be continuous and in mechanical equilibrium. The mechanical equilibrium transition line can be obtained by solving equations (1.14) to (1.27) and requiring that $\theta_1 - \theta_2 = \theta_3 = 0$.

1.5.3 Sonic Criterion

This transition criterion, which was also first introduced, as a possible transition criterion, by von Neumann (1943), is based on the argument that the RR⇔IR transition occurs when the corner-generated signals cannot catch-up with the reflection point, R, of an RR. Hence, as long as the flow Mach number behind the reflected shock wave is supersonic, the reflection point is isolated from the corner-generated signals, and they cannot reach it.

Consider figures 1.18a and 1.18b where two different I-R shock polar combinations are shown. While in figure 1.18a, the R-polar intersects the p-axis along its "weak" portion, in figure 1.18b the R-polar intersects the p-axis along its "strong" portion. Thus, while the flow behind the reflected shock wave is supersonic for the former case, it is subsonic for the latter. The limit between these two cases is shown in figure 1.18c where the R-polar intersects the p-axis exactly at its sonic point, s. This I-R shock polar combination is appropriate to the sonic criterion, or the catch-up condition, since this is the limit for which the corner-generated signals can catch-up with the reflection point, R, of the regular reflection. The transition line arising from the sonic criterion can be calculated by solving the two-shock theory, i.e., equations (1.5) to (1.13), and replacing θ_2 by θ_{2s}.

It is again worthwhile noting that the sonic criterion results in transition conditions which are very close to those resulting from the detachment criterion. In many cases the difference in the value of the reflecting wedge angle is only a fraction of a degree. For this reason it is almost impossible to distinguish experimentally between the sonic and detachment criteria.

It is important to note, however, that Lock & Dewey (1989) have developed an ingenious experimental set-up by which they were able to experimentally distinguish between the "sonic" and the "detachment" criteria. Their experimental investigation led to the conclusion that, in pseudo-steady flows, the RR⇔IR transition occurs when the corner-generated signals catch-up with the reflection point, R, i.e., at the sonic condition rather than the detachment one.

1.5.4 Length Scale Criterion

The length scale criterion was introduced by Hornung, Oertel & Sandeman (1979). The physical reasoning of this criterion is based

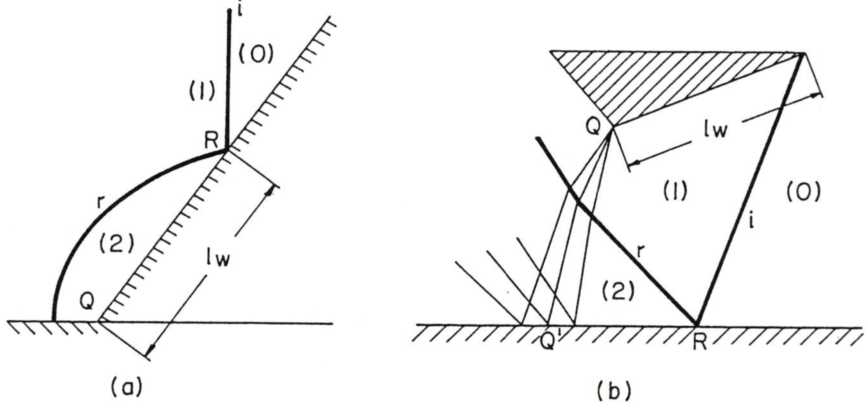

Figure 1.19 Definition of the physical length, l_w, which should be communicated to the reflection point, R, in order to have transition from regular to Mach reflection: a) pseudo-steady flows; b) steady flows.

on their argument that, unlike the wave configuration of an RR which is not associated with any length scale (both the incident and reflected shock waves extend to infinity), the wave configuration of an MR inherently includes a length scale, namely the finite length of the Mach stem which extends from the reflecting surface to the triple point. Thus they argued that in order for an MR to form, i.e., a shock wave with a finite length to be established, a physical length scale must be available at the reflection point, i.e., pressure signals must be communicated to the reflection point of the RR. This argument eventually led them to conclude that there are two different termination lines for RR depending on whether the flow under consideration is steady or pseudo-steady.

Consider the pseudo-steady RR in figure 1.19a and note that the length of the reflecting surface, l_w, can be communicated to the reflection point, R, only if a subsonic flow is established between Q and R (in a frame of reference attached to R). This requirement corresponds to the shock polar combination shown in figure 1.18c which, as discussed earlier, corresponds also to the sonic criterion. In a steady flow (figure 1.19b) the length, l_w, of the wedge used to generate the incident shock wave can be communicated to the reflection point R only if a propagation path exists between Q and R via the expansion wave at Q'. This is possible only if the flow between R and Q' is subsonic. According to Hornung, Oertel &

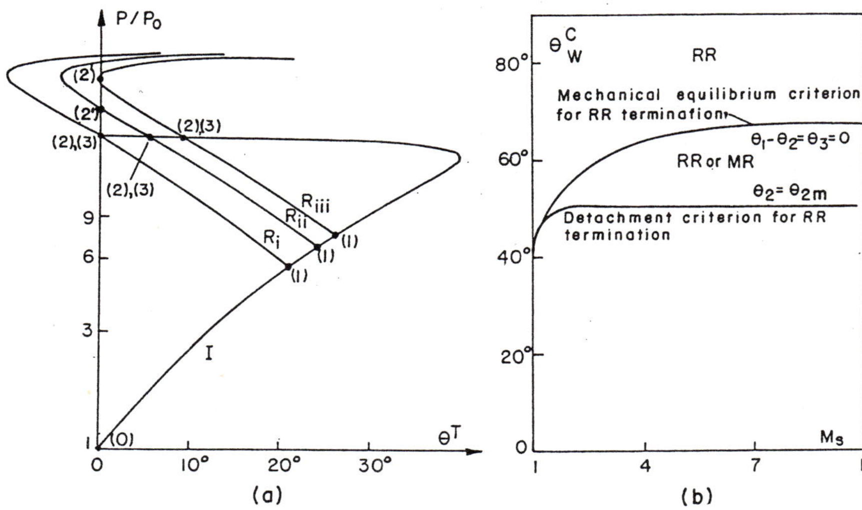

Figure 1.20 a) Various combinations of I-R shock polar; the I-R_I polar combination corresponds to the mechanical equilibrium criterion; the I-R_{III} polar combinations corresponds to the detachment (sonic) criterion; b) domains of RR and MR in the (M_s, θ_w^C)-plane as defined by the mechanical equilibrium and the detachment (sonic) criteria.

Sandeman (1979) this could happen if an MR existed, since the flow behind the Mach stem is always subsonic. Consequently, they argued that transition from RR to MR takes place the very first time MR becomes theoretically possible. This requirement corresponds to the shock polar combination shown in figure 1.17 which, as discussed earlier, corresponds also to the mechanical equilibrium criterion.

Thus, the physical arguments of Hornung, Oertel & Sandeman (1979) led to two different transition lines. In steady flows it predicts transition at the same point predicted by the mechanical equilibrium criterion, and in pseudo-steady flows it predicts transition at the same point predicted by the sonic criterion.

1.5.5 Summary, Critique and Discussion

The four foregoing suggested transition criteria yield three different RR⇔IR transition lines which can be calculated in the following ways:

- the transition line arising from the detachment criterion can be calculated using the two-shock theory while requiring that

$$\theta_2 = \theta_{2m} \tag{1.30}$$

- the transition line arising from the sonic criterion can be calculated using the two-shock theory while requiring that

$$\theta_2 = \theta_{2s} \tag{1.31}$$

- the transition line arising from the mechanical equilibrium criterion can be calculated using the three-shock theory while requiring that

$$\theta_1 - \theta_2 = \theta_3 = 0 \tag{1.32}$$

Recall that the transition lines arising from the length scale criterion are given by equation (1.31) for pseudo-steady flows and by equation (1.32) for steady flows. It should also be recalled that the transition lines as calculated by equations (1.30) and (1.31) are practically identical.

Figure 1.20a illustrates three different I-R polar combinations. The I-R_i polar combination corresponds to the mechanical equilibrium condition and the I-R_{iii} polar combination corresponds to the detachment condition. The I-R_{ii} polar combination corresponds to an intermediate situation. For this polar combination the mechanical equilibrium criterion predicts an MR at points (2), (3) where the R_{ii}-polar intersects the I-polar while the detachment criterion predicts an RR at point (2') where the R_{ii}-polar intersects the p-axis. For all the R-polars between R_i and R_{iii}, both RR and MR are theoretically possible.

Figure 1.20b illustrates the size of the dual-solution region in the (M_s,θ_w^C)-plane, where θ_w^C is the complementary angle of ϕ_1, i.e., $\theta_w^C = 90^\circ - \phi_1$. It is clearly seen that the area of disagreement between the mechanical equilibrium and the detachment criteria is very large. Note that if the transition line arising from the sonic criterion had been added to figure 1.20b it would have lain slightly above the detachment transition line.

Although Henderson & Lozzi (1975) reported that excellent agreement was obtained between the mechanical equilibrium criterion and their experiments in steady flows, i.e., wind tunnel experiments, there are unfortunately some difficulties associated with this criterion.

First of all, the mechanical equilibrium criterion does not apply over the entire range of incident shock wave Mach numbers, M_S, and angles of incidence, ϕ_1. There are combinations of M_S and ϕ_1 for which the condition given by equation (1.32) cannot be met.

Second, in their experiments in pseudo-steady flows, e.g., shock tube experiments, they observed the RR structure to persist not only inside the dual-solution region shown in figure 1.20b but also slightly below the detachment transition line, where RR is theoretically impossible. In the weak shock wave domain the persistence was up to 5° while in the strong shock wave domain RR prevailed to about 2° below its theoretical limit. Similar experimental results were obtained by other investigators who studied the RR⇔IR transition problem. Henderson & Lozzi (1975) attempted to resolve this anomaly by advancing a hypothesis that the RR-configurations observed beyond the limit predicted by the mechanical equilibrium criterion were undeveloped MR-configurations in which the Mach stem, the slipstream and the triple point were too close together to be resolved as in a well developed MR-configuration. However, in pseudo-steady flows, the shock wave configuration grows with time. Thus the triple point should eventually show up if a long enough reflecting wedge is used. This, unfortunately, did not occur even in experiments where the reflecting surface was very long.

Finally, Henderson & Lozzi's requirement of mechanical equilibrium is not justified when the flow under consideration is steady or pseudo-steady, since for these cases, depending upon the initial conditions, either an RR or an MR is established, and the requirement of a continuous pressure change during transition is unnecessary since transition does not take place. If, however, the flow under consideration is unsteady, and the reflection actually goes through a transition from RR to MR or from MR to RR, then their argument might apply. However, as will be shown in chapter 4, in the case of unsteady flows, the additional waves required by Henderson & Lozzi (1975) to be associated with a transition involving a sudden pressure drop indeed appear in the flow field.

In summary, experimental results in both steady and unsteady (including pseudo-steady) flows suggest that in steady flows the RR⇔IR transition generally agrees with the condition given by equation (1.32), while in unsteady flows the RR⇔IR transition seems to agree with the conditions given by either equation (1.30) or (1.31). Thus it can be concluded that the length scale criterion is an adequate criterion for the RR⇔IR transition for it results in the correct transition lines in both steady, pseudo-steady and unsteady flows.

As will be shown subsequently, the agreement between this transition criterion and careful experimental investigation was never satisfactory in the the vicinity of the transition lines. This fact has been motivating investigators to continue and search for the "correct" RR⇔IR transition criterion. However, one must recall that the transition criteria are based on the two- and three-shock theories which were developed under the assumption that all the discontinuities are straight in the vicinity of their intersection points and hence that the flow states are uniform. In addition to this assumption, which introduces inherent errors into predictions based on the two- and three-shock theories, it will be shown subsequently that the inclusion of viscous effects and real gas effects does improve the agreement between experiments and predictions based on these two fundamental theories.

1.6 References

Ben-Dor, G., "Relations Between First and Second Triple Point Trajectory Angles in Double Mach Reflection", AIAA J., Vol. 19, pp. 531-533, 1981.

Ben-Dor, G. & Takayama, K., "The Dynamics of the Transition from Mach to Regular Reflection over Concave Cylinders", Israel J. Tech., Vol. 23, pp. 71-74, 1986/7.

Colella, P. & Henderson, L.F., "The von Neumann Paradox for the Diffraction of Weak Shock Waves", J. Fluid Mech., Vol. 213, pp. 71-94, 1990.

Courant, R. & Freidrichs, K.O., *Hypersonic Flow and Shock Waves* , Wiley Interscience, New York 1948.

Henderson, L.F., "Exact Expressions for Shock Reflection Transition Criteria in a Perfect Gas", ZAMM, Vol. 62, pp. 258-261, 1982.

Henderson, L.F. & Lozzi, A., "Experiments on Transition of Mach Reflection", J. Fluid Mech., Vol. 68, pp. 139-155, 1975.

Hornung, H.G., Oertel, H. Jr. & Sandeman, R.J., "Transition to Mach Reflection of Shock Waves in Steady and Pseudo-Steady Flows with and without Relaxation", J. Fluid Mech., Vol. 90, pp. 541-560, 1979.

Kawamura, R. & Saito, H., "Reflection of Shock Waves - 1. Pseudo-Stationary Case", J. Phys. Soc. Japan, Vol. 11, pp. 584-592, 1956.

Krehl, P. & van der Geest, "The Discovery of the Mach Reflection Effect and Its Demonstration in an Auditorium", Shock Waves, Vol. 1, pp. 3-15, 1991.

Lee, J.-H. & Glass, I.I., "Pseudo-Stationary Oblique-Shock Wave Reflections in Frozen and Equilibrium Air", Prog. Aerospace Sci., Vol. 21, pp. 33-80, 1984.

Liepmann, H.W. & Roshko, A., *Elements of Gasdynamics* , John Wiley & Sons, New York, N.Y., U.S.A., 1957.

Lock, G. & Dewey, J.M., "An Experimental Investigation of the Sonic Criterion for Transition from Regular to Mach Reflection of Weak Shock Waves", Exp. in Fluids, Vol. 7, pp. 289-292, 1989.

Mach, E., "Über den Verlauf von Funkenwellen in der Ebene und im Raume", Sitzungsbr. Akad. Wiss. Wien, Vol. 78, pp. 819-838, 1878.

Neumann, J. von, "Oblique Reflection of Shocks", Explos. Res. Rep. 12, Navy Dept., Bureau of Ordinance, Washington, DC, U.S.A., 1943a.

Neumann, J. von, "Refraction, Intersection and Reflection of Shock Waves", NAVORD Rep. 203-45, Navy Dept., Bureau of Ordinance, Washington, DC, U.S.A., 1943b.

Reichenbach, H., "Contribution of Ernst Mach to Fluid Dynamics", Ann. Rev. Fluid Mech., Vol. 15, pp. 1-28, 1983.

Smith, L.G., "Photographic Investigation of the Reflection of Plane Shocks in Air", OSRD Rep. 6271, Off. Sci. Res. Dev., Washington, DC., U.S.A., or NDRC Rep. A-350, 1945.

White, D.R., "An Experimental Survey of the Mach Reflection of Shock Waves", Princeton Univ., Dept. Phys., Tech. Rep. II-10, Princeton, N.J., U.S.A., 1951.

2

Shock Wave Reflections in Pseudo-Steady Flows

List of Symbols

Latin Letters

a_i - local speed of sound in state (i)
A_{ij} = a_i/a_j
C_p - specific heat capacity at constant pressure
C_v - specific heat capacity at constant volume
h_i - enthalpy in state (i)
k - thermal conductivity
l_d - dissociational relaxation length
l_v - vibrational relaxation length
L = L_r/L_T
L_K - horizontal distance of the kink from the leading edge of the reflecting wedge
L_r - length of the reflected shock wave in TMR and DMR
L_T - horizontal distance of the triple point from the leading edge of the reflecting wedge
M_i^j - flow Mach number in state (i) with respect to point (j)
p_i - static pressure in state (i)
Pr - Prandtl number
P_{ij} = p_i/p_j

t - time
T_i - flow temperature in state (i)
u_i - flow velocity in state (i) with respect to R in RR or T in MR
u'_i - flow velocity in state (i) with respect to K in TMR or T' in DMR
u'_{ix} - x component of u'_i
u'_{iy} - y component of u'_i
V_a^b - velocity of point "a" with respect to point "b"
V_{ax}^b - x component of V_a^b
V_{ay}^b - y component of V_a^b
V_{ij} = V_i/a_j
V_i - flow velocity in state (i) in a laboratory frame of reference
V_s - incident shock wave velocity
x - coordinate
x_{char} - characteristic length
y - coordinate

Greek Letters

γ - specific heat capacities ratio
δ - kinematic boundary layer thickness
δ_T - thermal boundary layer thickness
δ^* - boundary layer displacement thickness
Δt - time interval
ε - roughness
ζ - boundary layer displaced angle
η = $\theta_w^{tr}(\varepsilon)/\theta_w^{tr}(0)$
θ_i - deflection angle of the flow while passing across an oblique shock wave into state (i) with respect to R in RR or T in MR
θ'_i - deflection angle of the flow while passing across an oblique shock wave into state (i) with respect to K in TMR or T' in DMR
θ_w - reflecting wedge angle
θ_w^C - complementary wedge angle
θ_w^{tr} - transition wedge angle

$\theta_w^{tr}(\varepsilon)$ - transition wedge angle over a surface having a roughness ε
λ - mean free path
μ - dynamic viscosity
ξ - inverse pressure ratio across the incident shock wave ($= P_{01}$)
ρ_i - flow density in state (i)
ϕ_i - angle of incidence between the flow and the oblique shock wave across which the flow enters state (i) with respect to R in RR or T in MR
ϕ_i' - angle of incidence between the flow and the oblique shock wave across which the flow enters state (i) with respect to K in TMR or T' in DMR
χ - first triple point trajectory angle in MR
χ' - kink trajectory angle in TMR or second triple point trajectory angle in DMR
χ_g - first triple point trajectory angle in MR at glancing incidence
ω_{ij} - angle between the discontinuities i and j

Subscripts

0 - flow state ahead of the incident shock wave, i.
1 - flow state behind the incident shock wave, i.
2 - flow state behind the reflected shock wave, r.
3 - flow state behind the Mach stem, m.
4 - flow state behind the secondary Mach stem, m'.
5 - flow state behind the secondary reflected shock wave, r'.

Shock waves having constant velocities with respect to an inertial frame of reference can be investigated using steady flow concepts by attaching a frame of reference to the shock wave. In such a frame of reference, the shock wave is stationary, and the entire flow field is known as either pseudo-stationary or pseudo-steady. The transformation, known as a Galilean transformation, is shown schematically in figure 2.1. In figure 2.1a a constant velocity shock wave, having a velocity of V_s, is propagating from left to right, towards a flow having a velocity V_i, and inducing behind it a flow velocity V_j. The velocities with respect to a frame of reference

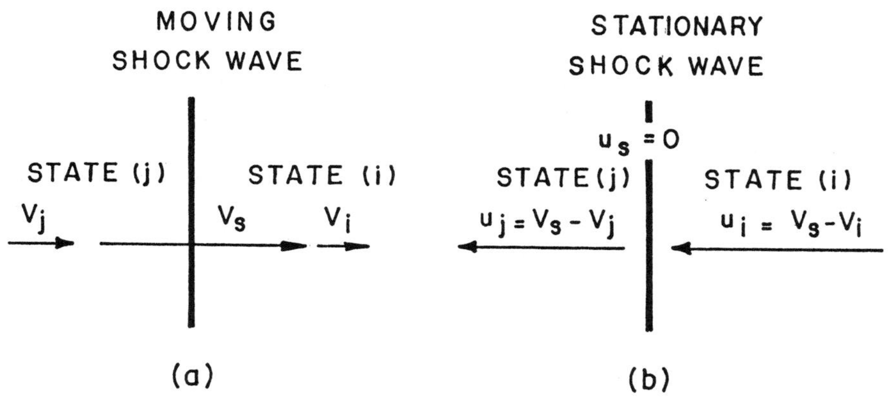

Figure 2.1 Schematic illustration of the Galilean transformation.

attached to the shock wave are shown in figure 2.1b. In this frame of reference the flow in state (i) propagates towards the stationary shock wave with the velocity $u_i = V_s - V_i$. Upon passing through the shock wave its velocity is reduced to $u_j = V_s - V_j$. The velocity field shown in figure 2.1b is actually obtained from the velocity field shown in figure 2.1a by superimposing on it a velocity equal to the shock wave velocity but in the opposite direction. The flow field of figure 2.1b is pseudo-steady, and hence can be treated using the steady flow theory for oblique shock waves.

2.1 Types of Reflections

When a planar moving incident shock wave encounters a sharp compressive straight corner in a shock tube, it is reflected by the wedge surface for the reasons discussed in section 1.2.2.

Depending upon the incident shock wave Mach number, M_s, and the reflecting wedge angle, θ_w, the three general types of reflection mentioned in section 1.1, i.e., RR, vNR and MR, can be obtained. As mentioned in section 1.1, the MR in pseudo-steady flows is always a direct-Mach reflection, DiMR, i.e., an MR in which the triple point, T, moves away from the reflecting surface. An intensive investigation at Princeton University under the supervision of Bleakney has led to the recognition that the DiMR in pseudo-steady flows can be further divided into three different types:

single-Mach reflection - SMR,
transitional-Mach reflection - TMR, and
double-Mach reflection - DMR.

The SMR was first observed by Mach (1878), the TMR was first reported by Smith (1945) and the DMR was discovered by White (1951). Furthermore, Ben-Dor (1981) has proved analytically that two different DMR-configurations are possible. In one of them the trajectory angle of the second triple point, χ', is larger than that of the first triple point, χ, and in the other one, $\chi' < \chi$. Lee & Glass (1983) verified this hypothesis, experimentally, and termed the former reflection as a positive double-Mach reflection, DMR^+, and the latter as a negative double-Mach reflection, DMR^-. Their experimental investigation further revealed that there are cases in which the second triple point of the double-Mach reflection can be located on the reflecting wedge surface, i.e., $\chi' = 0$. They termed this reflection as a terminal double-Mach reflection, TDMR.

Experimental observations of the wave configurations of RR, SMR, TMR and DMR in shock-tube experiments indicated that any point on these wave configurations, having a radius vector, $\bar{r}$, with the leading edge of the reflecting wedge as the origin, was transformed at a later time to a new point located at $C\bar{r}$, where C is a scalar constant. This observation means that instead of three independent variables, x, y and t, the reflection phenomenon is describable in terms of x/t and y/t, i.e., x and y may be measured relative to any point moving with constant velocity with respect to the reflecting wedge. In other words, the flow is self-similar and as such it can be treated as pseudo-steady. Further considerations regarding this issue can be found in Jones, Martin & Thornhill (1951). Some doubts regarding the accuracy of these observations have been raised in recent years. They will be further discussed in subsequent sections.

2.1.1 Regular Reflection - RR

The wave configuration and a holographic interferogram of a typical RR as obtained in shock tube experiments are shown in figures 2.2 and 2.3, respectively. The incident shock wave, i, propagates up the reflecting wedge surface, towards the quiescent gas in state (0). It touches the reflecting wedge surface at the reflection point, R. The reflected shock wave, r, unlike the incident shock wave, is straight only near the reflection point. It emanates from the reflection point and terminates perpendicularly on the shock tube bottom wall. The entire reflection grows linearly with time. The information that the incident shock wave collided with a

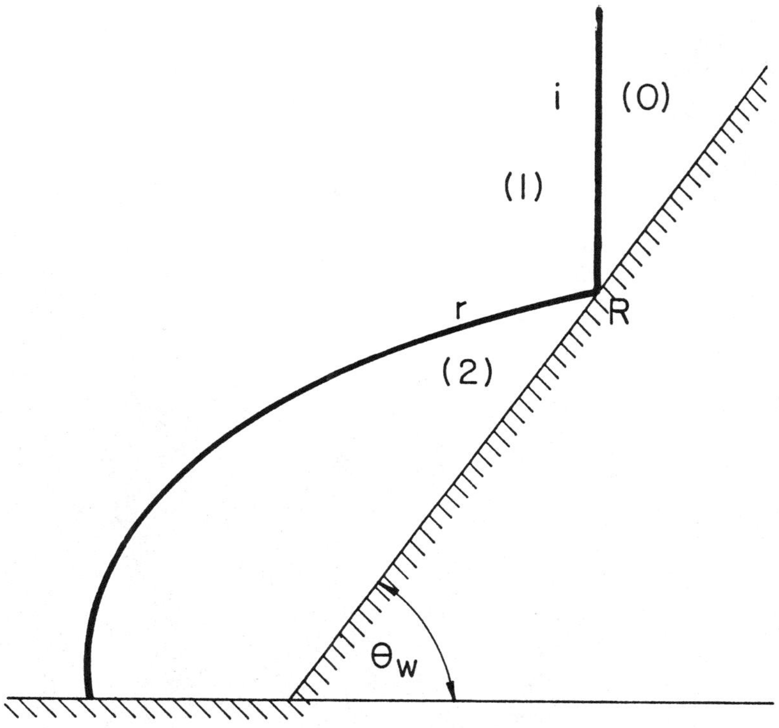

Figure 2.2 The wave configuration of a pseudo-steady RR.

compressive corner is available only to the flow field bounded by the reflected shock wave and the wedge surface.

If a frame of reference is attached to the reflection point, R, the RR becomes pseudo-steady as shown in figure 2.4. In this frame of reference the incident, i, and reflected, r, shock waves are stationary. The flow in state (0) which moves parallel to the reflecting wedge surface encounters the incident shock wave, i, obliquely, with an angle of incidence $\phi_1 = 90^\circ - \theta_W$. Upon passing through the incident shock wave, the flow is deflected towards the wedge surface by an angle θ_1, and assumes a new thermodynamic state, state (1). The reflected shock wave, r, which has an angle of incidence ϕ_2 with respect to the flow in state (1), deflects the flow in an opposite direction by an angle θ_2 to become again parallel to the reflecting wedge surface. The flow behind the reflected shock wave assumes a new thermodynamic state, state (2). Thus, in the vicinity of the reflection point of an RR $\theta_2 = \theta_1$.

Figure 2.3 Holographic interferogram of a regular reflection - RR with M_s = 1.17 and $\theta_w = 50^o$ in air at T_0 = 287.6K and $p_0 \approx 760$ torr (courtesy of Professor K. Takayama, Shock Wave Research Center, Institute of Fluid Science, Tohoku University, Sendai, Japan).

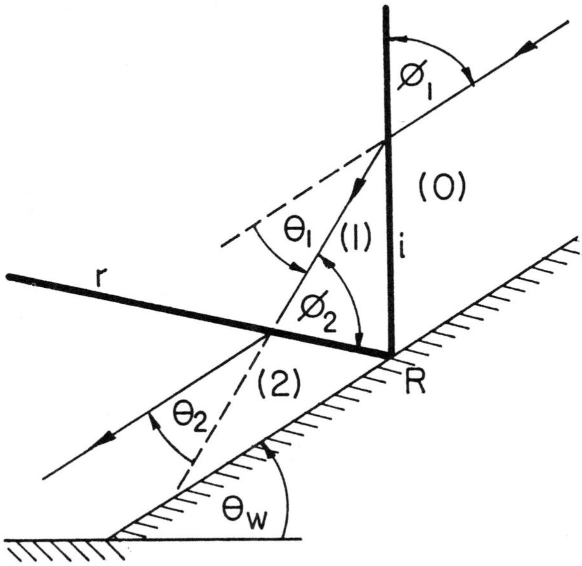

Figure 2.4 Definition of the flow parameters of a pseudo-steady RR.

Since, as mentioned in section 1.5.5, the RR⇔IR transition criterion in pseudo-steady flows is most likely the sonic criterion, the flow behind the reflection point, R, is always supersonic with respect to R, i.e., $M_2^R > 1$. The RR terminates when the flow in state (2) becomes sonic with respect to R, i.e., $M_2^R = 1$. This fact implies that the reflected shock wave is always straight in the vicinity of the reflection point, R, and hence one of the basic assumptions upon which the two-shock theory is based, namely, uniform flow regions in the vicinity of the reflection point, is satisfied.

It should be mentioned, however, that there are photographs of actual RR-configurations in which the reflected shock wave seems to be curved along its entire length. Although it is impossible, due to the lack of sufficient resolution capabilities, to categorically state whether the reflected shock wave indeed has a finite curvature at the reflection point, it is important to note that if RR-configurations with curved reflected shock waves do exist, then the sonic criterion is violated, and it should probably be replaced by the detachment criterion.

The governing equations for the flow field near the reflection point, R, of an RR are those arising from the two-shock theory, i.e., equations (1.5) to (1.13).

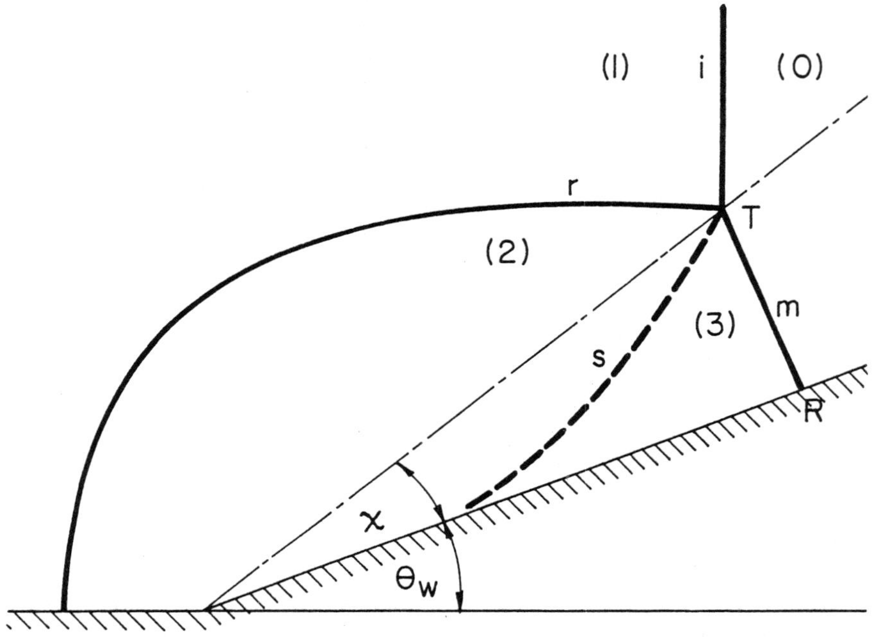

Figure 2.5 The wave configuration of a pseudo-steady SMR.

2.1.2 Single-Mach Reflection - SMR

The wave configuration and a holographic interferogram of a typical SMR as obtained in shock tube experiments are shown in figures 2.5 and 2.6, respectively. A comparison of the reflecting wedge angles in the photographs shown in figures 2.3 and 2.6 indicates that regular reflections are obtained, in general, at large reflecting wedge angles while Mach reflections are obtained, in general, at small reflecting wedge angles. The four discontinuities typical of an MR meet at the triple point, T, which moves along a straight line originating at the leading edge of the reflecting wedge and forming an angle χ with the reflecting surface. (Some remarks about the validity of this statement are given subsequently.) Unlike the incident shock wave, i, which is straight, the reflected shock, r, the Mach stem, m, and the slipstream, s, are curved. The information that the incident shock wave collided with a compressive corner is available only to the flow field bounded by the reflected shock wave, the Mach stem and the reflecting wedge surface.

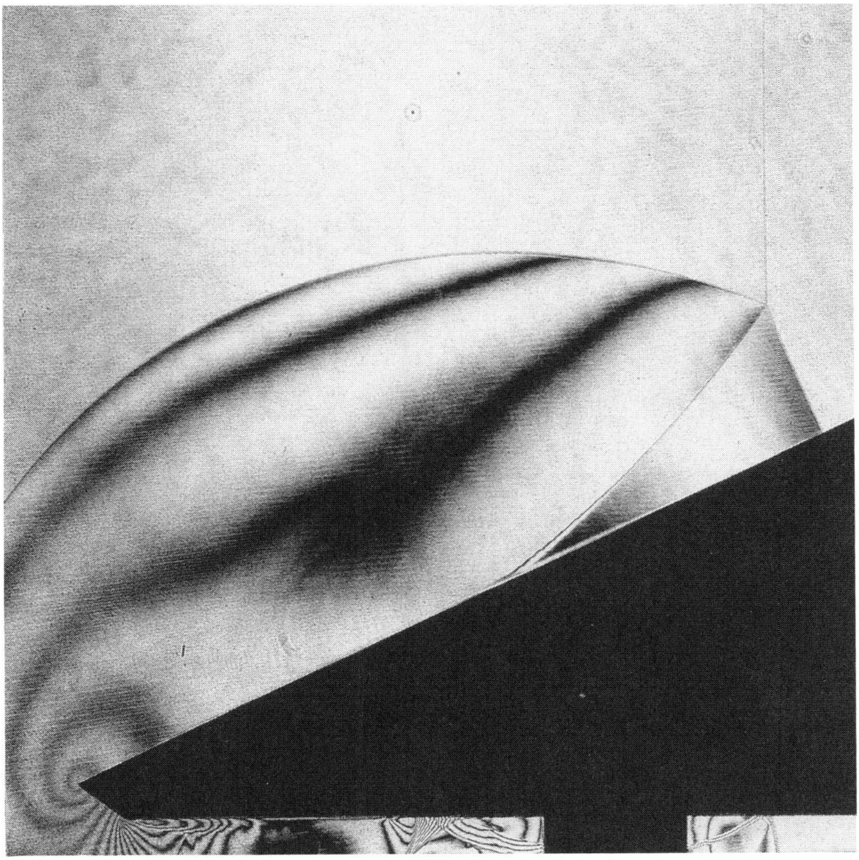

Figure 2.6 Holographic interferogram of a single-Mach reflection - SMR with $M_s = 1.70$ and $\theta_w = 25^o$ in air at $T_0 = 289.6K$ and $p_0 \approx 760$ torr (courtesy of Professor K. Takayama, Shock Wave Research Center, Institute of Fluid Science, Tohoku University, Sendai, Japan).

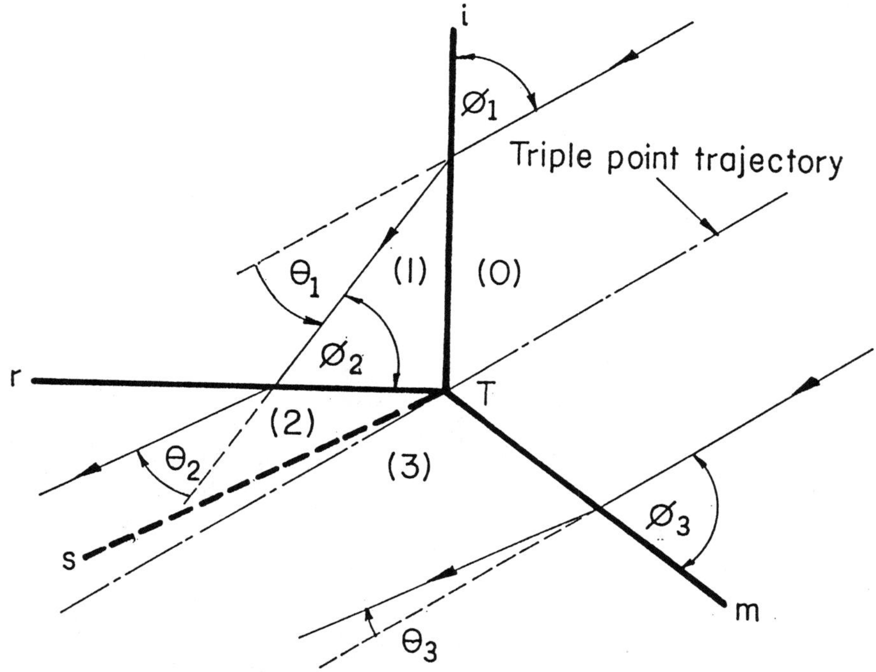

Figure 2.7 Definition of the flow parameter of a pseudo-steady SMR.

If a frame of reference is attached to the triple point, T, the SMR becomes pseudo-steady as shown in figure 2.7. In this frame of reference the four discontinuities, i.e., i, r, m and s are stationary. The flow in state (0), which was originally above the triple point trajectory, encounters the incident shock wave, i, obliquely, with an angle of incidence $\phi_1 = 90^\circ - (\theta_w + \chi)$. Upon passing through the incident shock wave, the flow is deflected towards the wedge by an angle θ_1, and assumes a new thermodynamic state, state (1). The reflected shock wave, r, which has an angle of incidence ϕ_2 with respect to the flow in state (1), deflects the flow in an opposite direction by an angle θ_2 to become parallel to the slipstream, s. The flow in state (0), which was originally below the triple point trajectory, encounters the Mach stem, m, obliquely, with an angle of incidence ϕ_3. Upon passing through the Mach stem it is deflected by an angle θ_3 to also become parallel to the slipstream. Thus, in the vicinity of the triple point of an MR $\theta_1 - \theta_2 = \theta_3$.

As will be shown subsequently, an SMR can exist as long as the flow behind the reflected shock wave is subsonic, in a frame of reference attached to the triple point, T, i.e., $M_2^T < 1$. For this reason only the incident shock wave, i, is straight in the vicinity of the triple point of an SMR. Therefore, one of the basic assumptions upon which the three-shock theory is based, namely, uniform flow regions in the vicinity of the triple point, is inherently violated. This fact might account for disagreements between experiment and predictions based on the three-shock theory regarding the SMR. The condition $M_2^T < 1$ is met when the R-polar intersects the I-polar along its strong portion, as shown, for example, in figure 1.15a. This can occur only in cases where the R-polar becomes tangent to the p-axis inside the I-polar. If, however, the R-polar intersects the I-polar along its weak portion, e.g., figure 1.18c, then $M_2^T > 1$, and an SMR is impossible.

The governing equations for the flow field in the vicinity of the triple point, T, of an SMR are those given by the three-shock theory, i.e., equations (1.14) to (1.27).

2.1.3 Transitional-Mach Reflection - TMR

As mentioned in the previous section, when the flow Mach number behind the reflected shock wave is supersonic with respect to the triple point, T, the reflection cannot be an SMR. Instead, the reflection is either a TMR or a DMR.

The wave configuration and a holographic interferogram of a typical TMR are shown in figures 2.8 and 2.9, respectively. In general, the wave configuration of a TMR is similar to that of an SMR (compare figures 2.5 and 2.8). The only difference is that unlike the case of an SMR where the reflected shock wave, r, is curved along its entire length, in a TMR, the reflected shock wave emanating from the triple point has a straight portion. The straight portion of the reflected shock wave extends from the triple point, T, to a point known as the kink, K. Beyond the kink the reflected shock wave (m' in figure 2.8) is curved along its entire length until it terminates perpendicularly on the shock tube bottom wall. The kink moves along a straight line originating from the leading edge of the reflecting wedge and forming an angle of χ' with the wedge surface.

Henderson & Lozzi (1975) hypothesized that the kink should be associated with a compression wave emanating from it and

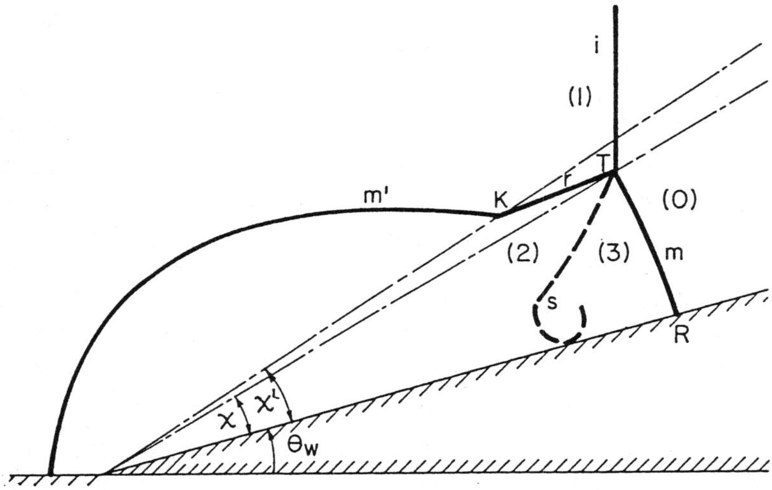

Figure 2.8 The wave configuration of a pseudo-steady TMR.

terminating on the slipstream. The existence of this compression wave was verified experimentally by Ben-Dor & Glass (1978) who evaluated finite fringe interferograms using a computerized technique [for details see Ben-Dor & Whitten (1979)] in order to obtain density contours in the flow field. This compression wave is probably responsible for the curled structure of the slipstream as shown in figure 2.8. Note that an SMR with a curled slipstream has never been observed, hence it is quite possible that, when the compression wave is strong enough, it imposes high enough pressures to force the slipstream to curl forward, as shown in figure 2.10a.

As the incident shock wave propagates up the wedge, the entire wave configuration grows linearly with time, hence the distance between points T and K also increases linearly with time. Thus in a frame of reference attached to the triple point, T, the kink, K, moves away from T and hence the flow Mach number in state (2) is smaller with respect to K than with respect to T. As mentioned earlier the condition for the existence of a TMR is $M_2^T > 1$. As will be shown subsequently, TMR exists as long as $M_2^K < 1$. When M_2^K becomes greater than unity, the TMR terminates and gives rise to a DMR.

The governing equations for the flow field in the vicinity of the triple point, T, of a TMR are those given by the three-shock theory, i.e., equations (1.14) to (1.27). Since the three-shock theory assumes straight shock waves, it is expected to predict the

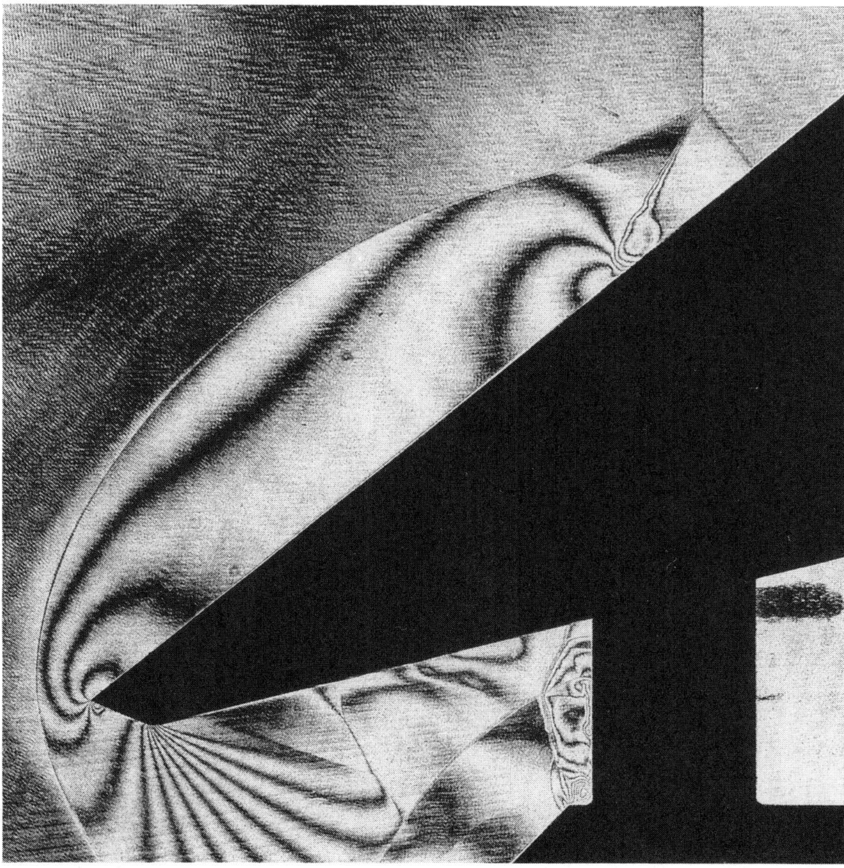

Figure 2.9 Holographic interferogram of a transitional-Mach reflection - TMR with M_s = 2.96 and $\theta_w = 40^o$ in air at T_0 = 290K and $p_0 \approx 760$ torr (courtesy of Professor K. Takayama, Shock Wave Research Center, Institute of Fluid Science, Tohoku University, Sendai, Japan).

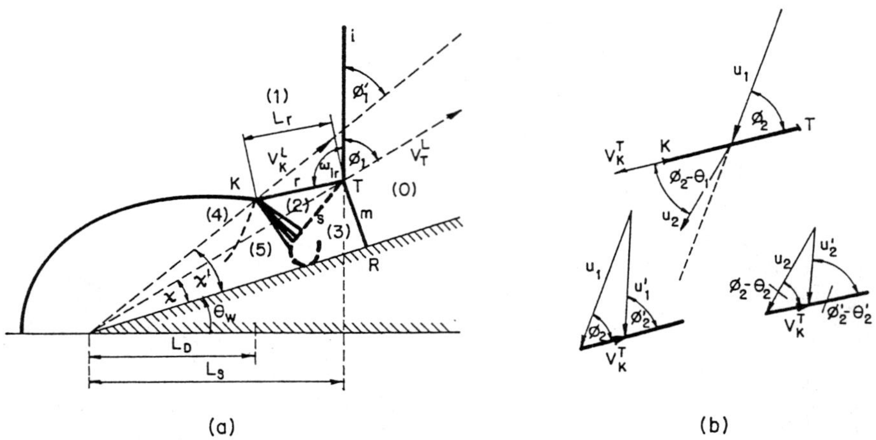

Figure 2.10 Definition of the flow parameters of a pseudo-steady TMR.

flow field in the vicinity of the triple point of a TMR better than in the vicinity of the triple point of an SMR, since in the case of a TMR the reflected shock wave is straight at T while in the case of an SMR it is curved there.

2.1.3.1 Analytical Formulation of a TMR

The solution of the flow field in the vicinity of the kink of a TMR must be done in a frame of reference in which the kink is stationary. Consequently, the relative motion of the flow in state (1) ahead of the reflected shock wave with respect to the kink must be known. This requirement is similar to the need of knowing the relative motion of the flow in state (0) ahead of the incident shock wave with respect to the triple point, if one wishes to solve the flow field in the vicinity of the triple point.

Consider figure 2.10a in which the triple point, T, moves with respect to a laboratory frame of reference with velocity V_T^L along a line inclined by an angle χ to the reflecting wedge surface, and the kink, K, moves with respect to the laboratory frame of reference with velocity V_K^L along a line inclined by an angle χ' to the reflecting wedge surface. Since the reflection is self-similar, the angles between the shock waves do not change as the TMR propagates along the reflecting surface, and hence, the kink, K, moves with respect to the triple point, T, along a line which coincides with the straight portion of the reflected shock wave, r. Simple vector analysis yields

$$V_T^L = V_S \operatorname{cosec} \phi_1 \tag{2.1}$$

Recall that $V_T^L = u_0$ by definition. Let us define by Δt the time from the moment the incident shock wave passed the leading edge of the reflecting wedge. Thus

$$\Delta t = \frac{L_T}{V_S} \tag{2.2}$$

where L_T is the distance travelled by the incident shock wave during the time interval Δt. Using this definition, the velocity of K with respect to T, i.e., V_K^T, is

$$V_K^T = \frac{L_r}{\Delta t} = \frac{L_r}{L_T} V_S = L V_S \tag{2.3}$$

where L_r is the length of the straight portion of the reflected shock wave, i.e., the distance from T to K, and L is defined as L_r/L_T.

By assuming that the Mach stem is straight and perpendicular to the reflecting wedge surface, and that the horizontal component of the kink velocity, in a laboratory frame of reference, is equal to the incident shock induced velocity, Law & Glass (1971) obtained

$$\frac{L_K}{L_S} = \left(1 - \frac{\rho_0}{\rho_1}\right) \tag{2.4}$$

where L_K is the horizontal distance of the kink from the leading edge of the reflecting wedge. It will be shown subsequently, that the above assumption regarding the horizontal velocity component of the kink, namely, the "Law-Glass assumption", was found to agree fairly well with experimental results. Expression (2.4) was proven by Bazhenova, Fokeev & Gvozdeva (1976) to be very good in the range $\theta_w < 40°$ and fairly good elsewhere.

Using equation (2.4) one can obtain

$$L = \frac{\rho_0}{\rho_1} \operatorname{cosec} (\phi_1 + \phi_2 - \theta_1) \tag{2.5}$$

Inserting equation (2.5) into equation (2.3) yields

$$V_K^T = \frac{\rho_0}{\rho_1} V_s \operatorname{cosec} (\phi_1 + \phi_2 - \theta_1) \tag{2.6}$$

Consider figure 2.10b, where the changes in the velocities as one moves from a frame of reference attached to the triple point, T, to a frame of reference attached to the kink, K, are schematically drawn to scale. If the direction of the reflected shock wave is x and a direction perpendicular to it is y, then

$$u'_{1x} = u_1 \cos \phi_2 - V_K^T$$

and

$$u'_{1y} = u_1 \sin \phi_2$$

where u_1 is the flow velocity in state (1) with respect to the triple point, and u'_{1x} and u'_{1y} are the x and y components of the flow velocity in state (1) with respect to the kink, i.e., the x and y components of u'_1 , where u'_1 is the flow velocity in state (1) with respect to the kink. The value of u'_1 then becomes

$$u'_1 = \left[u_1^2 + \left(V_K^T\right)^2 - 2u_1 V_K^T \cos\phi_2 \right]^{1/2} \tag{2.7}$$

The value of ϕ'_2 is

$$\phi'_2 = \tan^{-1}\left(\frac{u_1 \sin\phi_2}{u_1 \cos\phi_2 - V_K^T} \right) \tag{2.8}$$

Inserting the value of V_K^T from equation (2.6) into equations (2.7) and (2.8) finally results in

$$u'_1 = u_1 \left[1 + \frac{\sin^2 (\phi_1 - \theta_1)}{\sin^2 (\phi_2 + \phi_1 - \theta_1)} - 2 \frac{\sin (\phi_1 - \theta_1) \cos \phi_2}{\sin (\phi_2 + \phi_1 - \theta_2)} \right]^{1/2} \tag{2.9}$$

and

$$\phi_2' = \phi_2 + \phi_1 - \theta_1 \qquad (2.10)$$

where ϕ_2, ϕ_1 and θ_1 are defined in figure 2.7. The prime denotes that the properties are measured with respect to the kink, K. Note that the thermodynamic properties do not depend upon the frame of reference in which they are calculated, and hence in the following the prime will be used only to denote dynamic properties. The Mach number of the flow in state (1) with respect to the kink, M_1^K is

$$M_1^K = \frac{u_1'}{a_1} \qquad (2.11)$$

Once the flow Mach number ahead of the reflected shock wave M_1^K and the angle of incidence between the oncoming flow and the shock wave ϕ_1', in a frame of reference attached to the kink, are known, the flow Mach number in state (2) behind the reflected shock wave, M_2^K, can be calculated, using the oblique shock wave relations.

Figure 2.10a indicates that the flow in state (2), which was obtained from the flow in state (1) which was originally below the kink trajectory, passes through a compression wave, centered at the kink, across which the process is isentropic, and assumes a new thermodynamic state, state (5). At the same time, the flow in state (1) which originally was above the kink trajectory, passes through the curved part of the reflected shock wave and assumes a new thermodynamic state, state (4). The flows in states (4) and (5) have different entropies since they were both obtained from the same state, state (1), by passing through different wave systems, and reaching the same pressures, i.e., $p_4 = p_5$. Therefore states (4) and (5) must be separated by a slipstream, across which the flows are parallel and have the same pressure. Although a slipstream has never been observed experimentally, gas dynamic considerations require its existence. The fact that a slipstream has never been observed experimentally might suggest that the difference in the densities (as well as the other flow properties) between the two sides of the slipstream is too small to be resolved experimentally, i.e., $\rho_4 \approx \rho_5$. Note that since the reflection under consideration is a TMR, $M_2^K < 1$, and hence the Prandtl-Meyer function cannot be used to calculate the flow passing through the compression wave. Instead, a

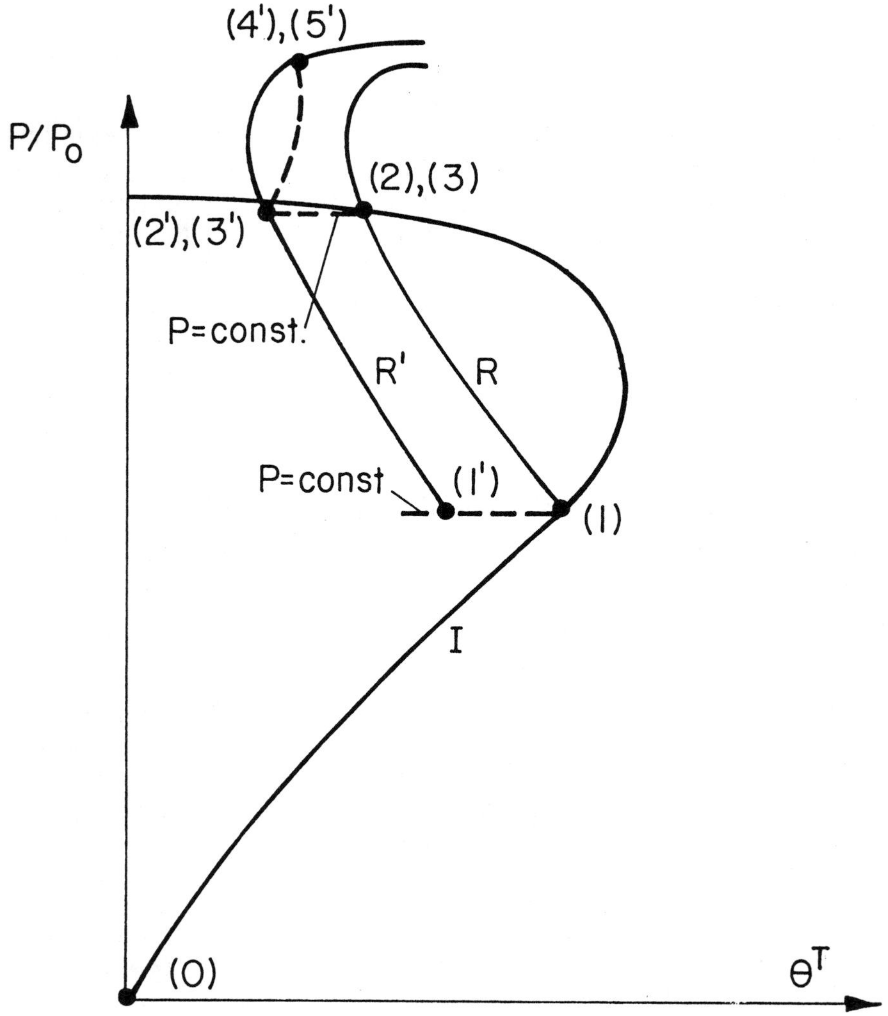

Figure 2.11 (p,θ)-solution of a TMR.

subsonic compression wave should be used to calculate the changes in the flow properties as it passes from state (2) to state (5).

2.1.3.2 Shock Polar Presentation of a TMR

Since states (4) and (5) of the TMR shown in figure 2.10a are separated by a slipstream, across which the pressures and flow directions are equal, they are mapped in the (p, θ)-plane at a single point.

Figure 2.11 presents the shock polar solution of a TMR. States (0), (1), (2) and (3) are obtained in the same way as for an MR (see section 1.4.2), which for the pseudo-steady case is identical to the shock polar solution of an SMR.

When moving over to a frame of reference attached to the kink, K, state (1) should be shifted along a constant pressure line (since $p_1 = p_1'$) to account for the fact that $\theta_1' \neq \theta_1$. The angle of incidence between the flow in state (0) and the incident shock wave, ϕ_1', in a frame of reference attached to the kink, K, can be obtained from

$$\phi_1' = \tan^{-1}\frac{V_{Kx}^L}{V_{Ky}^L} \tag{2.12}$$

where V_{Kx}^L and V_{Ky}^L are the x and y components of the kink velocity in a laboratory frame of reference. These velocity components can be calculated from

$$\begin{aligned} V_{Kx}^L &= V_K^L \sin\phi_1 - V_K^T \sin\omega_{ir} \\ V_{Ky}^L &= V_K^L \cos\phi_1 - V_K^T \cos\omega_{ir} \end{aligned} \tag{2.13}$$

where ω_{ir}, the angle between the incident and reflected shock waves, can be expressed as

$$\omega_{ir} = 180^\circ - (\phi_2 + \phi_1 - \theta_1) \tag{2.14}$$

Inserting equations (2.13) and (2.14) into equation (2.12) finally results in

$$\phi_1' = \tan^{-1}\left[\frac{1 - \rho_0/\rho_1}{\cot\phi_1 - \rho_0/\rho_1 \cot(\phi_2 + \phi_1 - \theta_1)}\right] \tag{2.15}$$

and the oncoming flow Mach number, M_0^K, with respect to the incident shock wave in a frame of reference attached to the kink is

$$M_0^K = \frac{M_S}{\sin\phi_1'} \tag{2.16}$$

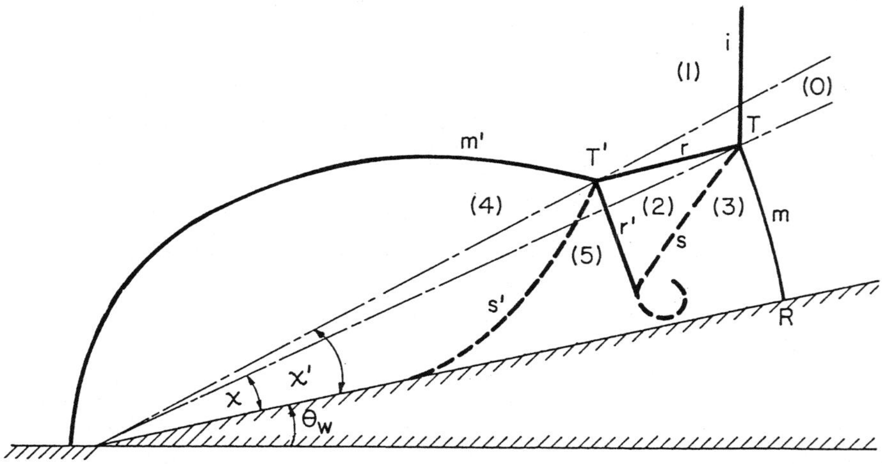

Figure 2.12 The wave configuration of a pseudo-steady DMR.

With the above calculated values of M_0^K and ϕ_1', the oblique shock wave equation can be used to calculate the flow deflection θ_1' in a frame of reference attached to the kink. Consequently, state (1') is shifted in the (p,θ)-plane by $\theta_1 - \theta_1'$ along the constant pressure line as shown in figure 2.11.

The R-polar with respect to the kink, i.e., R', originates from point (1') and is drawn for the value of M_1^K as calculated from equation (2.11).

Once the R'-polar is drawn, the location of states (2') and (3') is found simply by intersecting a constant pressure line, originating from state (2) and (3), with the R'-polar. Then the (p, θ)-relation appropriate to the centered compression wave, shown in figure 2.10a, is drawn (see dashed line in figure 2.11). The intersection of this dashed line with the R'-polar results in the flow states (4') and (5') on both sides of the slipstream emanating from the kink.

2.1.4 Double - Mach Reflection - DMR

As mentioned earlier, a TMR exists as long as the flow behind the reflected shock wave is supersonic with respect to the triple point, and subsonic with respect to the kink. When the flow Mach number behind the reflected shock wave becomes supersonic with respect to the kink, the TMR terminates and a DMR-configuration is formed.

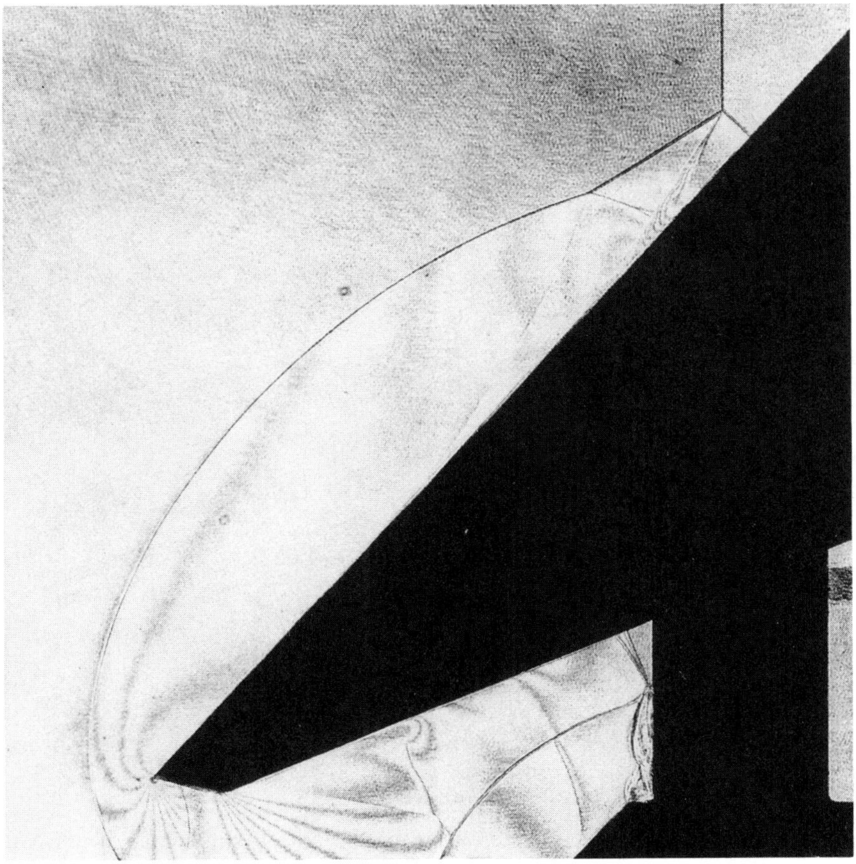

Figure 2.13 Holographic interferogram of a double-Mach reflection - DMR with $M_s = 2.74$ and $\theta_w = 47^o$ in air at $T_0 = 290K$ and $p_0 \approx 760$ torr (courtesy of Professor K. Takayama, Shock Wave Research Center, Institute of Fluid Science, Tohoku University, Sendai, Japan).

Figure 2.14 An enlarged interferogram of the curled slipstream of a DMR with M_s = 4.4 and $\theta_w = 30^o$ in CO_2 at T_0 = 294.3K and p_0 = 15 torr (courtesy of Professor I.I. Glass, Institute for Aerospace Studies, University of Toronto, Toronto, Ontario, Canada).

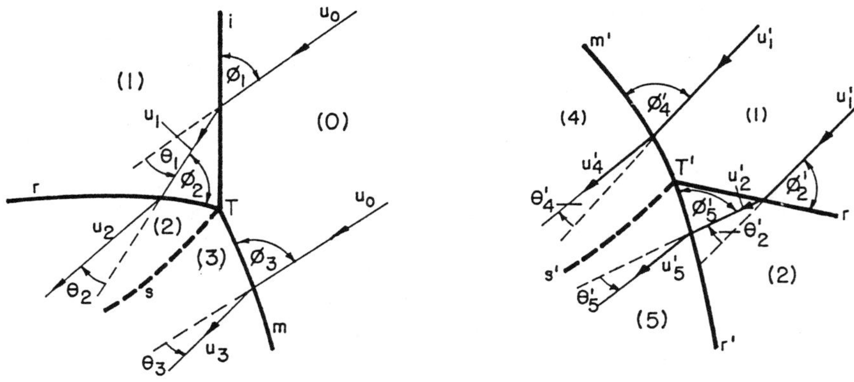

Figure 2.15 Definition of the flow parameters of a pseudo-steady DMR: a) the first triple point; b) the second triple point.

The wave configuration and a holographic interferogram of a typical DMR are shown in figures 2.12 and 2.13, respectively. The kink, K, of the TMR is replaced by a second triple point, T', and the centered compression wave is replaced by a shock wave, r'. The second triple point is complemented by a second slipstream, s'. The additional shock wave, r', emanates from the second triple point, T', and interacts with the first slipstream, s. The pressure jump across this shock wave forces the first slipstream to curl forward in a way similar to that of a TMR. As shown by Mirels (1985), the highest pressure in the flow field is obtained immediately behind the curled slipstream. An enlarged interferogram of the curled slipstream of a DMR is shown in figure 2.14. Note that the r' shock wave goes around the first slipstream, s, and breaks down to a compression wave. The nature of the interaction of the r' shock wave with the curled slipstream is yet to be investigated and understood.

The reflected shock wave, r, is straight between the two triple points, T and T'. Beyond the second triple point it is curved along its entire length until it terminates perpendicularly on the shock tube bottom wall. The curved part of the reflected shock wave is labeled, m', for it serves in the second triple point as the Mach stem, m, in the first triple point.

2.1.4.1 Analytical Formulation of a DMR

The solution of the flow field in the vicinity of the second triple point, T', of a DMR must be done in a frame of reference in which the second triple point is stationary. Since the second triple point of a DMR is identical in its dynamic behavior to the kink of a TMR, the relations given in sections 2.1.3.1 and 2.1.3.2 are valid for the case of a DMR. The governing equations of the flow field in the vicinity of the first triple point are identical to those given by the three-shock theory, [equations (1.14) to (1.27)].

Since the two triple points, T and T', are similar in their physical nature, the governing equations for the flow field in the vicinity of the second triple point can easily be obtained using the following analogy. Consider figures 2.15a and 2.15b which define the parameters of the flow fields in the vicinities of T and T', respectively. Let the notation b $\leftrightarrow$ a mean that symbol "b" in figure 2.15b is equivalent (analogous) to symbol "a" in figure 2.15a. Following this notation one can write:

state (1) $\leftrightarrow$ state (0)
state (2) $\leftrightarrow$ state (1)
state (5) $\leftrightarrow$ state (2)
state (4) $\leftrightarrow$ state (3)

$r \leftrightarrow i$
$r' \leftrightarrow r$
$m' \leftrightarrow m$
$s' \leftrightarrow s$

$u_1' \leftrightarrow u_0$
$u_2' \leftrightarrow u_1$
$u_5' \leftrightarrow u_2$
$u_4' \leftrightarrow u_3$

$\theta_2' \leftrightarrow \theta_1$
$\theta_5' \leftrightarrow \theta_2$
$\theta_4' \leftrightarrow \theta_3$

and finally:

$\phi_2' \leftrightarrow \phi_1$
$\phi_5' \leftrightarrow \phi_2$
$\phi_4' \leftrightarrow \phi_3$

Applying this analogy to equations (1.14) to (1.27) results in the governing equations describing the flow field in the vicinity of the second triple point, i.e.,

for r :

$$\rho_1 u_1' \sin \phi_2' = \rho_2 u_2' \sin (\phi_2' - \theta_2') \tag{2.17}$$

$$p_1 + \rho_1 (u_1')^2 \sin^2 \phi_2' = p_2 + \rho_2 (u_2')^2 \sin^2 (\phi_2' - \theta_2') \tag{2.18}$$

$$\rho_1 \tan \phi_2' = \rho_2 \tan (\phi_2' - \theta_2') \tag{2.19}$$

$$h_1 + \frac{1}{2}(u_1')^2 \sin^2 \phi_2' = h_2 + \frac{1}{2}(u_2')^2 \sin^2 (\phi_2' - \theta_2') \tag{2.20}$$

for r':

$$\rho_2 u_2' \sin \phi_5' = \rho_5 u_5' \sin (\phi_5' - \theta_5') \tag{2.21}$$

$$p_2 + \rho_2 (u_2')^2 \sin^2 \phi_5' = p_5 + \rho_5 (u_5)^2 \sin^2 (\phi_5' - \theta_5') \tag{2.22}$$

$$\rho_2 \tan \phi_5' = \rho_5 \tan (\phi_5' - \theta_5') \tag{2.23}$$

$$h_2 + \frac{1}{2}(u_2')^2 \sin^2 \phi_5' = h_5 + \frac{1}{2}(u_5')^2 \sin^2 (\phi_5' - \theta_5') \tag{2.24}$$

for m':

$$\rho_1 u_1' \sin \phi_4' = \rho_4 u_4' \sin (\phi_4' - \theta_4') \tag{2.25}$$

$$p_1 + \rho_1 (u_1')^2 \sin^2 \phi_4' = p_4 + \rho_4 (u_4')^2 \sin^2 (\phi_4' - \theta_4') \tag{2.26}$$

$$\rho_1 \tan \phi_4' = \rho_4 \tan (\phi_4' - \theta_4') \tag{2.27}$$

$$h_1 + \frac{1}{2}\left(u_1'\right)^2 \sin^2\phi_4' = h_4 + \frac{1}{2}\left(u_4'\right)^2 \sin^2\left(\phi_4' - \theta_4'\right) \tag{2.28}$$

and finally the boundary conditions across the slipstream, s':

$$p_5 = p_4 \tag{2.29}$$

$$\theta_2' - \theta_5' = \theta_4' \tag{2.30}$$

The prime denotes that the properties are measured with respect to the second triple point. However, it is omitted from the thermodynamic properties since they do not depend upon the frame of reference.

Since equations (2.17) to (2.30) are identical in their form to equations (1.14) to (1.27) they can be solved using the same procedure outlined in section 1.3.2. The procedure requires knowledge of p_1 ($\leftrightarrow p_0$), T_1 ($\leftrightarrow T_0$), u_1' ($\leftrightarrow u_0$) and ϕ_2' ($\leftrightarrow \phi_1$). The values of p_1 and T_1 are obtained from the solution of the flow field in the vicinity of the first triple point, and the values of u_1' and ϕ_2' can be calculated from equations (2.9) and (2.10), respectively.

Note that equations (2.17) to (2.20) and equations (2.25) to (2.30) are also valid for the flow field in the vicinity of the kink of a TMR. However, equations (2.21) to (2.24) should be replaced by the isentropic relations across a centered compression wave shown in figure 2.10a.

2.1.4.2 Shock Polar Presentation of a DMR

The shock polar presentation of a DMR is akin to that of a TMR which was described in section 2.1.3.2. The solution is again done in two steps, first in a frame of reference attached to the first triple point, and then in a frame of reference attached to the second triple point.

Figure 2.16 illustrates the shock polar presentation of a DMR. Once states (2') and (3') are found, in a way similar to that described for a TMR, the shock polar for the supersonic flow in state (2), i.e., $M_2^{T'}$, is drawn. This is shown as the R_1'-polar in figure 2.16. The intersection of the R_1'-polar with the R'-polar yields states (4) and (5), which lie on both sides of the second slipstream, s'. Note that $M_2^{T'}$ is calculated from

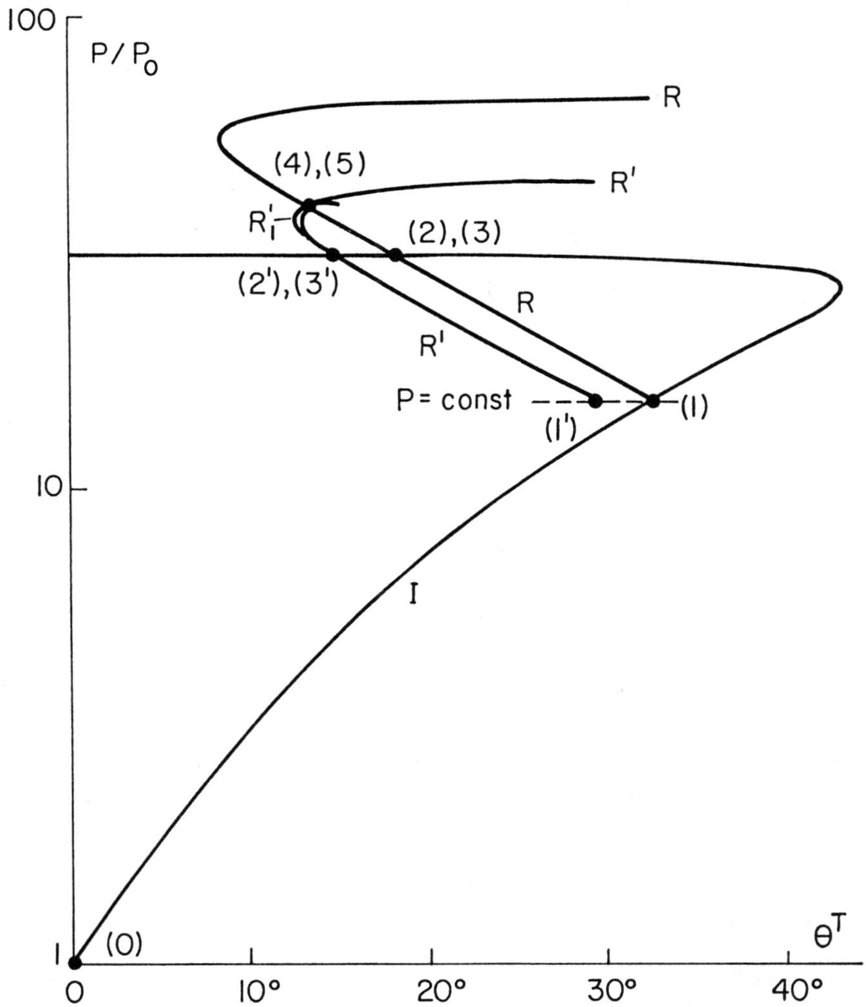

Figure 2.16 (p,θ)-solution of a DMR.

$$M_2^{T'} = \frac{u_2'}{a_2} \tag{2.31}$$

u_2' and a_2 are obtained from the solution of equations (1.14) to (1.27) and (2.17) to (2.30), respectively.

Note that states (2') and (3') are shifted from states (2) and (3) along a constant pressure line. Therefore, although states (2') and

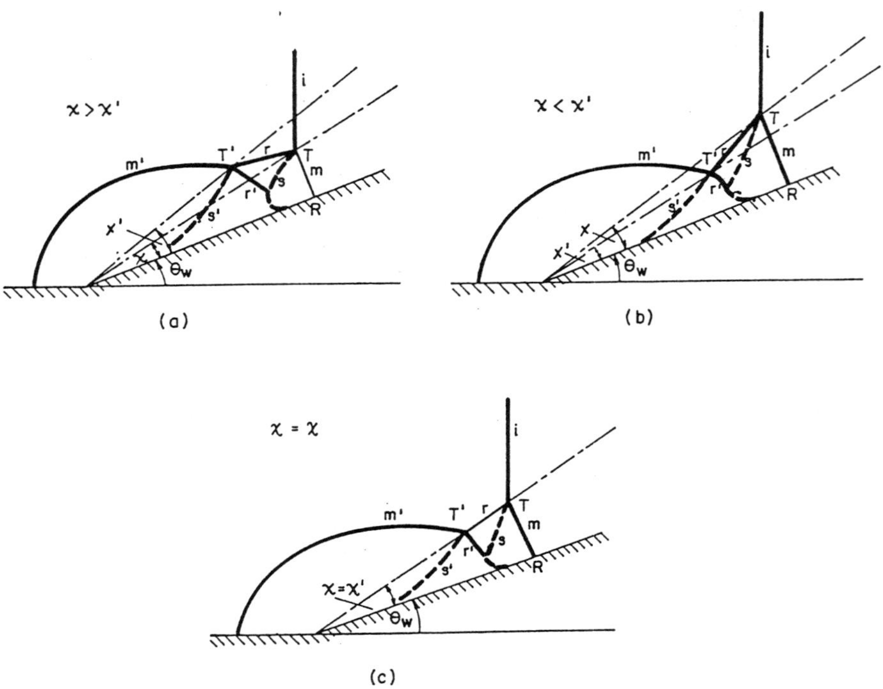

Figure 2.17 Three various wave configurations of possible pseudo-steady DMR: a) $\chi' > \chi$, i.e., DMR$^+$; b) $\chi' < \chi$, i.e., DMR$^-$; c) $\chi' = \chi$.

(3') seem to lie on the I-polar, they actually lie slightly below it as shown in figure 2.11.

2.1.5 Subtypes of DMR

Ben-Dor (1981) showed that the analytical solution of the governing equations of the first and second triple points [equations (1.14) to (1.27) and equations (2.17) to (2.30), respectively] together with the complementary equations given in section 2.2, which are needed to calculate the trajectory angles of the first and second triple points, result in situations in which the trajectory angle of the first triple point, χ, can be either larger or smaller than the trajectory angle of the second triple point, χ'.

This analytical prediction was later verified experimentally by Lee & Glass (1984). Their experimental investigation led to the recognition that the DMR can be further divided into subtypes of DMR.

a)

b)

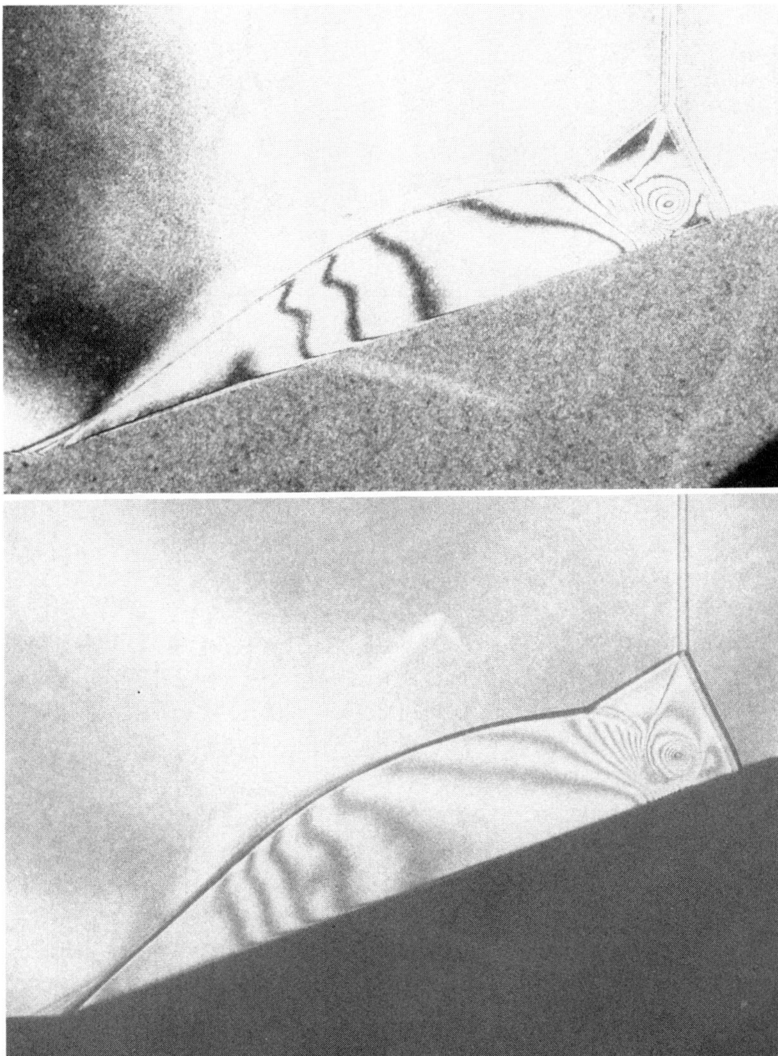

Figure 2.18 Interferograms of: a) a DMR in which $\chi' < \chi$ with $M_s = 3.76$ and $\theta_w = 20^o$ in SF_6 at $T_0 = 296.7K$ and $p_0 \approx 7.39$ torr; b) a DMR in which $\chi' = \chi$ with $M_s = 3.15$ and $\theta_w = 20^o$ in SF_6 at $T_0 = 297.1K$ and $p_0 = 107.9$ torr (courtesy of Professor I.I. Glass, Institute for Aerospace Studies, University of Toronto, Toronto, Ontario, Canada).

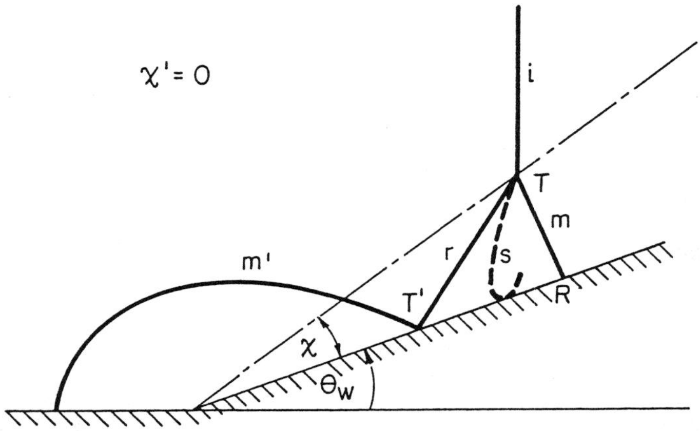

Figure 2.19 The wave configuration of a terminal double Mach reflection.

2.1.5.1 Positive Double-Mach Reflection - DMR+

A DMR in which $\chi' > \chi$ is shown in figure 2.17a. This DMR wave configuration was termed by Lee & Glass (1984) as positive double-Mach reflection - DMR+. An interferogram of a DMR+ is shown in figure 2.13.

2.1.5.2 Negative Double-Mach Reflection - DMR-

A DMR in which $\chi' < \chi$ is shown in figure 2.17b. This DMR wave configuration was termed by Lee & Glass (1984) as negative double-Mach reflection - DMR-. An interferogram of a DMR- is shown in figure 2.18a.

The limiting case between a DMR+ and a DMR-, i.e., a DMR wave configuration in which $\chi' = \chi$ is shown in figure 2.17c. An interferogram of this wave configuration is shown in figure 2.18b.

2.1.5.3 Terminal Double-Mach Reflection - TDMR

In their experimental study Lee & Glass (1984) discovered an additional DMR-configuration which is shown in figure 2.19. This wave configuration is a DMR- in which the second triple point, T', lies on the reflecting wedge surface and hence $\chi' = 0$, i.e., both the second reflected shock wave, r', and the second slipstream, s', typical to a DMR- wave configuration, disappeared, and the reflected shock wave, r, of the first triple point, T, reflects from the reflecting surface as a regular reflection. Lee & Glass (1984) termed this reflection as terminal double-Mach reflection - TDMR. A shadowgraph of a TDMR is shown in figure 2.20.

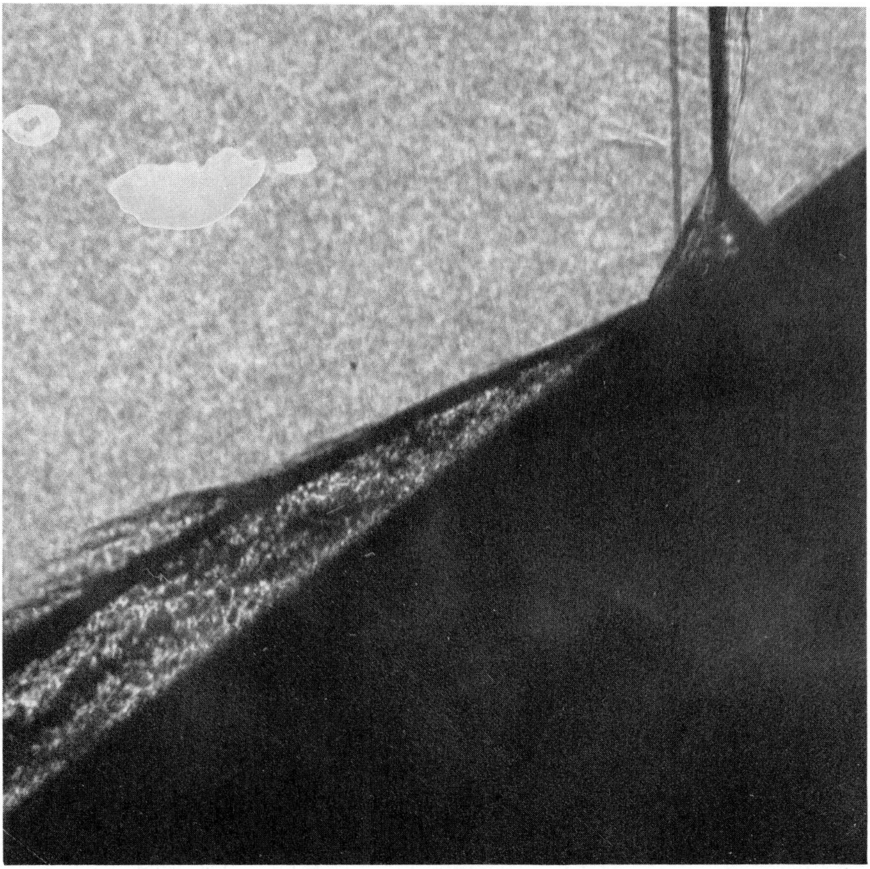

Figure 2.20 Shadowgraph of a terminal double Mach reflection - TDMR with M_s = 7.96 and $\theta_w = 37^o$ in SF_6 at T_0 = 296.5K and p_0 = 11.05 torr (courtesy of Professor I.I. Glass, Institute for Aerospace Studies, University of Toronto, Toronto, Ontario, Canada).

In additional experimental investigations Glass and his research group showed that TDMR can occur only in gases with very low specific heat capacities ratios - γ, such as Freon-12 or SF_6. TDMR cannot occur in monatomic, diatomic and triatomic gases when they behave as perfect gases. However, if these gases are forced into imperfect behavior (very high temperatures), then the occurrence of TDMR becomes possible. This should not be surprising since the excitation of real gas effects results in a reduction in the effective value of γ.

Some doubts regarding the possibility of the existence of a TDMR wave configuration were raised by investigators who argued that, due to viscous effects, the second triple point, T', cannot lie on the reflecting wedge surface. Unfortunately, however, the unclear photographs of TDMR-configurations (e.g., figure 2.20) which have been reported so far, only add to the uncertainty as to whether a TDMR is indeed possible.

2.1.6 von Neumann Reflection - vNR

The two- and three-shock theories which, as will be shown subsequently, are fairly good in predicting various aspects regarding the shock wave reflection phenomena for moderate and strong incident shock waves, fail completely for weak incident shock waves.

Figure 2.21 illustrates a (p, θ) presentation of such a case. The initial conditions for the case shown in figure 2.21 are $M_S = 1.2$, and $\phi_1 = 55^\circ$. As can be seen in figure 2.21 the R-polar does not intersect either the p-axis or the I-polar and hence, based on the two- or three-shock theories, neither an RR nor an MR are possible for this combination of M_S and ϕ_1. However, an actual experiment with these initial conditions results in a wave configuration which resembles a classical SMR. This fact, which has puzzled many investigators in the past four decades, was named the "von Neumann paradox" by Birkhoff (1950). The "paradox" was resolved only recently by Colella & Henderson (1990), who used a sophisticated numerical code by which the detailed structures of the discontinuities in the vicinity of the triple point could be resolved. Their numerical study established that the observed wave configuration was not an SMR at all, since the so-called "reflected shock wave" at the triple point was found not to be a shock wave, but a smoothly distributed self-similar band of compression waves of finite thickness, which evidently is too small to be resolved experimentally, but clear enough to be resolved numerically. As the compression waves retreat from the triple point they converge and

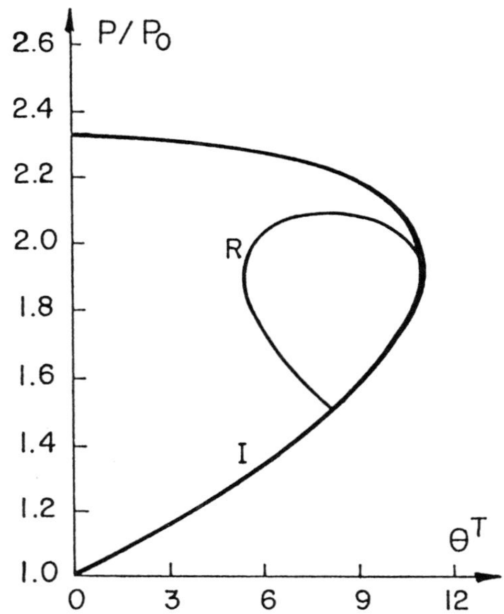

Figure 2.21 I-R shock polar combination in which the R-polar does not intercept either the p-axis or the I-polar.

steepen into a shock wave. The distance over which this happens is too small to be resolved experimentally. Thus, based on this recent finding, the fact that the three-shock theory failed to predict a reflection configuration in which there are only two shock waves, should not be surprizing. As a matter of fact, the three-shock theory was never meant to be applied under these circumstances. For this reason the "von Neumann paradox" is a misleading term. Furthermore, a successful prediction of this wave configuration using the three-shock theory would indeed be a "paradox". The above-mentioned wave configuration, which consists of an incident shock wave, i, a band of self-similar reflected compression waves, c, a Mach stem, m, and a slipstream, s, is shown schematically in figure 2.22. Colella & Henderson (1990) termed this wave configuration as a von Neumann reflection - vNR.

A comparison between the wave configurations of an SMR and a vNR reveals that, while in an SMR there is a slope discontinuity between the incident shock wave and the Mach stem, in a vNR the incident shock wave and the Mach stem appear to be a single wave with a smoothly turning tangent near the triple point. Furthermore, the slipstream has quite a sharp appearance in SMR-photographs, whereas it has a fuzzy appearance in vNR-photographs. As a matter of fact, the slipstream of a vNR reesembles a distributed shear layer

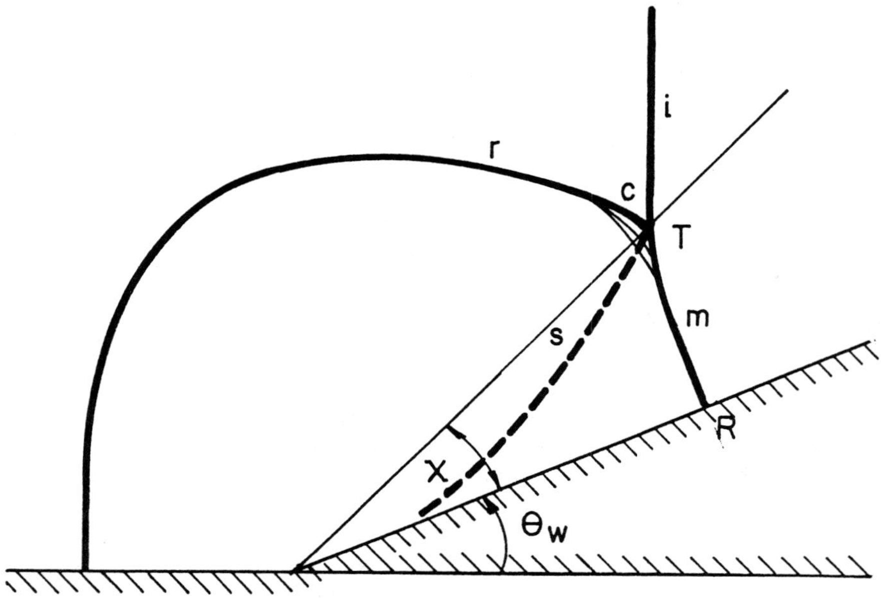

Figure 2.22 The wave configuration of a vNR.

rather than a shear discontinuity. Finally the so called "triple point" of a vNR is not a well defined single point, as in the case of an MR. Colella & Henderson (1990) suggested defining the triple point as the intersection between the leading compression wave and the incident shock wave (see figure 2.22).

Colella & Henderson's (1990) numerical study further revealed that the reflected shock wave of an SMR degenerates to a compression wave when it assumes an orientation normal to the flow behind the incident shock wave in a frame of reference attached to the triple point, T. The (p, θ)-polar combination for this case is shown in figure 1.15c. For this situation the flow is not deflected while passing through the reflected wave from state (1) to state (2). Consequently, the flow behind the reflected wave is also normal to it, and so is the slipstream to which the flow behind the reflected shock wave is parallel. Hence, all the (p, θ)-polar combinations which suggest that the flow deflection through the reflected shock wave is in the same direction as the flow deflection through the incident shock wave (e.g., figure 1.15b) are aphysical. Although, based on theoretical grounds, such MR-configurations are possible, in reality they do not materialize. Instead a vNR is obtained.

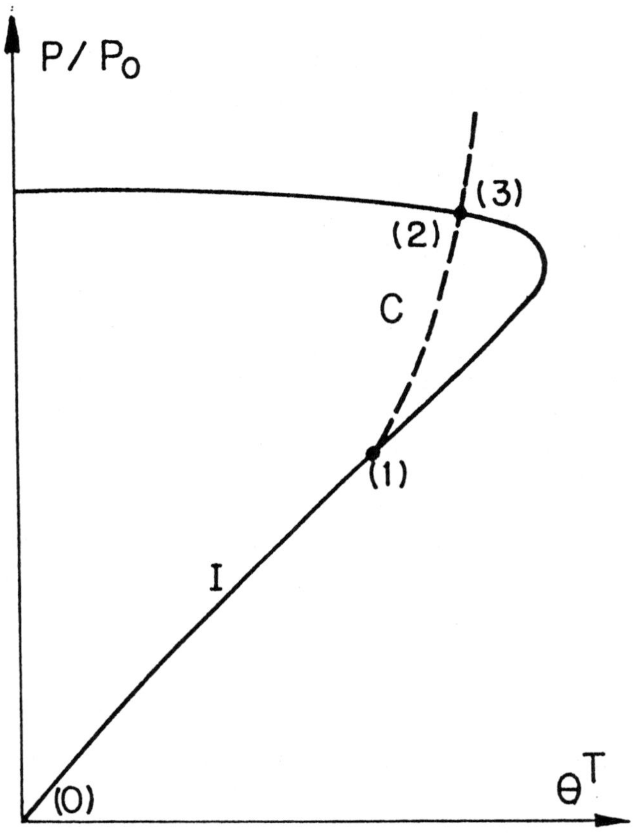

Figure 2.23 (p,θ)-solution of a vNR.

2.1.6.1 Shock Polar Presentation of a vNR

The presentation of a vNR using shock polars is, in general, quite similar to that of an SMR. The only difference is that the shock polar corresponding to the reflected shock wave, i.e. the R-polar, should be replaced by the (p,θ)-relation of a compression wave i.e., the C-polar. A schematic drawing of the (p,θ)-polar presentation of a vNR is shown in figure 2.23. The dashed line in figure 2.23 describes the (p,θ)-relation of a supersonic flow ($M_1 > 1$) through a compression wave.

Simple gas dynamic considerations imply that

$$\ln\frac{p_2}{p_1} = \frac{\gamma M_1^2}{\sqrt{M_1^2 - 1}}(\theta_2 - \theta_1) \tag{2.32}$$

2.2 Triple Point Trajectory Angles

Until recently it was believed that the triple points T and T' move along straight lines originating at the leading edge of the reflecting wedge. The angles between these lines and the reflecting wedge surface are the first and second triple point trajectory angles, χ and χ', respectively. Doubts concerning the fact that the first triple point originates at the leading edge of the reflecting wedge were first raised by Henderson & Gray (1981). Experimental support of their hypothesis was given recently by Reichenbach (1985) and by Schmidt (1989). Schmidt, in a careful experimental study, found that the first triple point is formed a few mean free paths downstream of the leading edge of the reflecting wedge. The reason for this was attributed by Schmidt to viscous effects. Based on his experiments, he concluded that the reflection is unsteady in the vicinity of the leading edge of the reflecting wedge. It becomes pseudo-steady only after the incident shock wave has propagated some distance from the leading edge. Using rarefied gases (initial pressures of a few torrs) the triple point was found to be formed a few millimeters downstream of the leading edge of the reflecting wedge.

Since the analysis so far is limited to inviscid flows, it is assumed in the following that both the first and the second triple points are generated at the leading edge of the reflecting wedge at the instant the incident shock wave strikes it. Even though this assumption might not be correct in view of the above-mentioned recent experimental findings, it is believed that the error introduced by using it is negligibly small.

2.2.1 First Triple Point Trajectory Angle - χ

A fairly good method for predicting the value of χ for a given set of initial conditions, i.e., M_S, ϕ_1, p_0 and T_0, was suggested by Law & Glass (1971). Their analytical model was based on the experimental observation that in most MR and vNR configurations the curvature of the Mach stem, m, is not too large. Based on this observation, they assumed a straight Mach stem which is normal to the reflecting wedge surface at the point where it touches it, i.e., the reflection point. Using this assumption together with the earlier assumption that the first triple point trajectory originates at the leading edge of the reflecting wedge, the following geometrical relation is self-explanatory (see figure 2.5):

$$\chi = 90^\circ - \phi_3 \qquad (2.33)$$

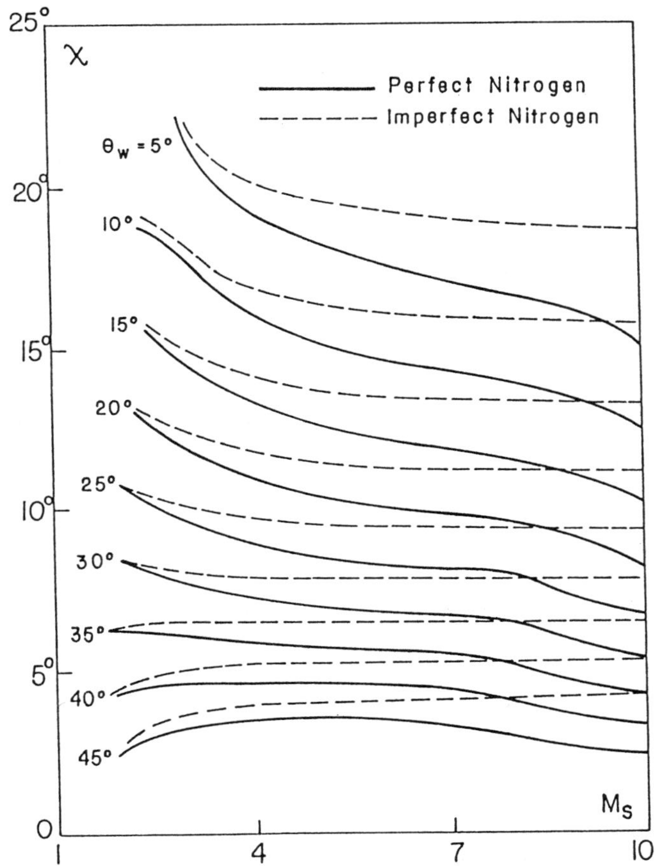

Figure 2.24 Variation of χ with M_s for given θ_w. Solid lines - perfect nitrogen with $\gamma = 1.4$, dashed lines - nitrogen in dissociational equilibrium with $p_0 = 15$ torr and $T_0 = 300$K.

This additional equation, together with the fourteen governing equations of the flow field in the vicinity of the triple point, i.e., the equations derived from the three-shock theory [equations (1.14) to (1.27)] can now be used to obtain a relation of the form

$$\chi = \chi\,(p_0, T_0, M_0, \phi_1) \tag{2.34}$$

Together with the relations

$$M_s = M_0 \sin \phi_1 \tag{2.35a}$$

and

$$\theta_w + \chi = 90^\circ - \phi_1 \tag{2.35b}$$

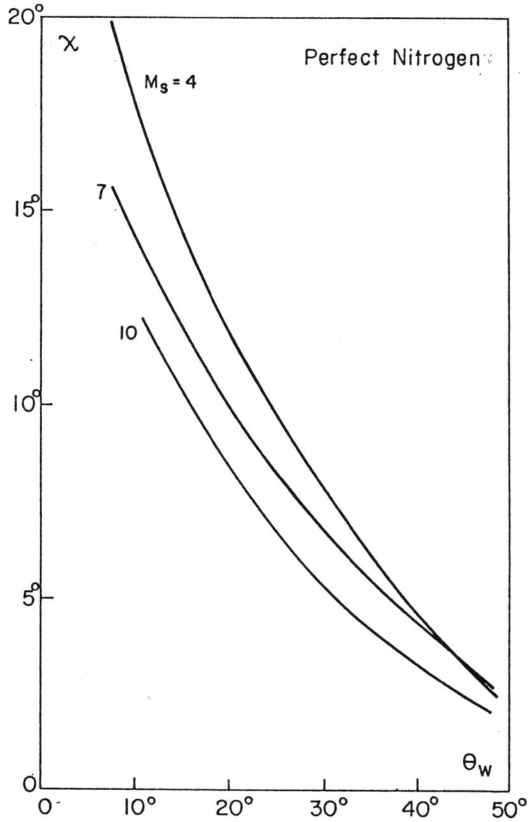

Figure 2.25 Variation of χ with θ_w for given M_s for a perfect nitrogen with $\gamma = 1.4$.

relation (2.34) can be transformed to the following form

$$\chi = \chi\,[\,p_0, T_0, M_S, (\theta_w + \chi)]$$

which can finally be reworked to obtain the following explicit form

$$\chi = \chi\,(\,p_0, T_0, M_S, \theta_w) \tag{2.36}$$

Figure 2.24 illustrates the dependence of the first triple point trajectory angle, χ, upon the incident shock wave Mach number, M_S, for nine different values of reflecting wedge angles, θ_w, which serves as a parameter. The solid lines are for a perfect nitrogen with $\gamma = 1.4$ and the dashed lines are for a real nitrogen in dissociational equilibrium. Similar plots, where the incident shock wave, M_S, is the parameter and the reflecting wedge angle, θ_W, is the variable are shown in figure 2.25 for a perfect nitrogen.

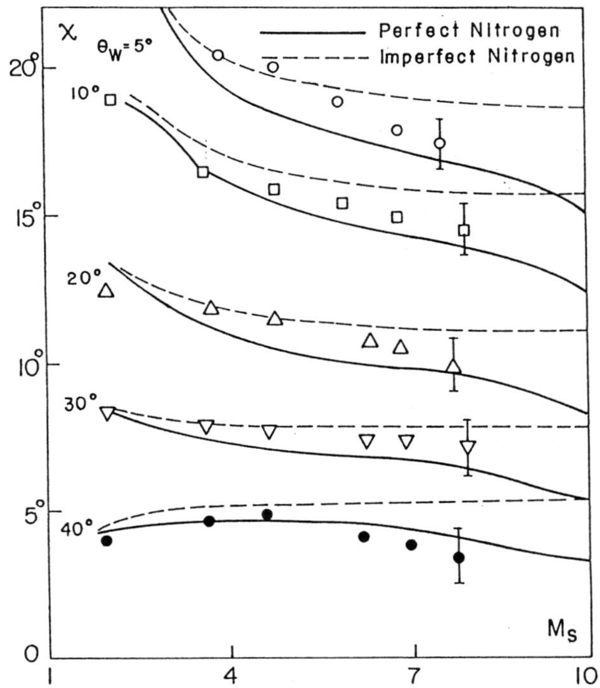

Figure 2.26 Variation of χ with M_s for given θ_w and comparison with Ben-Dor's (1978) experiments. Solid lines - perfect nitrogen with $\gamma = 1.4$, dashed lines - nitrogen in dissociational equilibrium with $p_0 = 15$ torr and $T_0 = 300K$.

Figure 2.24 indicates that for given initial conditions the value of χ is smaller when the gas is assumed to behave as a perfect gas. Real gas effects result in an increase in χ. Note that, in reality, the flow behind the triple point is probably neither perfect nor in dissociational equilibrium, but in an intermediate nonequilibrium thermodynamic state. Thus, the lines in figure 2.24 should most likely be regarded as the lower and upper limits of the actual value of χ.

2.2.1.1 Comparison with Experiments

Adding the experimental results of Ben-Dor (1978) to figure 2.24 results in figure 2.26. The comparison between theory and experiment indicates that the actual values of χ, as expected, lie in general between the values predicted by the perfect and imperfect gas models. As the incident shock wave Mach number, M_S, decreases, the experimental results approach the perfect gas line, as expected.

Clear disagreement between theory and experiments is seen for $\theta_w = 40^\circ$. This was explained by Bazhenova, Fokeev & Gvozdeva,

(1976) who showed that the assumption of a straight Mach stem is relatively good only for $\theta_W < 40^\circ$. As the value of θ_W approaches 40°, the assumption becomes progressively worse. Thus, in order to better predict the first triple point trajectory angles for $\theta_W \geq 40^\circ$, the Mach stem curvature should probably be accounted for.

The above mentioned approach failed also to predict with sufficient accuracy the value of χ for $M_s < 2$ when $\theta_W < 5^\circ$. The reason for the failure in this range lies probably in the following facts which contradict the assumptions upon which the analytical model is based, and hence could all contribute to the disagreement:

1) the Mach stem is not straight,
2) the triple point does not originate at the leading edge of the reflecting wedge,
3) the foot of the Mach stem is not perpendicular to the surface of the reflecting wedge [This was first observed and reported by Dewey & McMillin (1985)],
4) the flow is viscous, and
5) the reflection in this range is not always an MR. In many cases the reflection is a vNR for which the three-shock theory is not applicable.

2.2.1.2 First Triple Point Trajectory Angle for Small Wedge Angles

A technical approach to obtain a method by which the first triple point trajectory angle, χ, over small wedge angles could be predicted better than by the above-mentioned method of Law & Glass (1971) was suggested by Ben-Dor (1978) who made use of the fact that in shock tube experiments an MR or a vNR is obtained even for cases when the flow behind the incident shock wave is subsonic with respect to the reflection point, R, i.e., $M_1^R < 1$. The formation of the triple point, T, in this case results in a situation in which $M_1^T > 1$. Thus χ should assume a value large enough to force ϕ_1 to be sufficiently small so that M_1^T becomes supersonic, i.e., $M_1^T > 1$. Assuming that χ assumes a value so that ϕ_1 reaches the value appropriate to $M_1^T = 1$, one obtains:

$$\chi = \left[90^\circ - \phi_1\left(M_1^T = 1\right)\right] - \theta_W \tag{2.37}$$

Note that, unlike the method suggested by Law & Glass (1971) which is presented in section 2.2.1, the prediction of χ by equation (2.37) does not require the solution of the governing equations of the flow field in the vicinity of the first triple point.

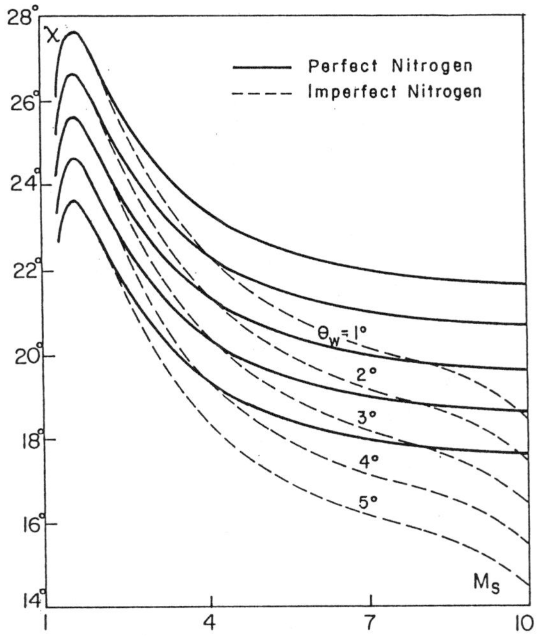

Figure 2.27 Variation of χ with M_s for small given θ_w. Solid lines - perfect nitrogen with $\gamma = 1.4$, dashed lines - nitrogen in dissociational equilibrium with $p_0 = 15$ torr and $T_0 = 300K$.

The above mentioned approach was applied to obtain figure 2.27 where the dependence of the first triple point trajectory angle upon the incident shock wave Mach number for small reflecting wedge angles is shown for both perfect and imperfect nitrogren. The comparison of the predictions based on this method with experiments is shown in figure 2.28. Although the prediction of χ, using this approach, is far from satisfactory, it is much better than the predictions obtained when the analytical method of Law & Glass (1971) is used. It should be noted that the agreement between the predicted values of χ and the experimental results improves as θ_w decreases. The agreement should be excellent for the case of glancing incidence, i.e., $\theta_w \rightarrow 0$. For this case Ben-Dor, Takayama & Dewey (1987) developed the following relation (for details see section 4.1.1.3.2)

$$\chi_g = \tan^{-1}\left\{\frac{1}{M_s}\left[A_{10}^2 - (M_i - V_{10})^2\right]^{1/2}\right\} \tag{2.38}$$

where $A_{10} = a_1/a_0$ and $V_{10} = V_1/a_0$. Here a_0 and a_1 are the local speeds of sound ahead of and behind the incident shock wave,

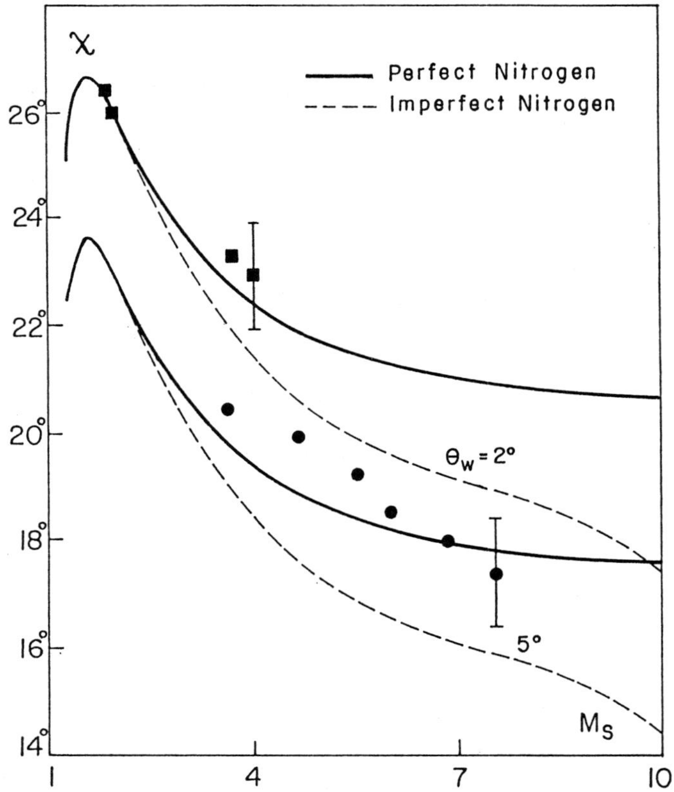

Figure 2.28 Variation of χ with M_s for small given θ_w. Solid lines - perfect nitrogen with $\gamma = 1.4$, dashed lines - nitrogen in dissociational equilibrium with $p_0 = 15$ torr and $T_0 = 300K$.

respectively, and V_1 is the shock induced flow velocity (V_j in figure 2.1a). For a given gas both A_{10} and V_{10} depend solely on the incident shock wave Mach number, M_S, through the following relations

$$V_{10} = \frac{2\left(M_s^2 - 1\right)}{(\gamma + 1)M_s} \tag{2.39}$$

and

$$A_{10} = \frac{\gamma - 1}{\gamma + 1} \frac{1}{M_S}\left[\left(\frac{2\gamma}{\gamma - 1} M_s^2 - 1\right)\left(M_s^2 + \frac{2}{\gamma - 1}\right)\right]^{1/2} \tag{2.40}$$

Using the Galilean transformation shown in figure 2.1, equation (2.38) becomes

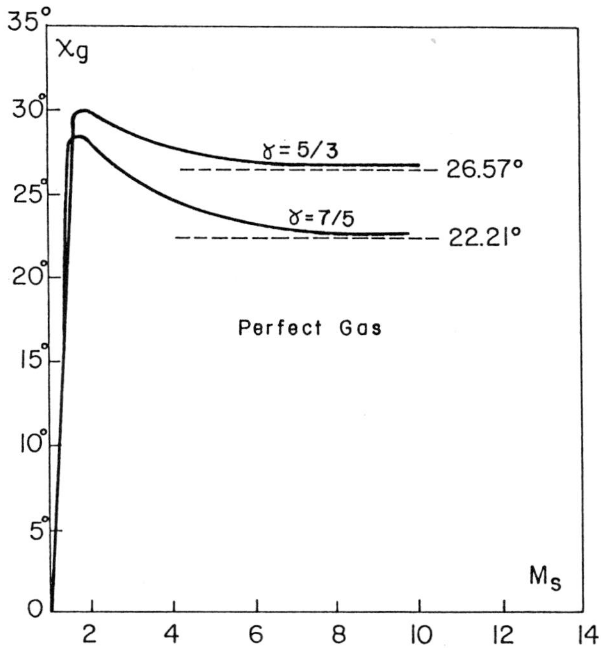

Figure 2.29 Variation of the triple point trajectory angle at glancing incidence with M_s.

$$\chi_g = \tan^{-1}\left\{\left[\frac{A_{10}^2}{M_s}\left(1 - M_1^2\right)\right]^{1/2}\right\} \tag{2.41}$$

where $M_1 = u_1/a_1$ and u_1 is the flow velocity behind the incident shock wave in a frame of reference attached to the incident shock wave (u_j in figure 2.1b). M_1 depends solely on M_S through the following relation

$$M_1 = \frac{\left(M_S - V_{10}\right)}{A_{10}} \tag{2.42}$$

When $M_S \rightarrow \infty$ the triple point trajectory angle at glancing incidence approaches a limit given by

$$\chi_g = \tan^{-1}\left(\frac{\gamma - 1}{\gamma + 1}\right)^{1/2} \tag{2.43}$$

which results in 22.21° and 26.57° for a diatomic ($\gamma = 7/5$) and a monatomic ($\gamma = 5/3$) gas, respectively.

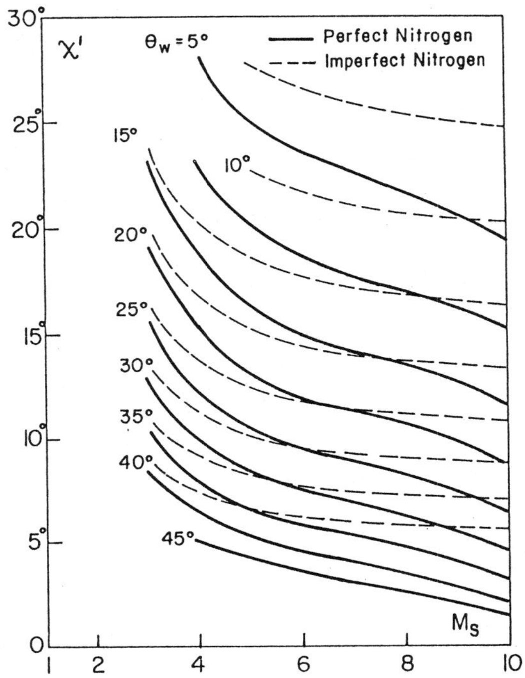

Figure 2.30 Variation of χ' with M_s for given θ_w. Solid lines - perfect nitrogen with $\gamma = 1.4$, dashed lines - nitrogen in dissociational equilibrium with $p_0 = 15$ torr and $T_0 = 300K$.

Using equation (2.41) the triple point trajectory angle at glancing incidence was calculated for both $\gamma = 7/5$ and $\gamma = 5/3$. The results are shown in figure 2.29. Figure 2.29 indicates that for a given gas, the triple point trajectory angle at glancing incidence has a maximum for a value of M_S in the range $1 < M_S < 2$. The precise value of M_S at which this maximum occurs can be calculated from equation (2.41) by setting $d\chi_g/dM_S = 0$. This yields

$$M_S = \frac{4\ (\gamma + 1)}{4 - (\gamma - 1)^2} \tag{2.44}$$

which results in $M_S = 1.58$ and $M_S = 1.73$ for a diatomic and a monatomic gas, respectively. The corresponding maximum triple point trajectory angles at glancing incidence are 28.69° and 30.04°.

2.2.2 Second Triple Point Trajectory Angle - χ'

Based on the assumptions that the first triple point originates at the leading edge of the reflecting wedge and that the Mach stem is straight and perpendicular to the reflecting wedge surface at its foot, a fairly good analytical method for predicting the value of the second triple point trajectory angle, χ', was suggested by Ben-Dor (1980). In addition to the assumptions upon which the method for predicting the first triple point trajectory angle is based, the method for predicting the second triple point trajectory angle is similarly based on the assumption that the second triple point trajectory also originates at the leading edge of the reflecting wedge. In view of the previously mentioned experimental findings regarding the origination of the first triple point trajectory angle, it is probable that, similarly to the first triple point, the second triple point also does not form at the leading edge of the reflecting wedge. Hence, the assumption that it does form at the leading edge of the reflecting wedge might introduce an inherent error into the predictions of the suggested method. However, it is believed that if such an error is introduced it is negligibly small.

The method for calculating the second triple point trajectory angle, χ', is based on the following geometrical relation which is self explanatory (see figure 2.10a):

$$\chi' = 90^\circ - \theta_w - \phi_1' \qquad (2.45)$$

where ϕ_1' is defined by equation (2.15).

The calculation of the second triple point trajectory angle is performed in the following way:

- the governing equations of the flow field in the vicinity of the first triple point which are given by the three-shock theory [equations (1.14) to (1.27)] are solved for a given set of initial conditions, i.e., M_S , ϕ_1, T_0 and p_0,
- the trajectory angle of the first triple point, χ, is calculated using equation (2.33),
- the reflecting wedge angle, θ_w, is calculated using equation (2.35b),
- the value of ϕ_1' is calculated using equation (2.15), and
- finally, the second triple point trajectory angle, χ', is calculated using equation (2.45).

Figure 2.30 illustrates the dependence of the second triple point trajectory angle on the incident shock wave Mach number and the reflecting wedge angle which serves as a parameter. The dashed lines are for real nitrogen in dissociational equilibrium and the solid lines are for a perfect nitrogen. Figures 2.31a and 2.31b illustrate the dependence of the second triple point trajectory angle on the reflecting wedge angle when the incident shock wave Mach number is the parameter, for a perfect and an imperfect nitrogen, respectively. Both figures indicate that χ' decreases as θ_w increases, and that the excitation of the internal degrees of freedom causes an increase in the value of χ' compared to its value when the gas is assumed to behave as a perfect gas. However, due to the fact that

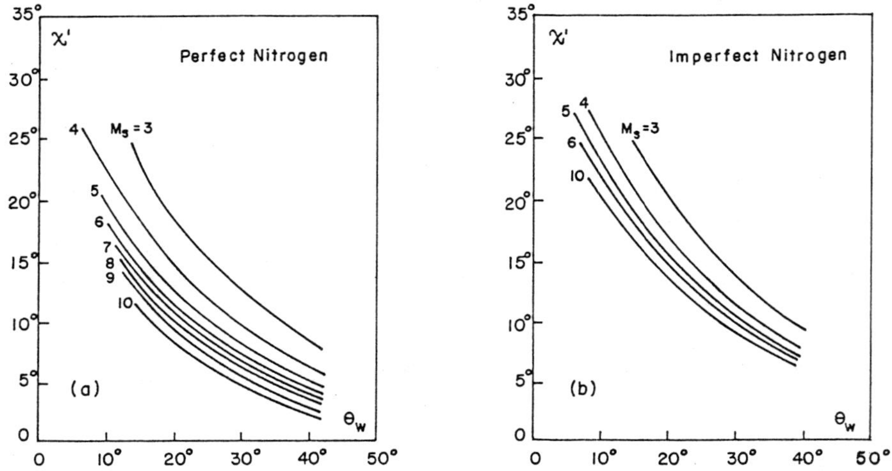

Figure 2.31 Variation of χ' with θ_w for given M_s: a) perfect nitrogen with $\gamma = 1.4$; b) nitrogen in dissociational equilibrium with $p_0 = 15$ torr and $T_0 = 300$K.

the flow is most likely neither perfect nor in dissociational equilibrium, the actual value of χ' is expected to be between the predictions based on perfect gas behavior and a gas in a dissociational equilibrium.

It should be noted that although the foregoing method for calculating the second triple point trajectory angle χ' results in values of χ' also for small wedge angles, TMR and DMR configurations, which have a kink or a second triple point, associated with a value of χ', do not exist at small wedge angles (see section 2.4). Therefore, this method should be used only for M_S and θ_W combinations which result in either a TMR or a DMR.

2.2.2.1 Comparison with Experiments

Figure 2.32 is a reproduction of figure 2.30 for $\theta_W = 30°$ and $40°$, which lie in the DMR-domain. While for $\theta_W = 30°$ the experimental results seem to agree excellently with the analytical predictions when the nitrogen gas is assumed to be in dissociational equilibrium, in the case of $\theta_W = 40°$ the experimental results lie, as expected, between the two limiting cases of perfect gas behavior and real gas in dissociational equilibrium behavior.

However, due to the fact that the suggested model is based on simplifying assumptions (see sections 2.2.1 and 2.2.2), which introduce inherent errors into the calculations of the first and the

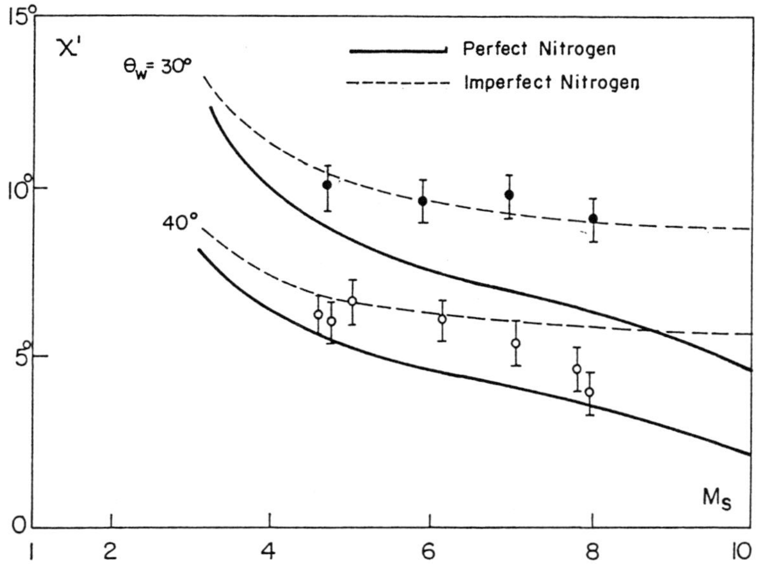

Figure 2.32 Variation of χ' with M_s for given θ_w and comparison with Ben-Dor's (1978) experiments. Solid lines - perfect nitrogen with $\gamma = 1.4$, dashed lines - nitrogen in dissociational equilibrium with $p_0 = 15$ torr and $T_0 = 300$K.

second triple point trajectory angles the good agreement which is seen in figure 2.32 should probably be considered accidental.

2.2.3 Comparison Between the First and Second Triple Point Trajectory Angles

The first and second triple point trajectory angles, χ and χ', respectively, as obtained using the analytical models presented in sections 2.2.1 and 2.2.2 are plotted as functions of the incident shock wave Mach number, M_S, for a perfect nitrogen and a real nitrogen in dissociational equilibrium in figures 2.33 and 2.34, respectively. In both figures the reflecting wedge angle, θ_w, serves as a parameter.

It is clearly seen in figure 2.33 that when the gas (nitrogen in this case) is assumed to behave as a perfect gas, there is a critical value of M_S, for each value of θ_w, at which $\chi = \chi'$. For values of M_S smaller than this value $\chi < \chi'$, while for values of M_S greater than this value $\chi > \chi'$. The locus of the critical values of M_S for which $\chi = \chi'$ is shown in figure 2.33 by a dotted line. This fact was used previously to distinguish between a positive double-Mach reflection, DMR^+, typified by $\chi' > \chi$, and a negative double-Mach reflection, DMR^-, typified by $\chi' < \chi$. The transition between these two types of DMR-configurations is the locus of all the points for which $\chi = \chi'$ (dotted line in figure 2.33).

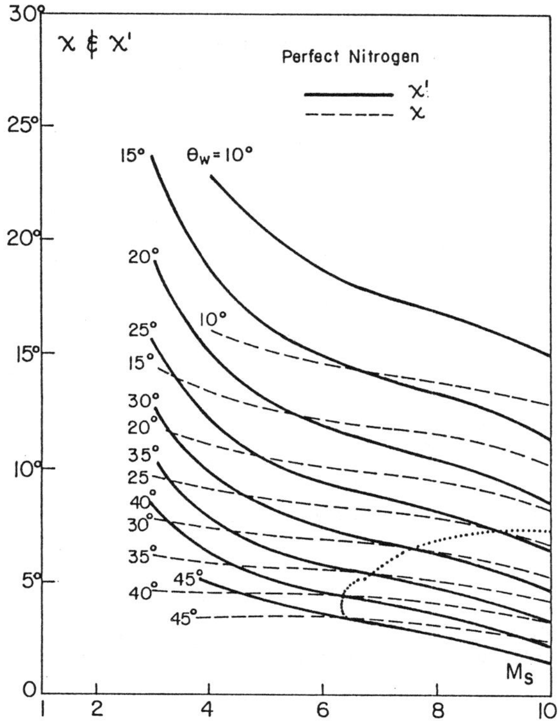

Figure 2.33 Comparison between χ and χ' for a perfect nitrogen with $\gamma = 1.4$.

If, however, real gas effects are accounted for, and the nitrogen gas is assumed to be in dissociational equilibrium immediately behind the shock waves (i.e., the dissociational relaxation length is assumed to be zero), then, as can be seen in figure 2.34, χ' approaches χ asymptotically as M_S increases, and a situation in which $\chi' < \chi$ is never reached.

As shown earlier, experimental results indicate that there are indeed DMR-configurations in which $\chi' < \chi$ (e.g., figure 2.18a). Based on these experimental observations it can be concluded, that the assumption of dissociational equilibrium immediately behind the shock fronts is wrong. Alternatively, if real gas effects are to be accounted for, then a non-equilibrium relaxation zone should be considered. Note, that the conclusion that the gas behaves as a perfect gas, cannot be reached from these results.

Figure 2.35 illustrates the dependence of χ and χ' on M_S for a perfect argon. Not only is χ' always greater than χ, i.e., $\chi' > \chi$, but it seems that as M_S increases the difference between χ' and χ approaches a constant value.

It can be concluded from comparing figures 2.33 and 2.35 that the value of γ plays a significant role in determining whether χ can

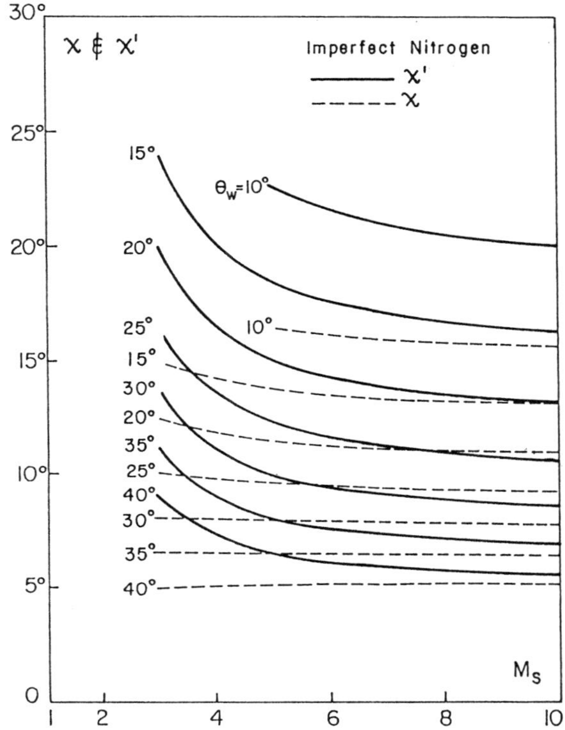

Figure 2.34 Comparison between χ and χ' for nitrogen in dissociational equilibrium with p_0 = 15 torr and T_0 = 300K.

exceed the value of χ'. It is evident from figures 2.33 and 2.35 that situations in which $\chi' < \chi$ are associated with gases having low values of γ. Thus it is not surprising that DMR^- wave configurations, i.e., DMR-configurations with $\chi' < \chi$, are obtained quite easily in heavy gases such as Freon-12, SF_6 and CO_2 which have low value of γ, even at room temperature.

2.3 Transition Criteria

As mentioned in section 1.5.5, of the various suggested criteria for the termination of RR, the one which best agrees with pseudo-steady shock tube experimental data, is the one arising from the length scale concept, which suggests that in pseudo-steady flow RR terminates when the flow behind the reflection point, R, of the RR becomes sonic in a frame of reference attached to R. This criterion implies that the RR⇔IR transition occurs when

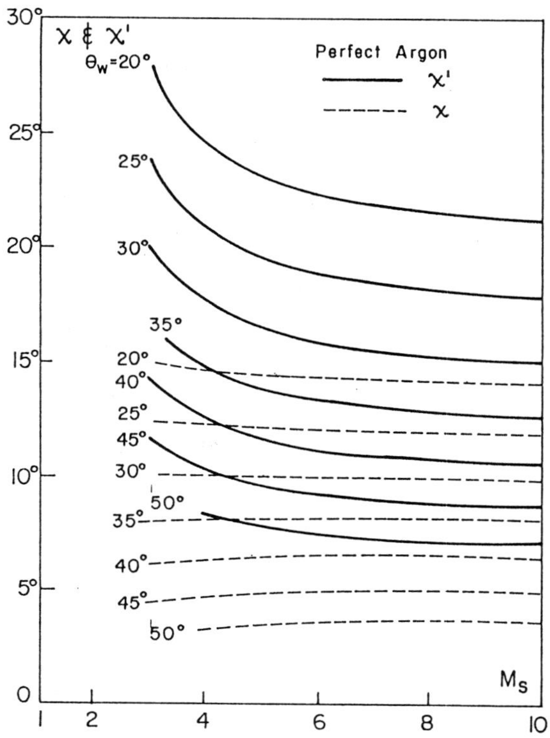

Figure 2.35 Comparison between χ and χ' for perfect argon with $\gamma = 5/3$.

$$M_2^R = 1 \qquad (2.46)$$

where M_2^R is the flow Mach number in state (2) behind the reflected shock wave of an RR with respect to the reflection point R. As long as $M_2^R > 1$, the corner-generated signals cannot catch-up with the reflection point and an IR-configuration is impossible.

Once the corner-generated signals have caught-up with the reflection point, an IR-configuration, typified by a shock wave with a finite length (the Mach stem) is formed. The IR can be either an MR or a vNR depending on the angle of incidence between the flow in state (1) and the reflected wave. The reflection is MR as long as $\phi_2 < 90^\circ$. Consequently, the MR⇔vNR transition occurs when

$$\phi_2 = 90^\circ \qquad (2.47a)$$

Note that when $\phi_2 = 90^\circ$, the flow passing through the reflected wave is not deflected. However, since the flow behind the reflected wave must be parallel to the slipstream, it is obvious that the above condition for the MR⇔vNR transition can also be written as

$$\omega_{rs} = 90^o \tag{2.47b}$$

where ω_{rs} is the angle between the reflected wave and the slipstream.

Since S_tMR and I_nMR cannot occur in pseudo-steady flows, the MR in pseudo-steady flows is always a D_iMR. Once the condition for the existence of a D_iMR is met, the value of the flow Mach number, in state (2), behind the reflected shock wave of a D_iMR with respect to the triple point, T, becomes the significant parameter in determining the particular type of the D_iMR.

As long as $M_2^T < 1$, the reflection is an SMR, typified by a curved reflected shock wave along its entire length. The fact that the reflected shock wave is curved along its entire length implies that a physical length scale is communicated to the triple point (from which the reflected shock wave emanates) through state (2). This communication path is possible only as long as $M_2^T < 1$. When the flow in state (2) becomes supersonic, i.e., $M_2^T > 1$, the communication path is blocked by a supersonic flow zone, and the reflected shock wave develops a straight portion, terminated by a kink in it, which most likely indicates the point along the reflected shock wave which has been reached by the corner-generated signals. Thus, the SMR terminates when

$$M_2^T = 1 \tag{2.48}$$

Once a kink is formed in the reflected shock wave, the value of the flow Mach number, in state (2), behind the reflected shock wave of a D_iMR with respect to the kink, K, becomes the significant parameter in determining whether the reflection remains a TMR or changes to a DMR.

As long as $M_2^K < 1$, the reflection is a TMR, typified by a centered compression wave at the kink. When the flow in state (2) becomes supersonic with respect to the kink, K, $M_2^K > 1$, the compression wave converges into a shock wave, and the kink changes into a triple point, the second triple point, T'. Thus the TMR⇔DMR transition occurs when

$$M_2^K = M_2^{T'} = 1 \tag{2.49}$$

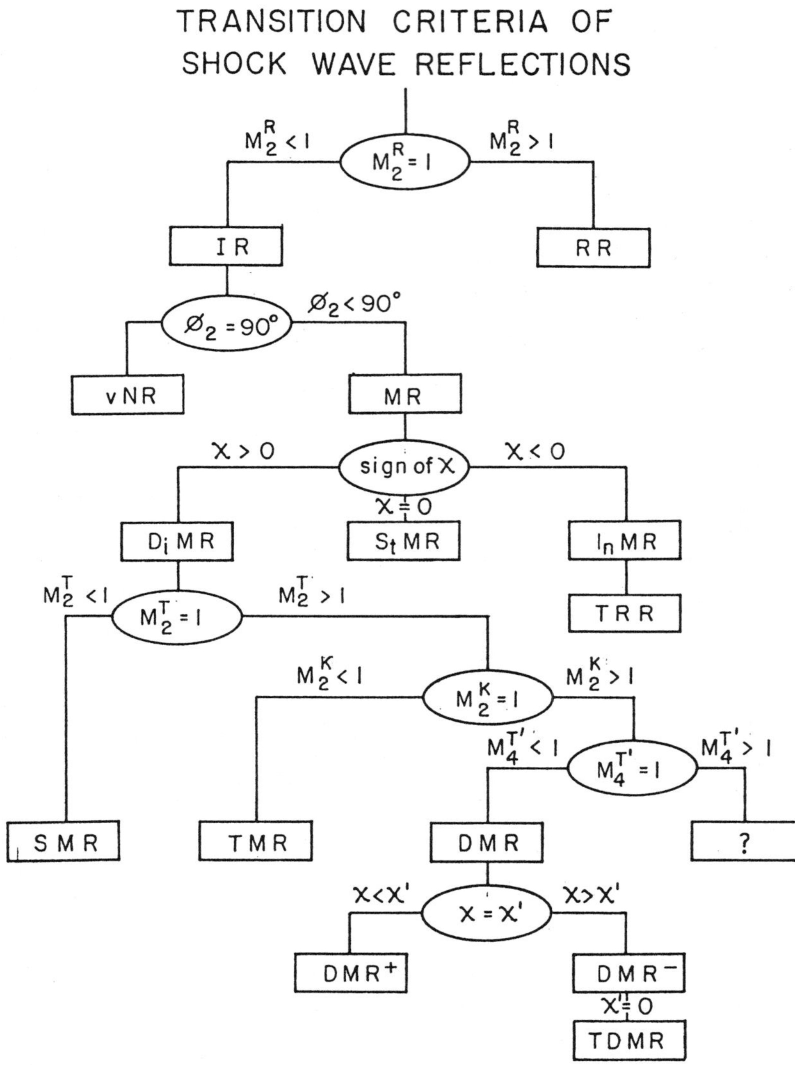

Figure 2.36 Schematic summary of the transition criteria between the various types of shock wave reflection configurations.

Following the logic of the above discussed sequence of transitions from SMR to TMR and then from TMR to DMR, it is hypothesized that when the flow behind m', in state (4) of a DMR (see figure 2.12) becomes supersonic with respect to the second triple point, T', the m' - shock wave will also develop a straight portion with a kink. Thus it is expected that the DMR will terminate when

$$M_4^{T'} = 1 \tag{2.50}$$

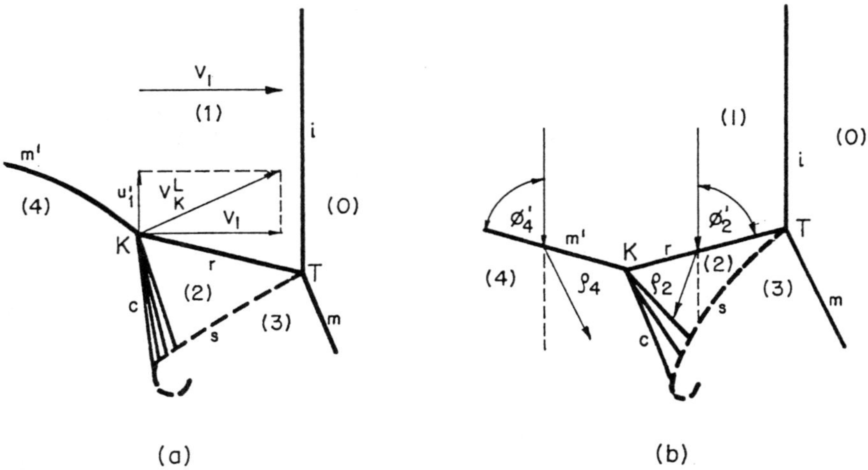

Figure 2.37 Definition of the various flow parameters in the vicinity of the kink of a TMR: a) in a laboratory frame of reference; b) in a frame of reference attached to the kink.

At this condition a new type of reflection is expected to be formed. Note that the condition given by equation (2.50) can be met only with extremely high incident shock wave Mach numbers, which in turn result in a situation in which real gas effects become significant, and therefore must be accounted for.

Once the condition for the existence of a DMR is met, the DMR wave configuration can be either a positive DMR or a negative DMR, depending on the values of the first and second triple point trajectory angles, χ and χ', respectively.

The reflection is a DMR$^+$ when $\chi > \chi'$ and it is a DMR$^-$ when $\chi < \chi'$. Thus the DMR$^+$⇔DMR$^-$ transition criterion is

$$\chi = \chi' \tag{2.51}$$

Finally, if the condition for the existence of a DMR$^-$ is met, i.e., $\chi < \chi'$, then a terminal double-Mach reflection, TDMR, can occur if

$$\chi' = 0 \tag{2.52}$$

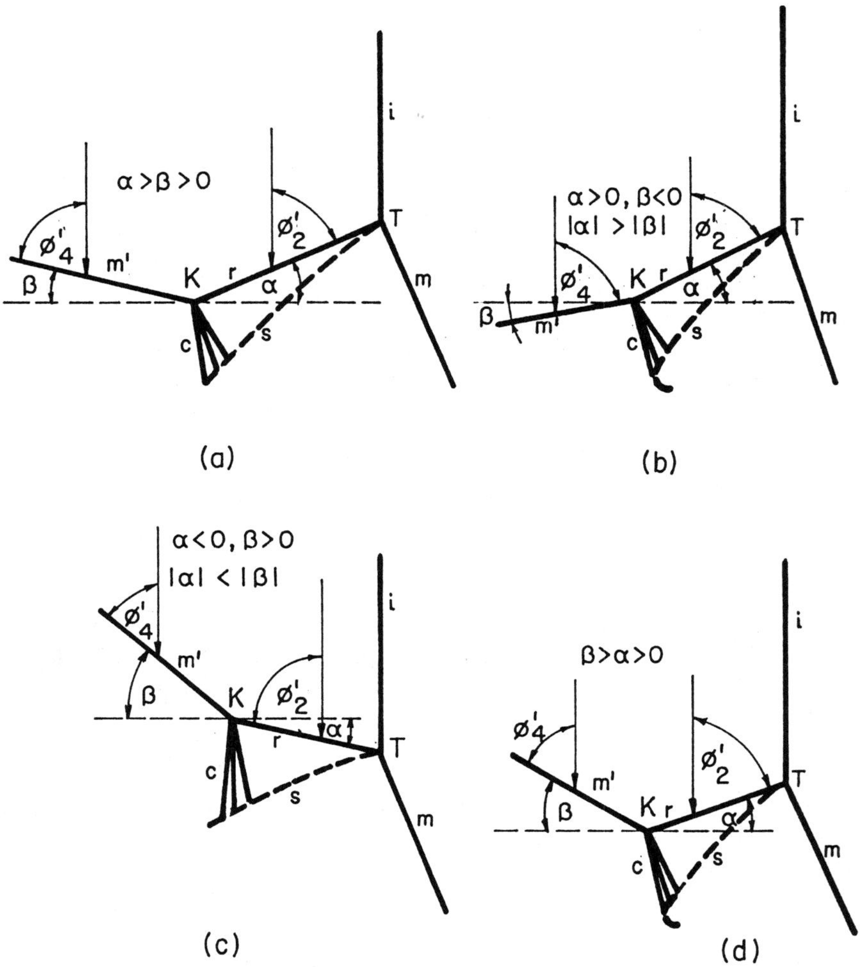

Figure 2.38 Schematic illustration of four different wave orientations in the vicinity of the kink in a TMR.

The foregoing discussion of the transition criteria between the various types of shock wave reflection configurations is summarized in figure 2.36.

2.3.1 Additional Requirement for the SMR⇔TMR Transition

Shirouzu & Glass (1986) proposed an additional necessary condition for the SMR⇔TMR transition which, as will be shown subsequently, has the property of slightly shifting the transition line based on the previously mentioned transition criterion [equation (2.48)]. They argued that the condition $M_2^T = 1$ is only necessary but not sufficient. Their motivation to search for additional conditions for the SMR⇔TMR transition originated from the failure of the above-mentioned transition criterion, i.e., $M_2^T = 1$, to sharply separate between the actual SMR- and TMR-domains.

Based on the previously mentioned Law-Glass assumption, i.e., that the horizontal velocity of the kink, K, of a TMR is equal to the shock induced velocity in region (1), they argued that the oncoming flow from region (1) towards the reflected shock wave, r, has only a vertical component in a frame of reference attached to the kink, K (see figure 2.37a). Thus, if one designates the angles of incidence to the reflected shock waves and the densities behind the reflected shock waves on either side of the kink as ϕ_2', ϕ_4', ρ_2 and ρ_4 (figure 2.37b) then the existence of a band of compression waves at the kink is equivalent to the requirement that $\rho_4 > \rho_2$. This, in turn, yields that $\phi_4' > \phi_2'$ since the flow in region (1) is uniform. Among the four possible orientations of the reflected shock wave in the wave configurations shown in figure 2.38, only configurations (a) and (b) satisfy the condition $\phi_4' > \phi_2'$. In both (a) and (b) the angle between the incident and the reflected shock waves, ω_{ir}, is greater than 90°. Therefore, Shirouzu & Glass (1986) concluded that

$$\omega_{ir} \geq 90^\circ \qquad (2.53)$$

is a necessary condition for the SMR⇔TMR transition. Note that the condition $\omega_{ir} > 90^\circ$ is not sufficient because configuration (d) in figure 2.38 also has $\omega_{ir} > 90^\circ$ but the fact that $\phi_4' < \phi_2'$ violates the requirement of $\rho_4 > \rho_2$ which is imposed by the compression wave at the kink of the TMR.

2.4 Domains of Different Types of Reflections

Once the various types of reflections which are possible in pseudo-steady flows, namely, regular reflection - RR, von

Neumann reflection - vNR, single-Mach reflection - SMR, transitional-Mach reflection - TMR, double-Mach reflection - DMR and its subtypes, i.e., positive double-Mach reflection - DMR$^+$, negative double-Mach reflection - DMR$^-$ and terminal double-Mach reflection - TDMR, have been identified, and the conditions for the existence (i.e., formation and termination) of each of these reflection configurations have been established, it is of interest to map the domains of these different types of reflections and their transition boundaries in the (M_S, θ_w)-plane. This plot is necessary if one is to predict, a priori, the resulting reflection for a given gas, its initial conditions, p_0 and T_0, the incident shock wave Mach number, M_S, and the reflecting wedge angle, θ_w.

2.4.1 Domains of Different Types of Reflections in the (M_S, θ_w^C)-plane

If the transition criteria, given in section 2.3, are used with the governing equations of the flow fields in the vicinities of the reflection point of an RR and the triple point of an MR (i.e., the two- and three-shock theories), it is possible to obtain a plot in the (M_0, ϕ_1)-plane delineating the domains of the various reflection configurations and their transition lines.

Then by using the following relations:

$$M_S = M_0 \sin \phi_1$$

$$\theta_w^C = 90^\circ - \phi_1$$

the domains and the transition lines can be transformed to the (M_S, θ_w^C)-plane, where θ_w^C, the complementary wedge angle, is equal to

$$\theta_w^C = \theta_w + \chi \tag{2.54}$$

and θ_w is the reflecting wedge angle; χ is the first triple point trajectory angle.

Figure 2.39 represents such a plot for perfect (solid lines) and for imperfect (dashed lines) air in vibrational, dissociational and ionizational equilibrium over a very large range of incident shock wave Mach numbers, $1 < M_S < 20$. Lines A, B, C and D represent the RR⇔IR, SMR⇔TMR, TMR⇔DMR and DMR$^+$⇔DMR$^-$

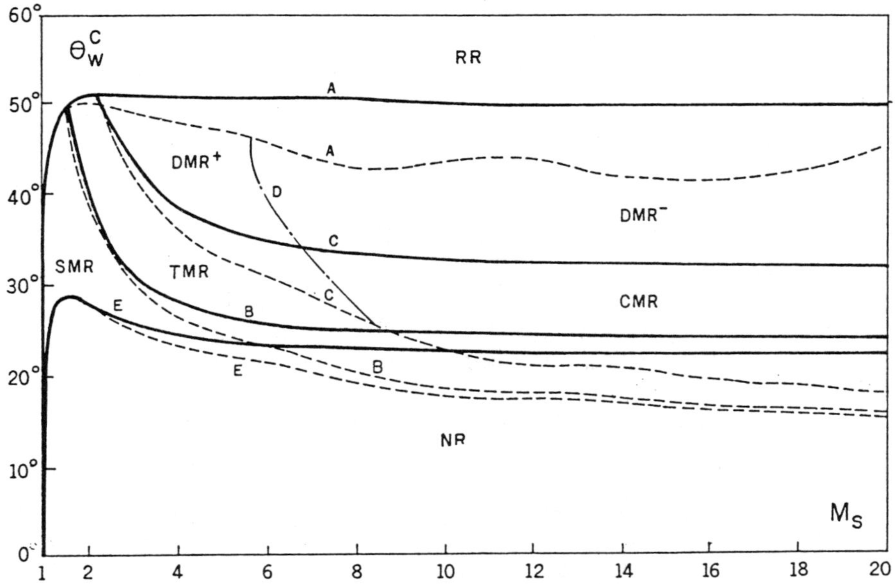

Figure 2.39 Domains and transition boundaries of the various types of shock wave reflections in the (M_S, θ_w^C)-plane. Solid lines are for perfect air with $\gamma = 1.4$, dashed lines are for imperfect air with $p_0 = 15$ torr and $T_0 = 300$K.

transition lines, respectively. Line E represents the boundary below which a reflection is impossible. Figure 2.39 indicates that for $M_S > 10$ the transition lines level out and do not change for a perfect air, while for equilibrium air they undulate as new degrees of freedom become excited.*

2.4.2 Domain of Different Types of Reflections in the (M_S, θ_w)-Plane

Once the transition lines between the various reflection configurations in the (M_S, θ_w^C)-plane are obtained, it is possible to transform them to the more physically meaningful plane, i.e., the (M_S, θ_w)-plane. This is done by analytically calculating the value of

---

* Similar plots for many other gases can be found in the various reports and technical notes of the Institute for Aerospace Studies of the University of Toronto (UTIAS). They are listed in section 5.2.

the first triple point trajectory angle, χ, and using equation (2.54) in order to transform the θ_w^C-axis in figure 2.39 to θ_w.

However, as mentioned earlier, the available analytical models for calculating χ are based on the following simplifying assumptions:
1) the Mach stem is straight,
2) the Mach stem is perpendicular to the reflecting wedge surface, and
3) the triple point is formed at the leading edge of the reflecting wedge.

Unfortunately, the Mach stem is usually not straight but curved. It is concave for low incident shock wave Mach numbers and significant Mach stem heights, and convex for high values of M_S. Furthermore, the other two assumptions were also found not to be entirely correct, probably due to viscous effects. Therefore, the transformation from θ_w^C to θ_w introduces an error which is inherent in the available methods for calculating the first triple point trajectory angle, χ.

Applying the above mentioned procedure to figure 2.39 results in figure 2.40, where the domains and transition boundaries of RR, SMR, TMR, DMR (DMR^+ and DMR^-) for a perfect (solid lines) and imperfect (dashed lines) air in vibrational, dissociational and ionizational equilibrium are shown in the (M_S, θ_w)-plane.*

2.4.3 Comparison of Analysis with Experiments

The following comparison of the foregoing analysis predicting the boundaries between the various types of reflections and their domains considers the experimental results which were obtained in the past two decades at the Institute for Aerospace Studies of the University of Toronto by Professor I. I. Glass** and his graduate students, and those reported by Ikui, Matsuo, Aoki & Kondoh (1981).

The results were obtained using the three well-known optical measurement techniques; interferometry, schlieren and shadowgraph photography.

The comparison of the foregoing analysis is done with experiments which were conducted using different gases in order to learn as much as possible about the limitations of the transition lines

---

* Similar plots for many other gases can be found in the various reports and technical notes of the Institute for Aerospace Studies of the University of Toronto (UTIAS). They are listed in section 5.2.

** The author is most thankful for the courtesy of Professor I. I. Glass and the Institute for Aerospace Studies of the University of Toronto for providing him with their valuable experimental results.

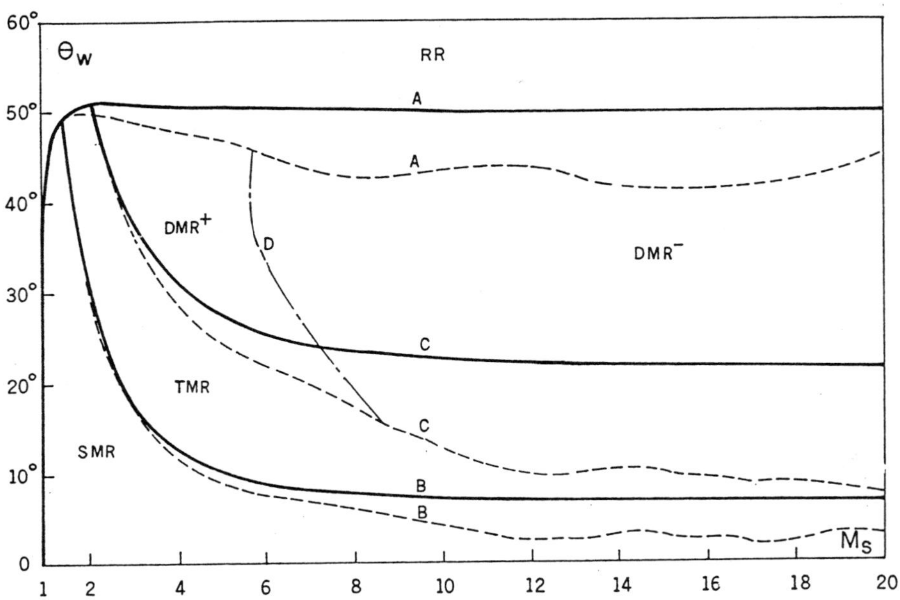

Figure 2.40 Domains and transition boundaries of the various types of shock wave reflections in the (M_s, θ_w)-plane. Solid lines are for perfect air with $\gamma =$ 1.4; dashed lines are for imperfect air with $p_0 = 15$ torr and $T_0 = 300$K.

as calculated using the above-mentioned criteria for the transition between the various types of reflections as well as the methods presented earlier for calculating the first and second triple point trajectory angles.

Figures 2.41 to 2.45 show comparisons of the analytical domains and transition boundaries for the various types of reflections with experiments in argon (Ar), oxygen (O_2), nitrogen (N_2), air, carbon dioxide (CO_2), sulfurhexaflouride (SF_6) and Freon-12.

Comparisons of the foregoing analysis and experimental results in Ar are shown in figures 2.41a and 2.41b. An examination of figure 2.41a shows results for RR and MR in the (M_S, θ_W)-plane. The various analytical domains are reasonably well validated. Three DMR-experiments (one at $\theta_W = 50°$ and two at $\theta_W = 40°$) lie in the TMR-domain. All the other experiments are in their appropriate domains.

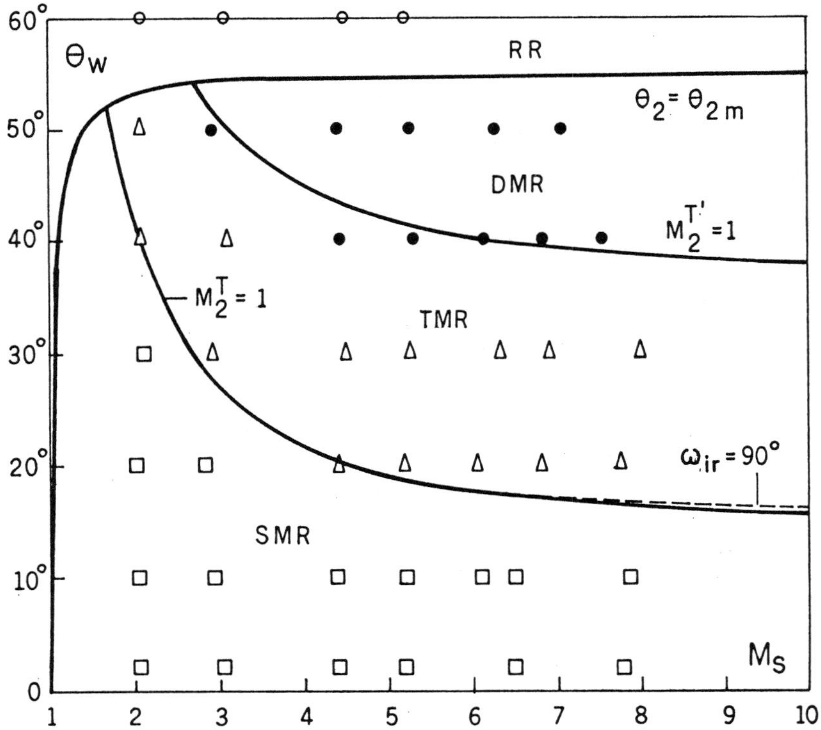

Figure 2.41 a) Domains and transition boundaries of the various types of shock wave reflections in the (M_S, θ_w)-plane for frozen argon with $\gamma = 5/3$ and experimental data of UTIAS (courtesy of Professor I.I. Glass).

The additional requirement for the formation of TMR, which was presented in section 2.3.1, is shown by a dashed line in this figure and in the following ones only in the range where $M_2^T \geq 1$.

When these results are plotted in the (M_S, θ_w^C)-plane (figure 2.41b), which should be more accurate since a relation for χ is not required to draw this figure, it is surprising that a poorer agreement is obtained. Now four DMR-experiments (one at $\theta_w^C \approx 50°$ and three at $\theta_w^C \approx 45°$) definitely lie in the TMR-domain, and one TMR-experiment (at $M_S \approx 4$) lies marginally in the SMR-domain. All the other experiments are in their appropriate domains. Since no experiments were made along the RR⇔IR transition line, little can be said about the well-known persistence of RR into the IR-domain. Note, that no real gas effects occur in Ar for the pressures and

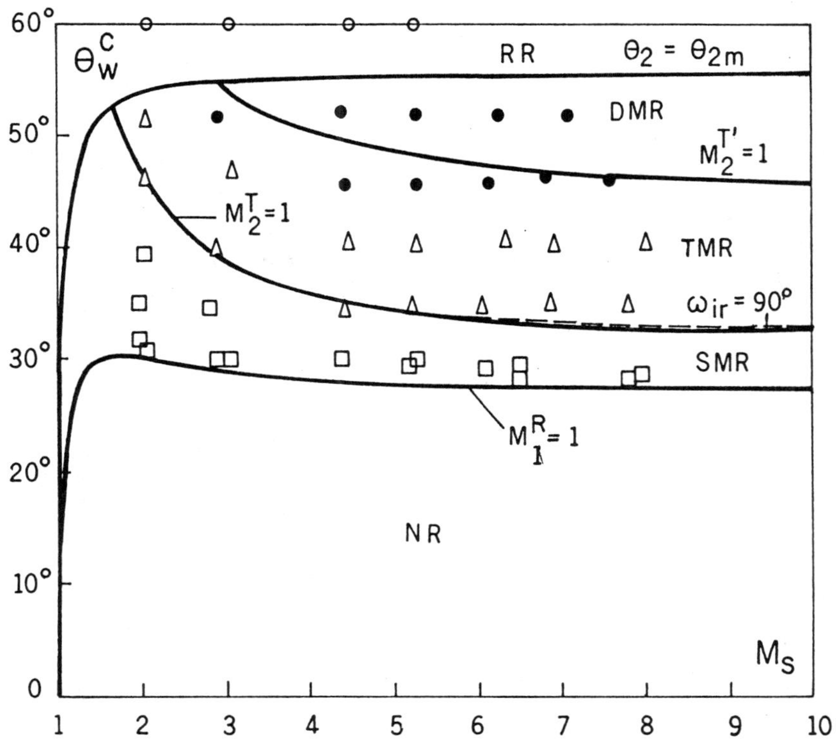

Figure 2.41 b) Domains and transition boundaries of the various types of shock wave reflections in the (M_s, θ_w^C)-plane for frozen argon with $\gamma = 5/3$ and experimental data of UTIAS (courtesy of Professor I.I. Glass).

Mach numbers of these experiments as the relaxation lengths are too large and the gas remains frozen with $\gamma = 5/3$. (For details on real gas effects see section 2.5.3.)

Figures 2.42a and 2.42b show the results for N_2 and O_2 in the (M_s, θ_w) and the (M_s, θ_w^C)-planes, respectively. In figure 2.42a five RR-experiments lie only marginally below the RR⇔IR transition line. All the other experiments, except one DMR at $M_s \approx 6$, lie in their appropriate domains. Note that the additional requirement for the formation of TMR is very helpful in this case, as it puts all the SMR-experiments in their appropriate domain while the $M_2^T = 1$ transition line fails with three SMR-experiments at $\theta_w = 10°$. In figure 2.42b, which as mentioned earlier should show a better agreement since θ_w^C is used instead of θ_w, the line $M_2^T = 1$ is indeed

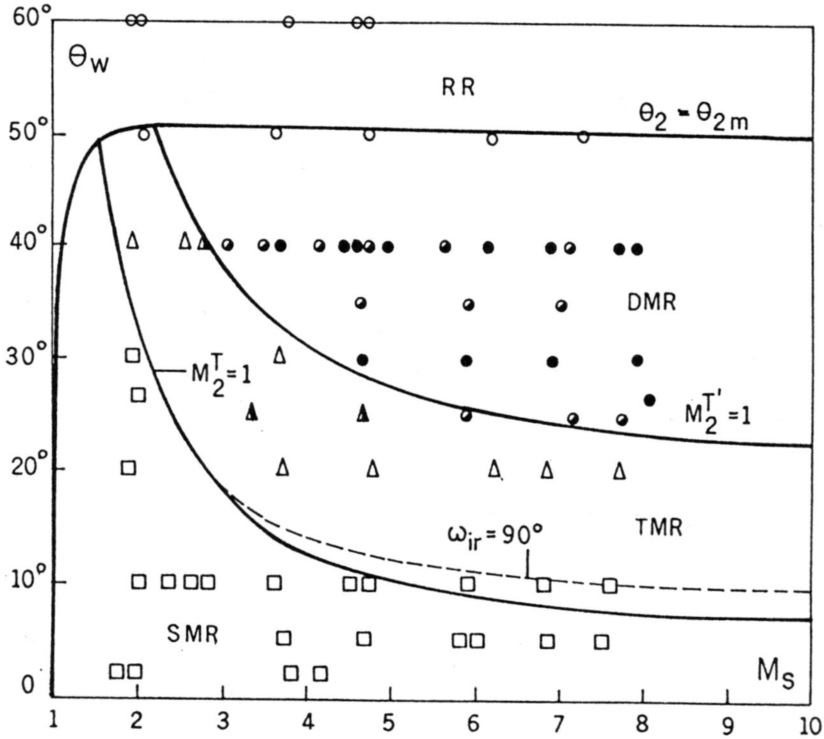

Figure 2.42 a) Domains and transition boundaries of the various types of shock wave reflections in the (M_s, θ_w)-plane for frozen oxygen and nitrogen with $\gamma = 1.4$ and experimental data of UTIAS (courtesy of Professor I.I. Glass).

seen to predict excellently the SMR⇔TMR transition. A comparison of figures 2.41b and 2.42b indicates that as γ decreases (5/3 in the former and 7/5 in the latter) the $M_1^R = 1$ and $M_2^T = 1$ lines, which bound the SMR-domain, lie closer together. This behavior unfortunately prevents a clear-cut distinction of the experiments.

Figures 2.42c and 2.42d show results for air in the (M_S, θ_W) and the (M_S, θ_W^C)-planes, respectively. It is seen from figure 2.42c that at the RR⇔IR transition line two RR-experiments, one at $M_S \approx 1.5$ and one at $M_S \approx 2$, lie in the SMR-domain, one RR-experiment at $M_S \approx 2.2$ lies slightly below the RR⇔IR transition line in the SMR-domain, one RR-experiment at $M_S \approx 2.5$ lies in the TMR-domain, and eight RR-experiments lie in the DMR-domain. As will be shown subsequently (section 2.5.4) this might be attributed to viscous effects along the reflecting wedge surface.

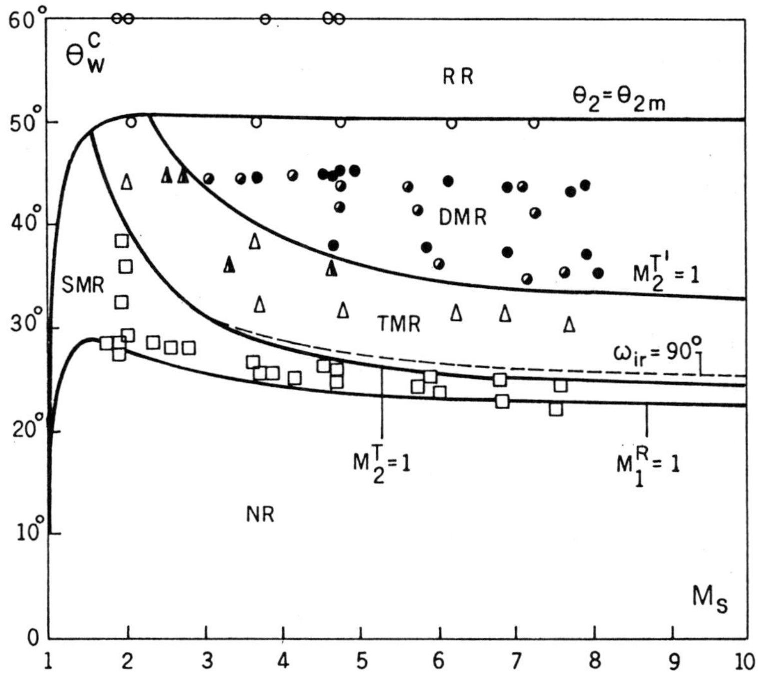

Figure 2.42 b) Domains and transition boundaries of the various types of shock wave reflections in the (M_S, θ_w^C)-plane for frozen oxygen and nitrogen with $\gamma = 1.4$ and experimental data of UTIAS (courtesy of Professor I.I. Glass).

Shirouzu & Glass (1982) have hypothesized that the TMR⇔DMR transition line must meet the SMR⇔TMR transition line at point Q (see figure 2.42c) on the RR⇔IR transition line, since the distance between the triple points T and T' should approach zero at that point. If the $M_2^{T'} = 1$ transition line is hand drawn to account for Shirouzu & Glass's (1982) hypothesis, then as can be seen in figure 2.42c, the two DMR-experiments in the TMR-domain are put into their appropriate DMR-domain. The reason that the calculated $M_2^{T'} = 1$ transition line fails to approach point Q as it approaches the RR⇔IR transition line, assuming that the hypothesis of Shirouzu & Glass (1982) is correct, is probably due to the simplifying assumptions upon which the model from which it is calculated is based. As mentioned in section 2.2.1.1, the assumptions were found by Bazhenova, Fokeev & Gvozdeva (1976) to become

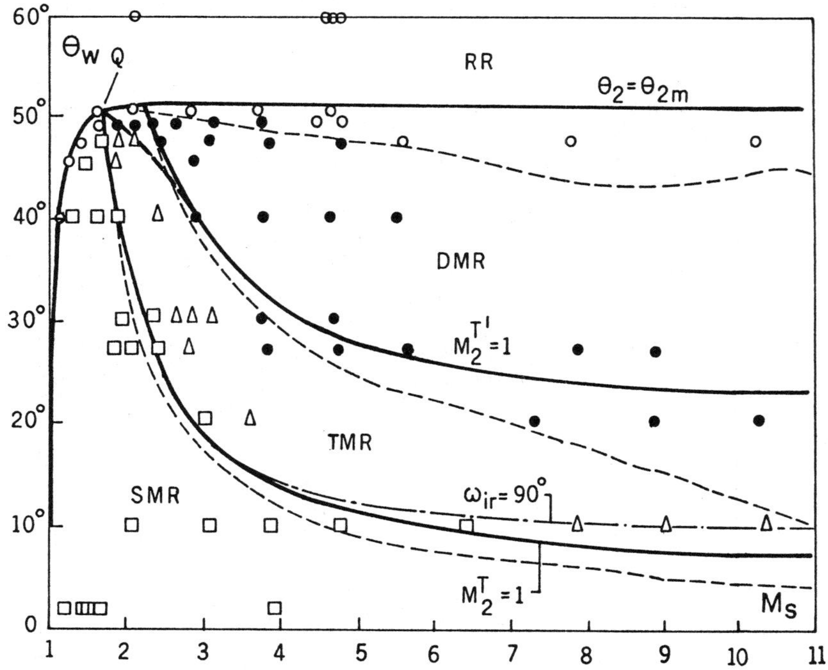

Figure 2.42 c) Domains and transition boundaries of the various types of shock wave reflections in the (M_s,θ_w)-plane and experimental data of UTIAS. Solid lines are for perfect air with $\gamma = 1.4$, dashed lines are for imperfect air with $p_0 = 15$ torr and $T_0 = 300$ (courtesy of Professor I.I. Glass).

progressively poorer as θ_W increased beyond a value of about 40°. Note also that behavior of the $M_2^{T'} = 1$ transition line as hypothesized by Shirouzu & Glass (1982) would also put the DMR-experiments at $\theta_W = 50°$ in figures 2.41a and 2.42a into their appropriate domains.

Further inspection of figure 2.42c indicates that eight DMR-experiments lie in the TMR-domain and three SMR-experiments, at $\theta_W = 10°, 20°$ and $30°$, lie in the TMR-domain very close to the SMR⇔TMR transition line. The $\omega_{ir} = 90°$ line puts the SMR-experiment at $\theta_W = 10°$ into its appropriate SMR-domain.

The RR⇔IR, SMR⇔TMR and TMR⇔DMR transition lines were also drawn (dashed lines) in figure 2.42c for an imperfect air in equilibrium (initially at $p_0 = 15$ torr and $T_0 = 300$K). It is seen, in general, that when real gas effects are accounted for, all the RR-

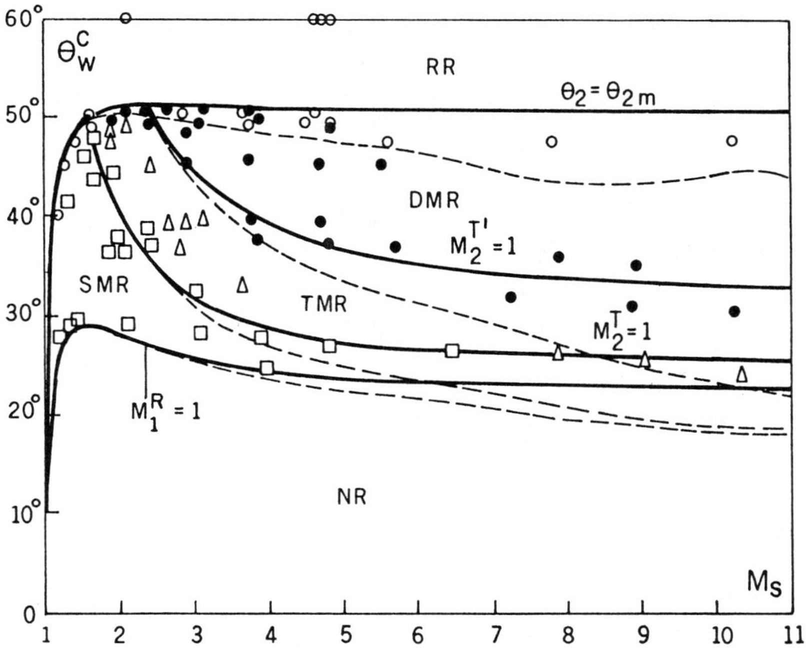

Figure 2.42 d) Domains and transition boundaries of the various types of shock wave reflections in the (M_s, θ_w^C)-plane and experimental data of UTIAS. Solid lines are for perfect air with $\gamma = 1.4$, dashed lines are for imperfect air with $p_0 = 15$ torr and $T_0 = 300$K (courtesy of Professor I.I. Glass).

experiments, except one at $M_S \approx 2$, which were outside their perfect gas domain now lie inside their imperfect gas domain. Unfortunately, however, one DMR-experiment at $M_S \approx 4$ is put into the RR-domain when the imperfect gas transition line is considered. The $M_2^{T'} = 1$ transition line with real gas effects is also able to put five DMR-experiments which were in the TMR perfect gas domain into their appropriate DMR imperfect gas domain. Only one DMR-experiment at $M_S \approx 4$ lies in the TMR-domain. The $M_2^T = 1$ transition line with real gas effects does not change the overall picture as far as the SMR-experiments are concerned. It does, however, put the SMR-experiment at $M_S \approx 6.5$ further into the TMR-domain.

It should be mentioned, however, that it is impossible to draw conclusions from figure 2.42c about the effectiveness of accounting for real gas effects since all the imperfect gas transition lines in

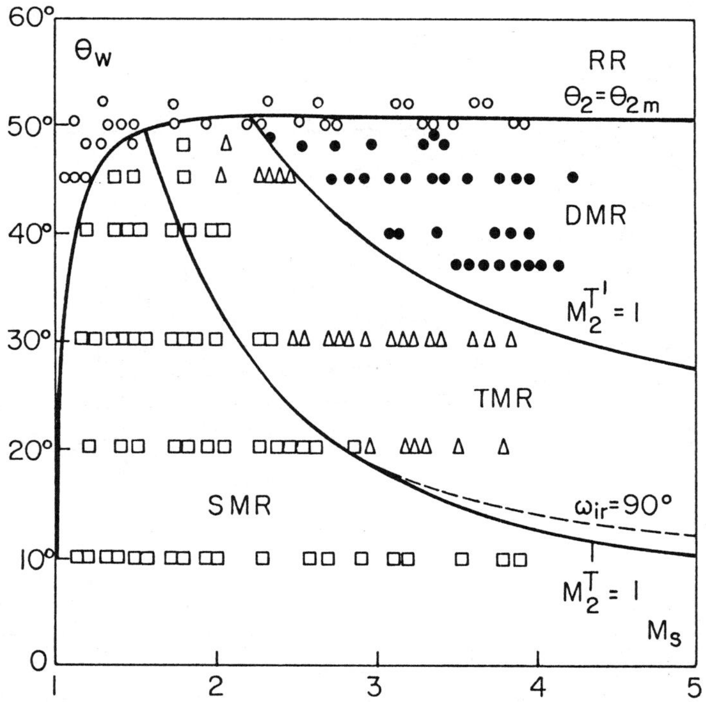

Figure 2.42 e) Domains and transition boundaries of the various types of shock wave reflections in the (M_s,θ_w)-plane for perfect air with $\gamma = 1.4$ and experimental data of Ikui et al. (1981).

figure 2.42c are drawn for $p_0 = 15$ torr and $T_0 = 300K$, while the experiments were conducted in a wide range of initial pressures. (T_0, the room temperature, was probably close to 288K in all the experiments.) As will be shown in section 2.5.3, real gas effects strongly depend on the initial pressure which, in turn, is very dominant in shifting the transition lines. Furthermore, as will be discussed in section 2.5.3 the use of real gas effects in calculating some of the transition lines is not fully justified. The question whether real gas effects should or should not be accounted for depends on the relaxation lengths. Thus, the fact that their inclusion improves the agreement between the analysis and the experiment might be accidental.

A detailed inspection of figure 2.42d clearly indicates that the agreement of the experiments with the analytical transition lines in the (M_s, θ_w^C)-plane is not better than that in the (M_s, θ_w)-plane (figure 2.42c) even though it should be. Nine RR-experiments lie in

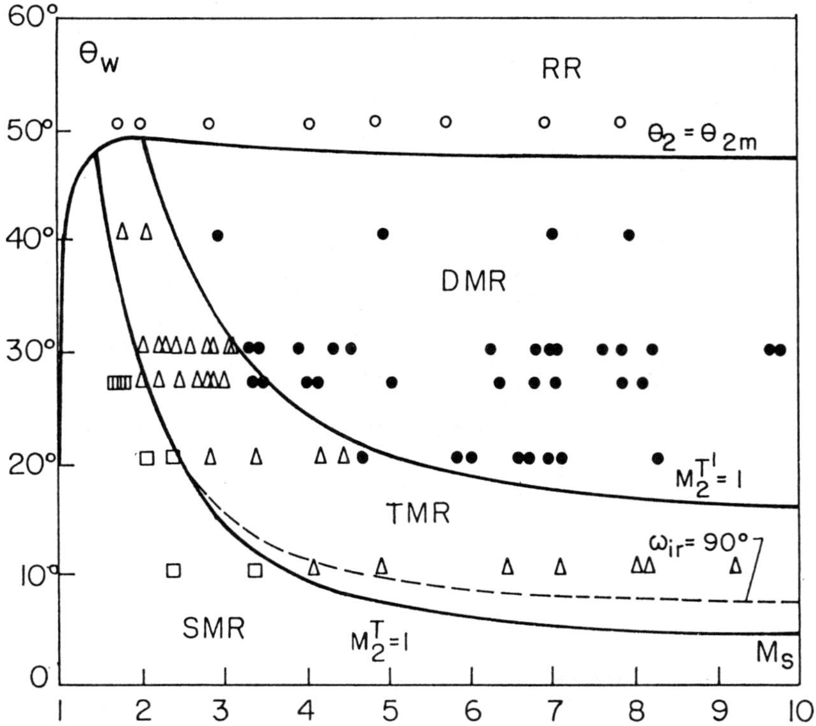

Figure 2.43 a) Domains and transition boundaries of the various types of shock wave reflections in the (M_s,θ_w)-plane for perfect CO_2 with $\gamma = 1.29$ and experimental data of UTIAS (courtesy of Professor I.I. Glass).

the DMR-domain, nine DMR-experiments lie in the TMR-domain, one TMR-experiment lies in the SMR-domain, and four SMR-experiments lie in the TMR-domain. The distribution of the RR- and DMR-experiments at $\theta_w \approx 50^\circ$ and $M_s = 4$ and 5 in figure 2.42d is such that there is no way of separating them and locating them in their appropriate domains by a physically realistic boundary. Thus, one can only wonder how accurately the initial conditions of experimental measurements are recorded and reported, and how good is the subjective interpretation of the experiments by experimentalists.

Figure 2.42e includes the detailed results of Ikui, Matsuo, Aoki & Kondoh (1981) for air. It is seen that excellent agreement is obtained for the RR-experiments at the RR⇔IR transition line. No evidence of the real persistence of RR into the IR-domain can be seen. For $M_s > 2$ the RR-experiments lie marginally below the RR⇔IR transition

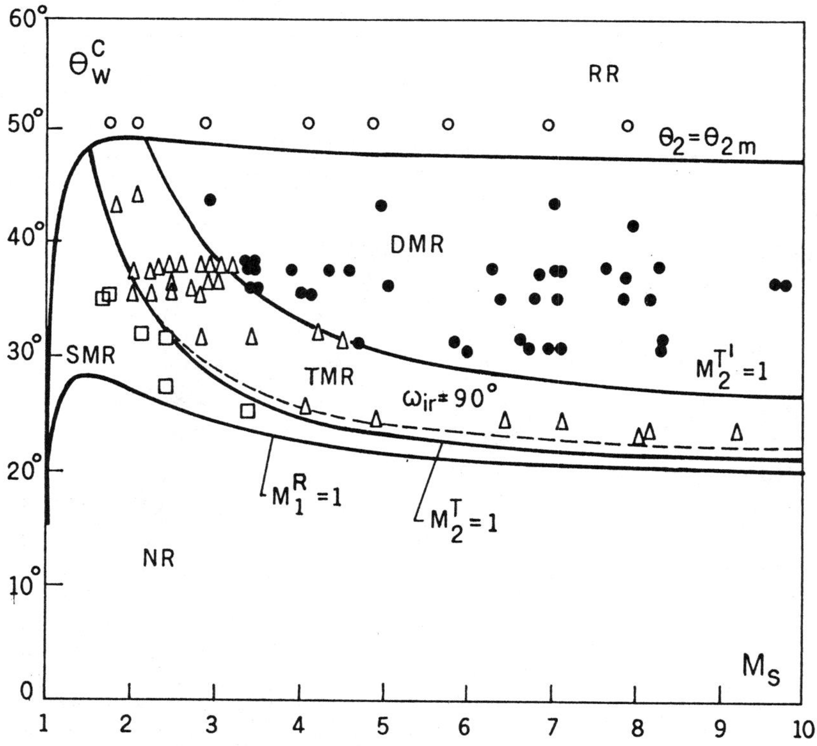

Figure 2.43 b) Domains and transition boundaries of the various types of shock wave reflections in the (M_s,θ_w^C)-plane for perfect CO_2 with $\gamma = 1.29$ and experimental data of UTIAS (courtesy of Professor I.I. Glass).

line. Eight SMR-experiments lie in the TMR-domain. All the TMR- and the DMR-experiments lie in their appropriate domains. Thus the perfect gas transition lines appear quite appropriate for these results. The dashed line for $\omega_{ir} = 90°$ is not helpful, but a new SMR⇔TMR transition line, which would lie closer to the TMR⇔DMR transition line, would place all the SMR-experiments in their appropriate domain.

It should also be noted that while Ikui, Matsuo, Aoki & Kondoh (1981) reported SMR results at $\theta_w \approx 48°$ and $M_s \approx 1.8$ (see figure 2.42e), Shirouzu & Glass (1982) reports TMR and DMR results in the vicinity of this point (see figure 2.42c). Thus, it is possible that either wrong interpretation of similar experimental results adds to the disagreement between theory and experiments, or perhaps the shock reflection phenomenon depends on some additional factors which

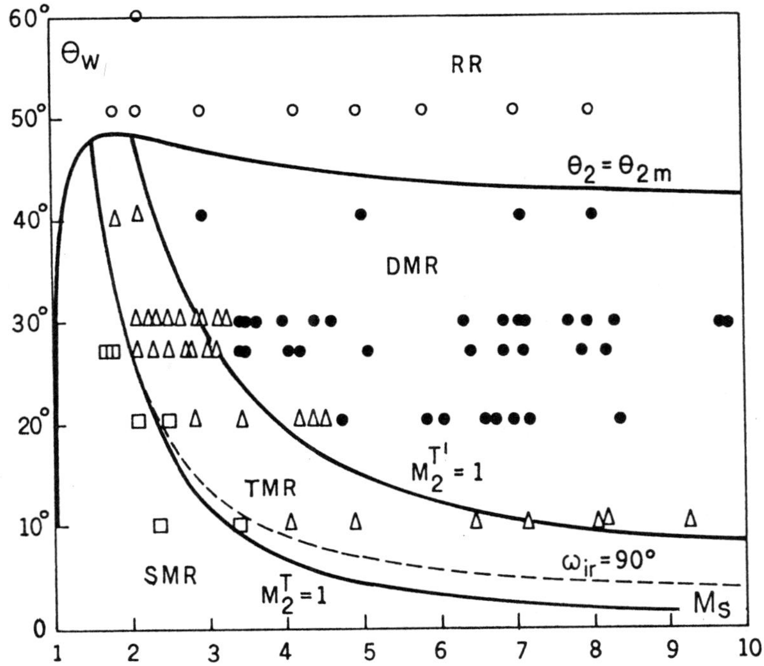

Figure 2.43 c) Domains and transition boundaries of the various types of shock wave reflections in the (M_s, θ_w)-plane for CO_2 in vibrational equilibrium and experimental data of UTIAS (courtesy of Professor I.I. Glass).

have not as yet been identified, e.g., acceleration or attenuation of the incident shock wave, size of the shock tube, size of the reflecting wedge, roughness of the reflecting wedge surface, etc.

Figures 2.43a to 2.43f deal with CO_2 in the (M_S, θ_w) and (M_S, θ_w^C)-planes. Figure 2.43a shows a comparison of the experiments with the analytically predicted transition lines for a perfect CO_2 with $\gamma = 1.29$. Although the use of a perfect gas model is not justified in the case of CO_2 even at room temperature, since the vibrational relaxation lengths are short and hence a vibrational equilibrium model would be more appropriate, excellent agreement is obtained, by and large, for all the experimental points except two DMR-experiments at $M_S \approx 3.5$ and one DMR-experiment at $M_S \approx 4.7$ which lie in the TMR-domain. One TMR-experiment at $M_S \approx 4$ agrees better with the $M_2^T = 1$ criterion than with the $\omega_{ir} = 90^\circ$. The RR-experiments at $\theta_w = 50^\circ$ are not sufficient to test the RR⇔IR transition line, as they do not lie close enough to it. As

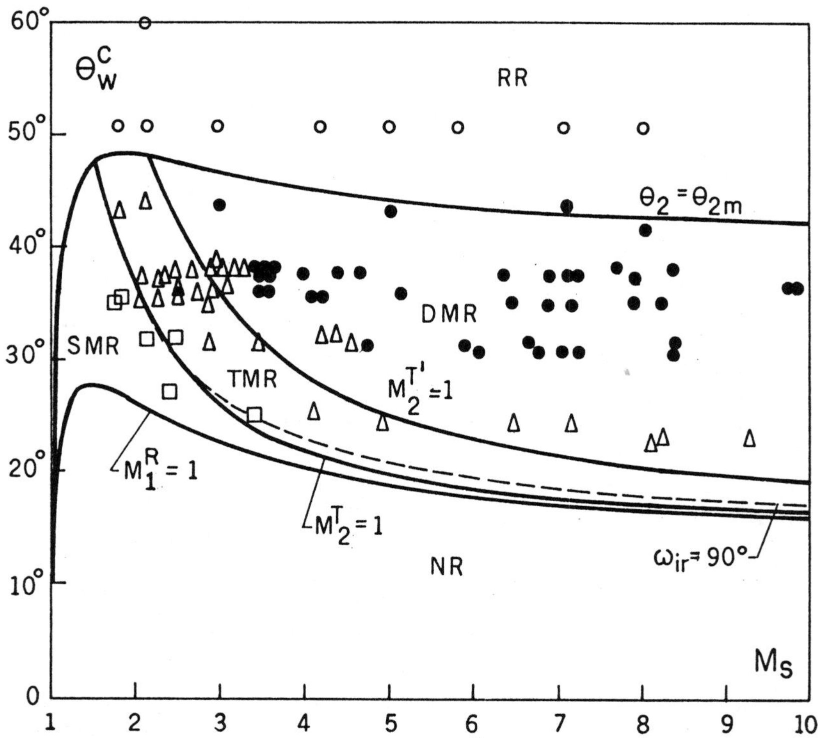

Figure 2.43 d) Domains and transition boundaries of the various types of shock wave reflections in the (M_s, θ_w^C)-plane in vibrational equilibrium and experimental data of UTIAS (courtesy of Professor I.I. Glass).

expected, when the above experiments are plotted in the (M_S, θ_w^C)-plane better agreement is obtained. This is shown in figure 2.43b, where only one TMR-experiment at $\theta_w^C \approx 35°$ lies in the SMR-domain. Note that this is the first evidence of the failure of the $M_2^T = 1$ transition line to accurately bound the TMR-domain. In all the experiments shown in figures 2.41 and 2.42 not even one TMR-experiment lay beyond the $M_2^T = 1$ transition line for O_2, N_2 and air. The fact that the $M_2^T = 1$ transition line failed to accurately place this TMR-experiment in its predicted domain, in spite of its excellent performance for diatomic gases, might be due to the above-mentioned unjustified use of a perfect gas model for CO_2 even at room temperature.

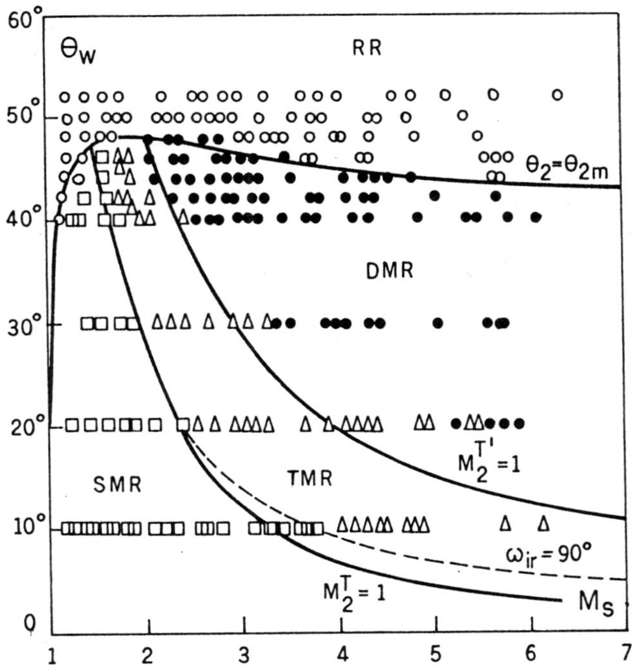

Figure 2.43 e) Domains and transition boundaries of the various types of shock wave reflections in the (M_s, θ_w)-plane for CO_2 in vibrational equilibrium and experimental data of Ikui et al. (1981).

The transition lines for CO_2 in vibrational equilibrium are shown in figure 2.43c. The above-mentioned TMR-experiment which did not agree with the $M_2^T = 1$ transition line, now lies in its appropriate TMR-domain. Eight TMR-experiments lie in the equilibrium DMR-domain (three of them at $M_s > 8$ only marginally) and two SMR-experiments lie in the equilibrium TMR-domain. One of them, at $\theta_w = 10°$, which lies only slightly above the $M_2^T = 1$ transition line, is below the $\omega_{ir} = 90°$ inside its appropriate SMR-domain as defined by this additional requirement for the SMR⇔TMR transition.

Figure 2.43d shows the transition lines for a CO_2 in vibrational equilibrium in the (M_s, θ_w^C)-plot together with the above-mentioned CO_2 experiments. Although the agreement between theory and experiment should be better, since the value of χ is not needed to draw these transition lines, the results are poorer. Fourteen TMR-

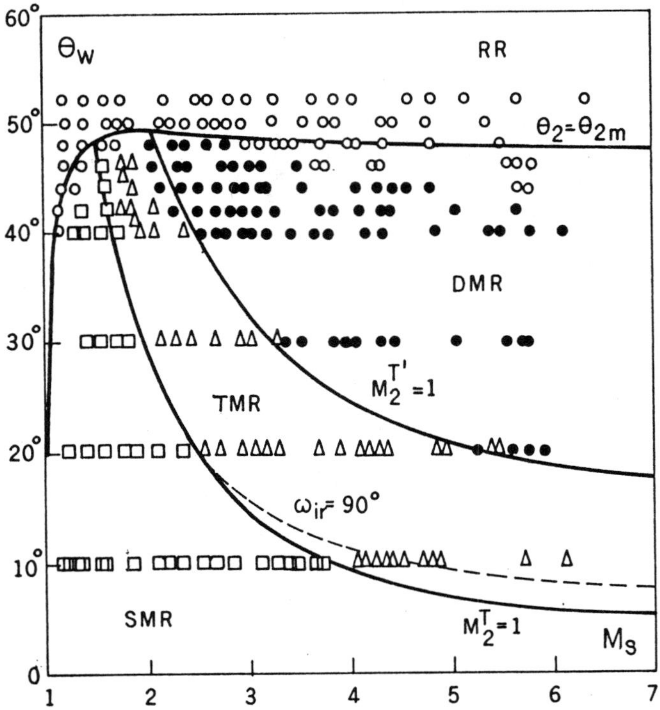

Figure 2.43 f) Domains and transition boundaries of the various types of shock wave reflections in the (M_s,θ_w)-plane for perfect CO_2 with $\gamma = 1.31$ and experimental data of Ikui et al. (1981).

experiments lie in the DMR-domain and two SMR-experiments lie in the TMR-domain as defined by the $M_2^T = 1$ transition line. The $\omega_{ir} = 90^\circ$ transition line is successful in relocating only one of these two experiments into its appropriate SMR-domain. All the RR- and DMR-experiments are correctly located in their appropriate domains. Note how narrow the SMR-domain becomes as M_s increases beyond 5.

Figure 2.43e shows the experimental results of Ikui, Matsuo, Aoki & Kondoh (1981) for CO_2 in the vibrational equilibrium (M_s, θ_w)-plane. In general, good agreement over the entire incident shock wave Mach number range is seen. Only two RR-experiments lie below the RR⇔IR transition line inside the SMR-domain. Three DMR-experiments lie in the RR-domain slightly above the RR⇔IR transition line. There is no evidence of RR persistence into the IR-domain at $M_s > 1.7$. Four SMR-experiments at $M_s \approx 2$ lie marginally in the TMR-domain, and eleven TMR-experiments are placed in the DMR-domain. Note that two of these TMR-experiments, at $\theta_W = 20^\circ$ and $M_S \approx 5.5$, lie betweeen DMR-experiments with $\theta_W = 20^\circ$ and M_S greater and smaller than 5.5. Thus, one can only wonder how accurately investigators interpret their experimental results. The results for $\theta_W = 10^\circ$ clearly indicate that sometimes it is difficult to distinguish TMR-configurations from DMR-configurations, the latter having weak discontinuities at their second triple point. Perhaps this might account for some of the discrepancy along the TMR⇔DMR transition line. It is again evident from figure 2.43e that not even one TMR-experiment lies beyond the $M_2^T = 1$ transition line. The $\omega_{ir} = 90^\circ$ puts the four SMR-experiments at $\theta_W = 10^\circ$ into their proper SMR-domain.

Figure 2.43f again shows the CO_2 experimental results of Ikui, Matsuo, Aoki & Kondoh (1981) in the (M_S, θ_W)-plane with analytical transition lines calculated using a perfect gas model with $\gamma = 1.31$. It is seen that, by and large, good agreement is obtained with all the experiments. The RR⇔IR transition line does not fit the experimental results as well as the transition lines for the CO_2 in vibrational equilibrium (figure 2.43e) since fourteen RR-experiments lie in the DMR-domain. Only two RR-experiments, at $M_S \approx 1.5$, lie in the SMR-domain and two RR-experiments at $M_S \approx 2$ lie in the TMR-domain. Four borderline DMR-experiments at $M_S \approx 2.2$ lie in the TMR-domain. Again not even one TMR-experiment disagrees with the $M_2^T = 1$ transition line, while four of them clearly lie outside the TMR-domain as defined by the additional $\omega_{ir} = 90^\circ$ requirement.

Figure 2.44a shows the experimental results of Ikui, Matsuo, Aoki & Kondoh (1981) for Freon-12 in the (M_S, θ_W)-plane where the transition lines are drawn for a perfect gas with $\gamma = 1.141$. The perfect gas lines exhibit, in general, reasonably good agreement with the experiments. Four RR-experiments lie in the TMR-domain. Many DMR-experiments lie in the TMR-domain. Two TMR-experiments at $\theta_W = 10^\circ$ lie in the DMR-domain. The $M_2^T = 1$ transition line separates perfectly between the SMR- and TMR-experiments. All the TDMR-experiments lie above the RR⇔IR transition line in the RR-domain. The dashed line is an experimental best fit for the RR⇔TDMR transition. Note that this transition could be predicted analytically, as is done in figure 2.44b, by requiring $\chi' = 0$. The only evidence of the persistence of RR beyond its theoretical limit are four RR-experiments which lie in the TMR-domain.

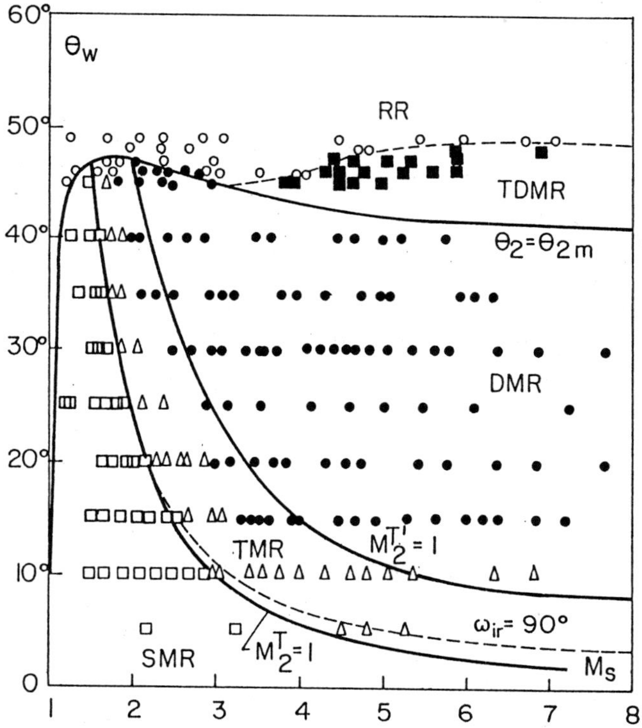

Figure 2.44 a) Domains and transition boundaries of the various types of shock wave reflections in the (M_s,θ_w)-plane for perfect Freon-12 with $\gamma = 1.141$ and experimental data of Ikui et al. (1981).

An (M_S, θ_W)-plot for Freon-12 in equilibrium (solid lines) is shown in figure 2.44b. The transition lines for a perfect gas model (dashed lines) with $\gamma = 1.333$ are also added to figure 2.44b. A comparison of figures 2.44a and 2.44b (solid lines in figure 2.44a and dashed line in figure 2.44b) suggests that the SMR⇔TMR and TMR⇔DMR transition lines are better for a perfect gas model with $\gamma = 1.141$ than with $\gamma = 1.333$. The RR⇔IR transition line, however, is better for a perfect gas model with $\gamma = 1.333$ than with $\gamma = 1.141$ as it places all the TDMR-experiments in the DMR-domain. Thus it is possible that different values of γ (which is an indication of real gas effects) should be used for calculating the various transition lines if a perfect gas model is to be used for calculating the transition boundaries of Freon-12. When the equilibrium transition lines are considered, it is again seen that the

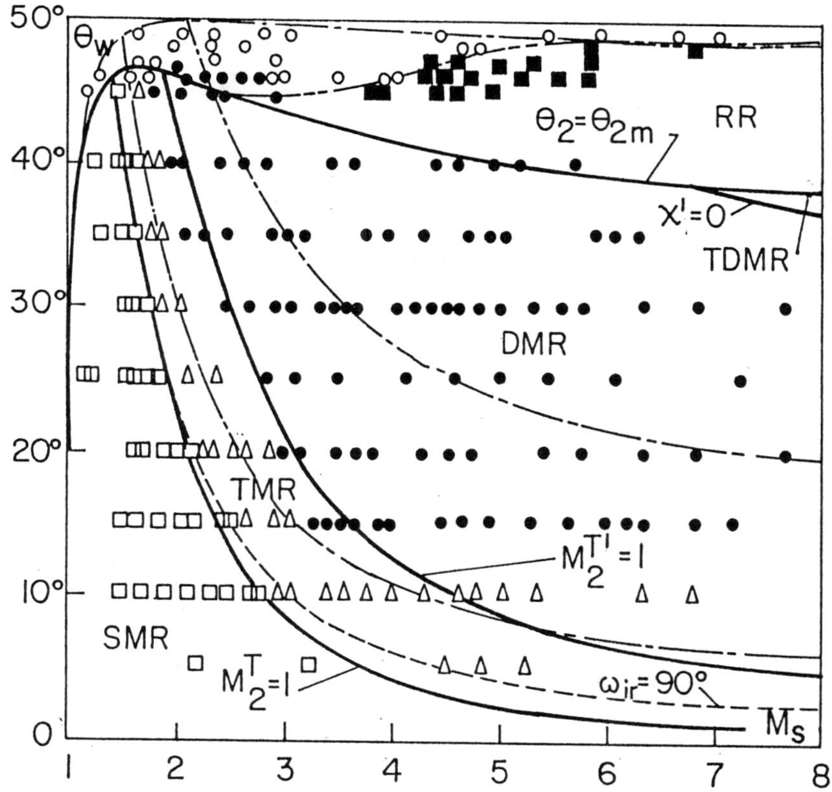

Figure 2.44 b) Domains and transition boundaries of the various types of shock wave reflections in the (M_s,θ_w)-plane. Dashed-dotted lines are for perfect Freon-12 with γ = 1.333. Solid lines are for Freon-12 in equilibrium. Experimental data of Ikui et al. (1981).

$M_2^T = 1$ transition line is very accurate in separating between the SMR- and the TMR-experiments. Only two SMR-experiments at $\theta_w = 15^\circ$ and two at $\theta_w = 40^\circ$ lie beyond this line in the TMR-domain. When the $\omega_{ir} = 90^\circ$ requirement is added, the SMR-experiments at $\theta_w = 15^\circ$ are placed in their proper SMR-domain, but one TMR-experiment at $\theta_w = 10^\circ$ is pushed outside its TMR-domain into the SMR-domain. In addition to the TDMR-experiments which lie in the RR-domain, seven DMR-experiments also lie above the RR$\Leftrightarrow$IR transition line. The $M_2^{T'} = 1$ transition line is better in the case of the equilibrium model than in the case of a perfect gas

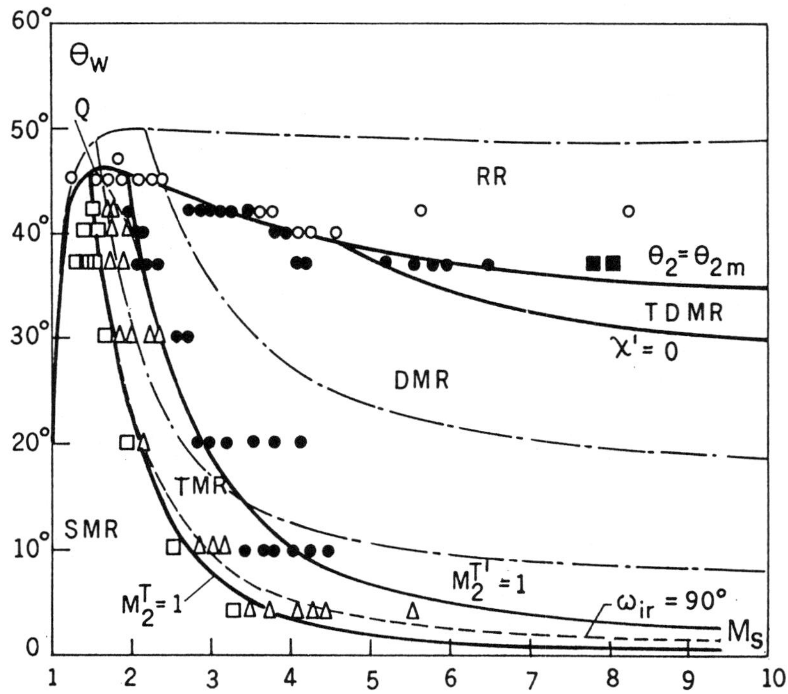

Figure 2.45 a) Domains and transition boundaries of the various types of shock wave reflections in the (M_s,θ_w)-plane and experimental data of UTIAS. Solid lines are for SF_6 in vibrational equilibrium, dashed-dotted lines are for perfect SF_6 with $\gamma = 1.333$ (courtesy of Professor I.I. Glass).

with $\gamma = 1.141$ for $M_S < 5$ and poorer for $M_S > 5$ (compare the solid lines in figures 2.44a and 2.44b).

Note that the line arising from the condition $\chi' = 0$ yields a small TDMR-domain below the RR⇔IR transition line. However, all the TDMR-experiments lie outside this predicted domain inside the RR-domain. The location of the TDMR-experiments in figure 2.44b raises serious doubts about the validity of the calculation of the RR⇔IR transition line in the case of Freon-12.

Figures 2.45a and 2.45b show the results for SF_6 in the (M_S, θ_W) and the (M_S, θ_W^C)-planes, respectively. Considerable efforts were made in this case to check the transition lines themselves. Figure 2.45a shows the transition lines for an equilibrium gas model (solid lines) and a perfect gas model (dashed-dotted lines) with $\gamma = 1.333$. The perfect gas transition lines do not agree with the experiments.

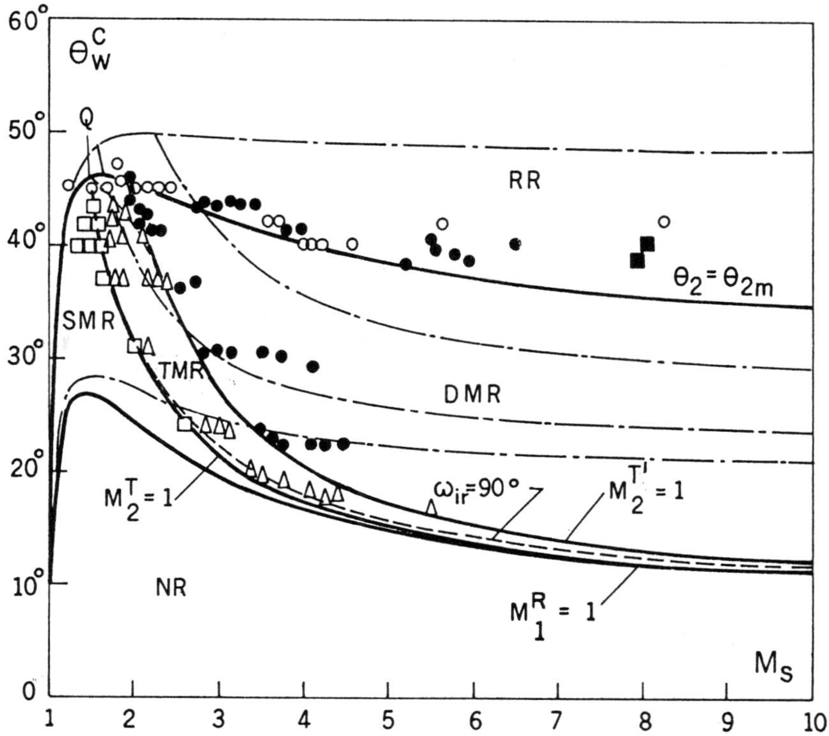

Figure 2.45 b) Domains and transition boundaries of the various types of shock wave reflections in the (M_s,θ_w^C)-plane and experimental data of UTIAS. Solid lines are for SF_6 in vibrational equilibrium, dashed-dotted lines are for perfect SF_6 with $\gamma = 1.333$ (courtesy of Professor I.I. Glass).

The equilibrium transition lines are far better. Three RR- and three DMR-experiments lie in the TMR-domain below the RR⇔IR transition line. There is no serious evidence of the RR persistence beyond its theoretical limit. Only one TMR-experiment at $M_S \approx 3.5$ lies outside its domain as defined by the the $M_2^T = 1$ transition line. The additional requirement for the TMR formation, however, puts three more TMR-experiments into the SMR-domain. Two TDMR-experiments at $M_S \approx 8$ lie above the RR⇔IR transition in the RR-domain. They lie only slightly outside their predicted TDMR-domain, inside which there are four DMR-experiments. In spite of the above-mentioned disagreements, it is quite clear from figure 2.45a that, by and large, the equilibrium transition lines map the experimental results of SF_6 quite well.

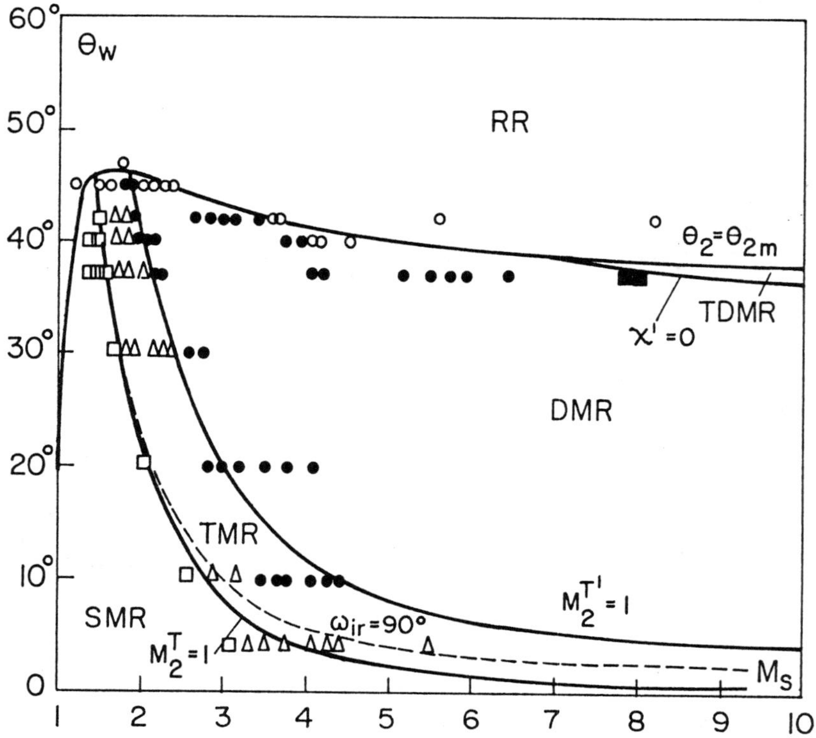

Figure 2.45 c) Domains and transition boundaries of the various types of shock wave reflections in the (M_s,θ_w)-plane and experimental data of UTIAS for perfect SF_6 with $\gamma = 1.093$ (courtesy of Professor I.I. Glass).

When the above-mentioned results are plotted in the (M_S, θ_w^C)-plane (figure 2.45b) many DMR-experiments shift into the RR-domain over the entire range of M_S. Consequently, the (M_S, θ_w^C)-plot again gives poorer, rather than better, agreement with the RR⇔IR transition line. It should be noted that if the three-shock theory were used to calculate the RR⇔IR transition line, rather than the two-shock theory, then χ would have a value of 1° to 3° along the RR⇔IR transition line. In such a case the RR⇔IR transition line would have been pushed upwards for better agreement with the experiments. The small dashed line [hypothesized by Shirouzu & Glass (1982)], connecting the $M_2^T = 1$ transition line to point Q on RR⇔IR transition line, indicates the fact that in actuality this takes place. Note also that the SMR-domain practically vanishes for

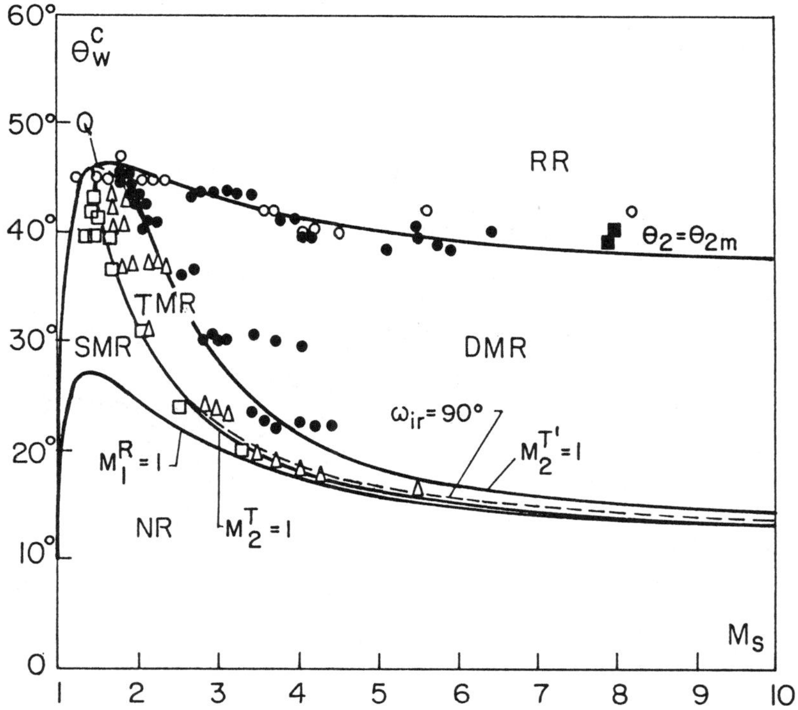

Figure 2.45 d) Domains and transition boundaries of the various types of shock wave reflections in the (M_s,θ_w^C)-plane and experimental data of UTIAS for perfect SF_6 with $\gamma = 1.093$ (courtesy of Professor I.I. Glass).

$M_s > 5$ as the $M_2^T = 1$ transition line approaches the $M_1^R = 1$ boundary of the no-reflection domain.

Figure 2.45c shows the above discussed SF_6 experimental results in the (M_s, θ_w)-plane with transition lines calculated using a perfect gas model with $\gamma = 1.093$. The experiments are represented quite well in this plot. Two RR-experiments lie below the RR⇔IR transition line in the TMR-domain. Two RR-experiments at $M_s \approx 2$, and three at $M_s \approx 4.5$ lie marginally below the RR⇔IR transition line in the DMR-domain, clearly showing the persistence of RR beyond its theoretical limit. Several DMR-experiments lie in the TMR-domain. Those at $\theta_w = 20°$, 40°, 43° and 45° are quite close to the transition line; however those at $\theta_w = 10°$ disagree far too much with the $M_2^{T'} = 1$ transition line. Note that all these DMR-experiments were perfectly located in their appropriate domain in the

(M_S, θ_w^C)-plot of figure 2.45b, and hence one must conclude that the poor agreement seen in figure 2.45c is due to the use of perfect gas transition lines for a gas which has some of its internal degrees of freedom already excited, as well as the use of χ for obtaining the less accurate (M_S, θ_w)-plots. The equilibrium plot in figure 2.45a is superior in providing the RR⇔IR transition line. Note that the SMR⇔TMR and the TMR⇔DMR transition lines are also better in figure 2.45a. The two TDMR-experiments, which in figures 2.45a and 2.45b lie outside the MR-domain, are placed in figure 2.45c in the MR-domain, slightly below their appropriate TDMR-domain.

Figure 2.45d provides the (M_S, θ_w^C)-plot for SF_6 when it is assumed to behave as a perfect gas with $\gamma = 1.093$. The few TMR-experiments which were located in the SMR-domain in figure 2.45c are now in their proper TMR-domain. Similarly, all the SMR-experiments are excellently placed in their SMR-domain. Three DMR-experiments at $M_S \approx 3.5$ lie in the TMR-domain. The several DMR-experiments at $M_S \approx 2$, which lie in the TMR-domain, would lie correctly in the DMR-domain if the TMR⇔DMR transition line were arbitrarily changed to join the SMR⇔TMR transition line on the RR⇔IR transition line (point Q). Note also that several DMR- and the two TDMR-experiments lie slightly above the RR⇔IR transition line in the RR-domain. Overall, the equilibrium transition lines, shown in figure 2.45a, provide the best agreement between theory and experiments for SF_6.

2.4.4 Summary, Critique and Discussion

It can be summarized from the foregoing comparison between the analytical predictions of the various transition lines and the experimental results that, in general, the (M_S, θ_w)-plots for the various gases provide the experimenter with good engineering results. Comparing the results for Ar, N_2, O_2 and air with those for CO_2, Freon-12 and SF_6 clearly indicates that the agreement between the analysis and the experiments was, in general, much better for the former gases which, unlike the latter gases, could be assumed to obey the perfect gas equation of state under the experimental conditions.

Inspecting figures 2.41b, 2.42b, 2.42d, 2.43b, 2.43d, 2.45b and 2.45d which represent the (M_S, θ_w^C)-plots for the various gases, indicates that the boundary beyond which reflection cannot occur, i.e. $M_1^R = 1$, is excellent. Not even one experiment resulted in a reflection in the NR-domain.

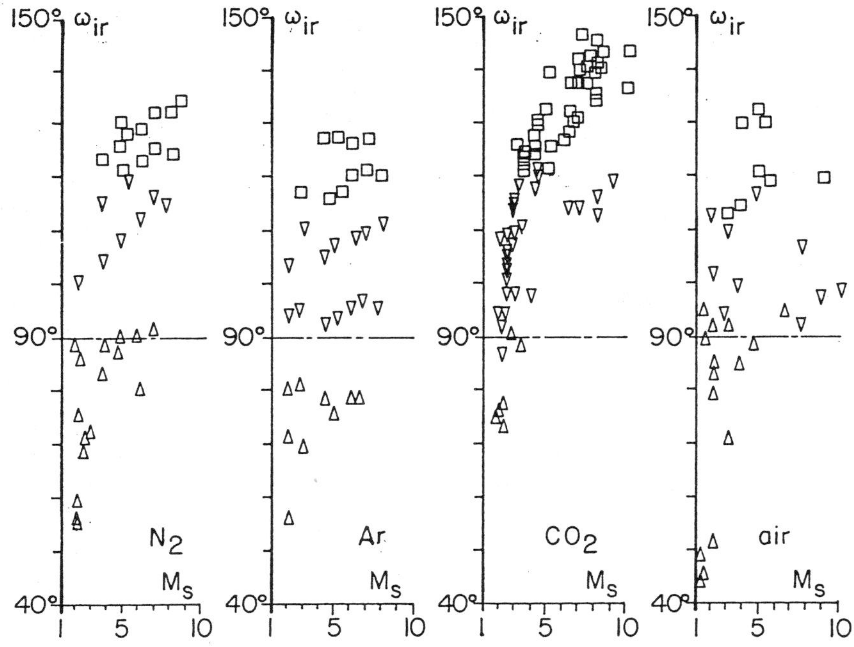

Figure 2.46 The value of ω_{ir} as measured in actual SMR and TMR experiments for N_2, Ar, CO_2 and air. The experimental data is from UTIAS (courtesy of Professor I.I. Glass).

The same figures indicate that almost no TMR-experiments lie in the SMR-domain on the wrong side of the $M_2^T = 1$ transition line.

Unfortunately, however, SMR-experiments do lie on the wrong side of the $M_2^T = 1$ transition line, in the TMR-domain. Thus it can safely be concluded that TMR cannot occur for $M_2^T < 1$.

Figure 2.46 illustrates the distribution of SMR- and TMR-experiments for N_2, Ar, CO_2, and air in the (ω_{ir}, M_S)-plane. All the TMR-experiments except one for CO_2 satisfy the condition $\omega_{ir} > 90^\circ$. These results provide very good evidence that $\omega_{ir} > 90^\circ$ is a necessary condition for the SMR⇔TMR transition. Several SMR-experiments which lie above the $\omega_{ir} = 90^\circ$, in the case of N_2 and air, suggest that $\omega_{ir} \geq 90^\circ$ is not a sufficient condition. In spite of this clear evidence, which is completely based on experimental measurements and hence, includes neither calculations nor assumptions, the behavior of the calculated $\omega_{ir} = 90^\circ$ transition line in figures 2.41 to 2.45 does not help in clearly separating the SMR- and the TMR-domains. As is evident from figures 2.42a, 2.42c, 2.43c, 2.43e and 2.44b the calculated $\omega_{ir} = 90^\circ$ line has the potential for sometimes placing SMR-experiments, which violated the $M_2^T = 1$ transition line, in their appropriate domain. Not even one SMR-experiment which was in its appropriate domain as defined by the $M_2^T = 1$ transition line was moved into the TMR-domain due to the additional calculated $\omega_{ir} \geq 90^\circ$ requirement. This is not the case for many TMR-experiments which originally were in their appropriate TMR-domain as defined by the $M_2^T = 1$ transition line and were pushed into the SMR-domain as defined by the calculated $\omega_{ir} = 90^\circ$ boundary (see figures 2.43a, 2.43f, 2.44a, 2.44b, 2.45a and 2.45c). This contradictory result clearly suggests that an error is introduced during the analytical calculation of ω_{ir}. As will be shown in section 2.5.4.2 the inviscid three-shock theory does indeed fail to accurately predict the angles between the various discontinuities near the triple point, T. Thus, it is expected that the $\omega_{ir} = 90^\circ$ transition line would better predict the experimental results if it is calculated using a three-shock theory which is modified to account for viscous effects.

Figures 2.41 to 2.45 also indicate that the $M_2^{T'} = 1$ transition line is not accurate enough in clearly and sharply distinguishing between the TMR- and DMR-domains. This is probably due to the oversimplifying assumptions used in calculating the triple point trajectory angles χ and χ' as well as the motion of the second triple point with respect to the first triple point. It is expected that more realistic models would result in a much better agreement.

The earlier-mentioned hypothesis of Shirouzu & Glass (1982) about the need for the TMR⇔DMR transition line to approach the SMR⇔TMR transition line as both of them reach the RR⇔IR transition line is verified by the experimental results. The failure of the $M_2^T = 1$ transition line to do so, is also probably due to the oversimplifying assumptions upon which its calculation is based. Unfortunately, these assumptions were found by Bazhenova, Fokeev & Gvozdeva (1976) to be very poor, especially in the reflection wedge angles range, where these two transition lines are expected to merge.

Unlike the fairly good agreement between the analysis and the experiments for monatomic and diatomic gases, which as mentioned earlier can be safely assumed to behave as perfect gases under the experimental conditions, the experimental results for the more complex gases, e.g., CO_2, Freon-12 and SF_6, do not indicate good agreement with the analysis. The main reason is probably due to the fact that these gases, which are already excited at room temperature, result in nonequilibrium flows behind the shock waves. Thus neither a perfect gas solution nor an equilibrium solution could be expected to reproduce the experimental results. At most, these solutions can be regarded as upper and lower limits of the actual solution. The obvious persistence of DMR-experiments far into the RR-domain suggests that the RR⇔IR transition line should be calculated with more realistic assumptions. Since this transition line is calculated by solving the flow field at the reflection point, R, of an RR, which lies on the reflecting wedge surface, perhaps heat transfer as well as viscosity are extremely important in calculating this transition line. However, real gas effects (see figures 2.41 to 2.45) and viscosity (see figure 2.53) cause the RR⇔IR transition line to shift downwards in a direction opposite to that required by the dashed lines in figures 2.44a and 2.44b. Thus, it might be that the heat transfer mechanism is the additional factor needed for calculating a more accurate RR⇔IR transition line.

The fact that all the TDMR-experiments (see figures 2.44 and 2.45) lie outside their predicted domain clearly suggests that the $\chi' = 0$ transition line is not calculated accurately enough. This again is probably due to the inadequate way in which real gas effects were accounted for, i.e., equilibrium rather than nonequilibrium, as well as the oversimplifying assumptions used to calculate χ'. In addition, the fact that $\chi' = 0$ forces the second triple point to be on the reflecting wedge surface probably means that viscosity and heat transfer effects cannot be neglected if one is to calculate a more realistic $\chi' = 0$ transition line.

In summary, the fact that the two- and three-shock theories fail to accurately predict the two- and three-shock wave reflection configurations in pseudo-steady flows might be attributed to two major causes:

1. some of the assumptions upon which the two- and three-shock theories are based are not justified, and
2. there are some additional factors in the experimental set-up used to generate shock wave reflections in shock tubes which influence the actual two- and three-shock wave configurations.

In the following, both of these major causes are examined. Wherever possible, modifications are introduced into the two- and three-shock theories.

2.5 Modifications of the Perfect Inviscid Two- and Three-Shock Theories

The foregoing comparisons between the analytical predictions based on the two- and three-shock theories and the experimental results clearly indicate that the two- and three-shock theories must be modified in order to improve the analytical predictions and obtain better agreement with the experimental results.

The major assumptions upon which the two- and three-shock theories are based are:

1. the flow field is steady,
2. the discontinuities at the reflection point of an RR and at the triple points of an MR are straight, i.e., the flow regions bounded by each pair of neighboring discontinuities is uniform,
3. the flow obeys the equations of state of a perfect gas ($p = \rho RT$),
4. the flow is inviscid ($\mu = 0$),
5. the flow is thermally non-conductive ($k = 0$), and
6. the contact discontinuity of the triple point is a slipstream, i.e., it is infinitely thin.

In the following, the validity of each of these assumptions is discussed. Whenever possible, a modified model in which the assumption under consideration is relaxed is presented.

2.5.1 Non-Steady Effects

The two- and three-shock theories assume that the flow fields in the vicinity of the reflection point of an RR, and in the vicinity of the triple points of an MR, are steady. Hence, strickly speaking, if these theories are used to investigate shock wave reflection configurations in steady flows (e.g., reflections which are obtained in wind tunnels) then the requirement of steady flow is fulfilled.

However, in shock tube experiments, the flow is not steady. The use of the two- and three-shock theories for investigating shock-wave reflection configurations in this case was justified by experimental observations in the early 1940's, which reported that the reflection configurations were self-similar, and hence, the flow field could be considered as pseudo-steady.

Unfortunately, however, recent experimental investigations regarding the point of formation of the first triple point raise some doubts about the validity of the self-similar behavior of the MR configuration. As mentioned earlier, Reichenbach (1985) and

Schmidt (1989) provided experimental evidence that the triple point does not form at the leading edge of the reflecting wedge, but at some distance up the reflecting wedge surface. Furthermore, they showed that the triple point trajectory is not straight near the point along the reflecting wedge surface where the triple point is formed. Consequently, an MR in its early stages, cannot have a self-similar structure. It is possible, however, that it approaches a self-similar structure after the triple point trajectory straightens out.

Further comments regarding the self-similarity of the reflection configurations in pseudo-steady flows are given in section 2.5.7.

2.5.2 Non-Straight Discontinuities

The two- and three-shock theories assume that the discontinuities at the reflection point of an RR and at the triple point of an MR are straight. This assumption implies that the flow region bounded by any two neighboring discontinuities is uniform.

In the case of an RR (see figures 2.2 and 2.3), which exists as long as the flow behind the reflection point, R, is supersonic with respect to the reflection point, i.e., $M_2^R > 1$, this assumption is justified since both the incident shock wave, i, and the reflected shock wave, r, are separating supersonic flow regions. Thus, the assumption of straight discontinuities in the vicinity of the reflection point of an RR should not introduce any error into predictions based on the two-shock theory when it is used to calculate the flow properties in the vicinity of the reflection point.

Unfortunately, however, this is not the case with the three-shock theory. While the incident shock wave, i, of the triple point, T, of an MR is always straight, both the Mach stem, m, and the slipstream, s, are always curved in the vicinity of the triple point. Thus, the assumption of straight discontinuities in the vicinity of the triple point introduces an inherent error into the predictions based on the three-shock theory.

Furthermore, the reflected shock wave, r, of the first triple point, is straight in the vicinity of the triple point, T, only for TMR (see figures 2.8 and 2.9) and DMR (see figures 2.12 and 2.13) which exist only for $M_2^T > 1$. In the case of an SMR (see figures 2.5 and 2.6), the reflected shock wave is curved in the vicinity of T since an SMR is limited to the $M_2^T < 1$ domain. This fact introduces further inherent errors into predictions based on the three-shock theory when it is used to calculate the flow properties in the vicinity of the triple point of an SMR.

Similarly, the discontinuities in the vicinity of the second triple point, T', of a DMR (see figures 2.12 and 2.13) are not all straight.

Both the second Mach stem, m', and the second slipstream, s', are curved. Thus, the use of the assumption of straight discontinuities introduces an inherent error into predictions based on the three-shock theory when it is used to calculate the flow properties in the vicinity of the second triple point of a DMR, or the kink of a TMR.

2.5.3 Real Gas Effects

When a shock wave propagates through a gas, the translational and rotational degrees of freedom of the gas molecules are excited to a new state of equilibrium. The distance along which this occurs is known as the relaxation length. Its length is equal to a few mean-free-paths, which is of the order of the thickness of the shock wave front. The other internal degrees of freedom require a longer time (or distance) to reach equilibrium. Thus, in the analysis of gasdynamic shock wave phenomena the role of the relaxation length is very important. If the relaxation length of an internal degree of freedom is much longer than a characteristic length of the phenomenon, then the internal degree of freedom can be treated as frozen at its pre-shock state. If, however, the relaxation length of an internal degree of freedom is considerably shorter than a characteristic length of the phenomenon, it can be assumed to be in equilibrium immediately behind the shock wave front. Note that this is only a simplifying assumption, since equilibrium is approached only at a distance of the order of the relaxation length. The gas is in non-equilibrium when the relaxation length is of the order of the characteristic length of the phenomenon. In such a case, the frozen and the equilibrium solutions might be considered as two extreme cases bounding the real non-equilibrium solution.

Although there are no straight-forward rules for choosing a characteristic length of a given phenomenon, there are cases where the choice is quite simple. In the case of the TMR⇔DMR transition which is based on the flow Mach number behind the reflected shock wave with respect to the kink or the second triple point, M_2^K or $M_2^{T'}$, the distance between the kink K or the first, T, and the second, T', triple points can be considered as a characteristic length. This is because the flow state in the vicinity of the second triple point depends on the length of the relaxation zone behind the incident shock wave.

Consider figure 2.47 where a DMR reflection at three different positions along a given reflecting wedge is schematically drawn. The vibrational and the dissociational relaxation lengths, l_v and l_d, respectively, are also drawn behind the incident shock wave. These lengths, which depend solely on the incident shock wave Mach number and the flow properties ahead of it, are constant, as is

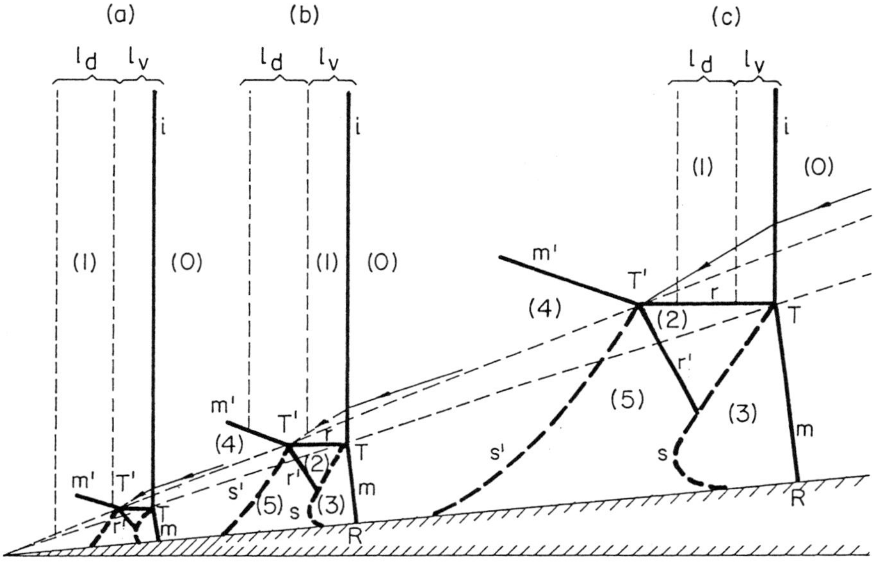

Figure 2.47 Schematic illustration of the way by which real gas effects might affect a DMR.

shown in figure 2.48 for N_2, O_2 and CO_2. The reflection configuration, however, grows with time. When the incident shock wave is at position "a" the second triple point, T', is still inside the zone where the vibrational degree of freedom has not yet reached its equilibrium. The dissociational degree of freedom is still frozen at that position, as its excitation can be assumed to start only after vibrational equilibrium is reached. Thus, if the flow field in the vicinity of the second triple point is to be solved, the flow approaching the second triple point from state (1) should be treated as a non-equilibrium flow. At a later time, the incident shock wave reaches position "b". At this position the second triple point, T', lies beyond the vibrational relaxation zone, and hence the flow is already in vibrational equilibrium. However, the dissociational degree of freedom which is already excited at this position is not in equilibrium yet. Thus, the treatment of the flow field in the vicinity of the second triple point, T', in this situation is different from that in the previous situation. When the incident shock wave reaches position "c", the second triple point, T', is already located beyond

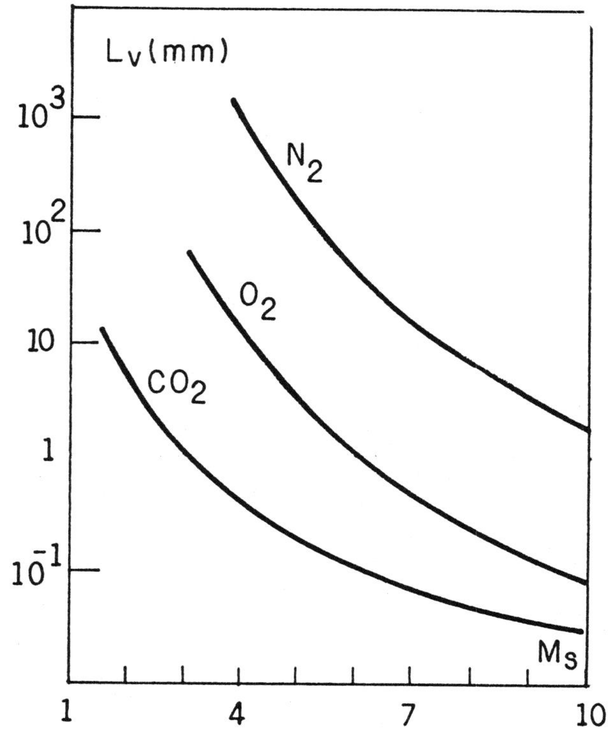

Figure 2.48 Variation of the vibrational relaxation lengths of N_2, O_2 and CO_2 behind the incident shock wave for $p_0 = 15$ torr and $T_0 = 300K$.

the dissociational relaxation zone. Thus, the flow approaching the second triple point from state (1) is in dissociational equilibrium.

The foregoing illustration and discussion clearly indicate the important role of the various relaxation lengths in accurately solving the flow field in the vicinity of the second triple point. It is clear that the flow field in the vicinity of the second triple point depends on the continuously changing distance between the second triple point, T', and the incident shock wave, i, or alternatively on the location of the incident shock wave along the reflecting wedge surface.

Note, that unlike the second triple point, which moves backwards with respect to the incident shock wave, the first triple point is a part of the incident shock wave and hence all the internal degrees of freedom are frozen in its vicinity and retain their pre-shock states. Therefore, while real gas effects must be accounted for when the flow field in the vicinity of the second triple point is solved, a frozen gas solution should be carried out for the flow in the vicinity of the first triple point. This implies that the SMR⇔TMR transition criterion, i.e., $M_2^T = 1$, should be based on a frozen flow in states (1), (2) and (3) while the TMR⇔DMR transition criterion, i.e., $M_2^{T'} = 1$, depends on the thermodynamic state of the flow in state (1), which, as shown earlier, depends on the location of the incident shock wave along the reflecting wedge.

It is evident from the foregoing discussion that the choice of a characteristic length in the case of a DMR is quite obvious if the flow in the vicinity of the second triple point is to be solved. In other cases, such as RR or SMR, the choice of a characteristic length is much more difficult. However, as long as the flow fields in the vicinity of the reflection point, R, of an RR, or the first triple point, T, in an MR, are to be solved, it seems justified to assume that the internal degrees of freedom of the flow remain frozen and hence retain their pre-shock values. Thus gases, such as Ar, N_2, O_2 and air, which at room temperature can be assumed to behave as perfect gases, should be treated as perfect gases near the reflection point of an RR or the first triple point of an MR. Similarly, gases which are already excited at room temperature, such as Freon-12, SF_6 and CO_2 should be treated as frozen at their excited level in the vicinities of R and T.

The foregoing discussion also suggests that even in the case of a single reflection configuration, it is not possible to determine one single characteristic length. In the case of DMR, for example, two different characteristic lengths should be considered for treating each of the two triple points. The two characteristic lengths might differ by orders of magnitude. In addition, the internal degrees of freedom do not relax in the simplified manner shown in figure 2.48. The dissociational relaxation does not start after vibrational equilibrium is reached, but earlier. Therefore there are positions in the flow field where both the vibrational and dissociational degrees of freedom are not fully excited. When the temperatures are high enough electronic excitation and ionization further complicate the phenomenon. The situation is even further complicated if one recalls that the relaxation processes themselves do not have a precise length, for it takes a considerably longer distance than the relaxation length [usually defined as the distance where the flow properties reach (1 - 1/e) of their equilibrium value which is reached theoretically at infinity] to achieve the final equilibrium state.

The foregoing discussion suggests that obtaining a real non-equilibrium solution might be too complicated. Therefore, it is a common practice to model real gas behavior by assuming an equilibrium of the most likely activated degree of freedom, i.e.,

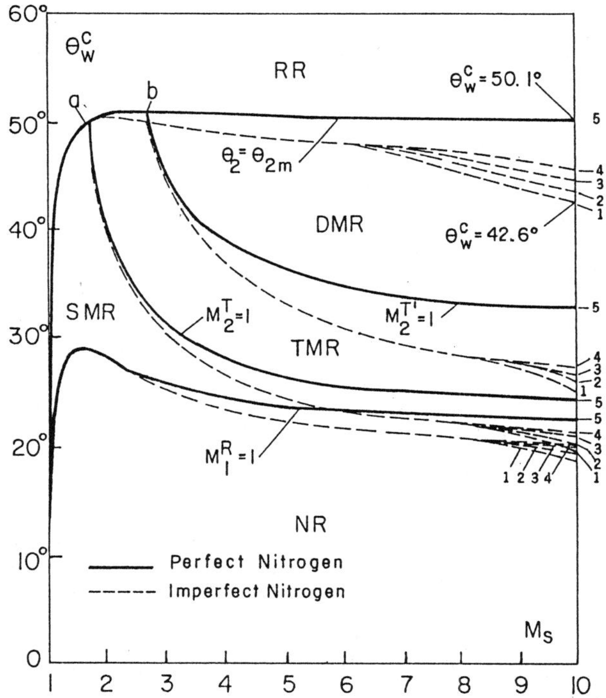

Figure 2.49 Domains and transition boundaries of different types of shock wave reflections in the (M_s, θ_w^C)-plane. Solid lines are for perfect nitrogen with $\gamma = 1.4$. Dashed lines are for imperfect nitrogen in dissociational equilibrium with $T_0 = 300K$ and $p_0 = 1, 10, 100, 1000$ torr for lines 1 to 4, respectively.

vibrational, dissociational, electronic excitational or ionizational equilibrium. Such a model should therefore be regarded as an upper limit on the phenomenon, where the perfect gas model most likely results in the lower limit. Further discussions concerning the assumptions regarding the excitation of the internal degrees of freedom can be found in Shirouzu & Glass (1986) and Glaz, Colella, Collins & Ferguson (1988).

Figure 2.49 illustrates the domains of different types of reflections in the (M_s, θ_w^C)-plane for perfect nitrogen (solid lines) and nitrogen in dissociational equilibrium (dashed lines). The real gas model assumes a dissociational equilibrium everywhere. At relatively low incident shock wave Mach numbers, the transition lines for the imperfect nitrogen diverge from their appropriate perfect gas transition lines due to the excitation of the vibrational degree of

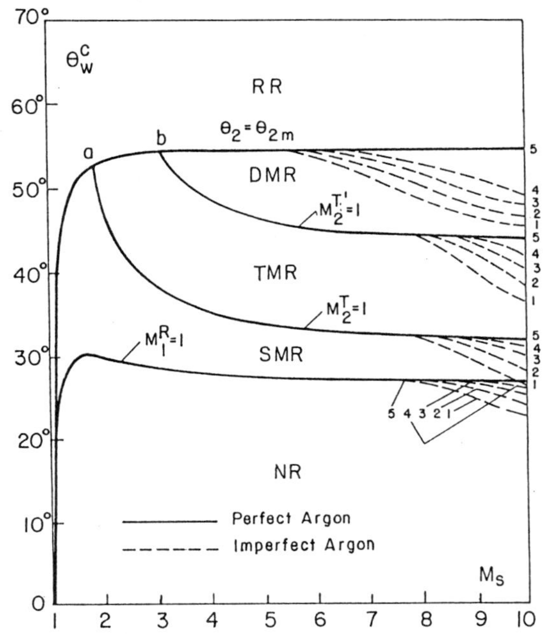

Figure 2.50 Domains and transition boundaries of different types of shock wave reflections in the (M_S, θ_w^C)-plane. Solid lines are for perfect argon with $\gamma = 5/3$. Dashed lines are for imperfect argon in ionizational equilibrium with $T_0 = 300K$ and $p_0 = 1, 10, 100, 1000$ torr for lines 1 to 4, respectively.

freedom. Each of the real gas boundary lines splits up, at a higher incident shock wave Mach number, into four lines (numbered 1 to 4). This split is due to the excitation of the dissociational degree of freedom. Each of these four lines corresponds to a different initial pressure, p_0. The lower is the initial pressure, the greater is the shift from the appropriate perfect gas line. Unlike the dissociational relaxation, which is pressure dependent, the vibrational relaxation does not depend on the pressure. Therefore the transition lines for different initial pressures, but identical initial temperatures, converge to a single line.

Figure 2.49 clearly indicates that, for a given incident shock wave Mach number and a given reflecting wedge angle, the resulting reflection depends on the thermodynamic state of the gas. If the gas can be assumed to be in dissociational equilibrium, then at high incident shock wave Mach numbers the phenomenon depends also on the initial pressure. For example, at $M_S = 10$ and $\theta_w^C = 44°$ an RR should be expected if $p_0 < 1$ torr, while a DMR would take place if $p_0 > 1000$ torr. Since, however, the real flow is most likely not in dissociational equilibrium, one can conclude from figure 2.49 that for $M_S = 10$ and $p_0 = 1$ torr an RR cannot be obtained for $\theta_w^C < 42.6°$ and an MR is impossible for $\theta_w^C > 50.1°$. In the intermediate range $42.6° < \theta_w^C < 50.1°$ either an RR or a DMR could be obtained, depending upon the actual thermodynamic state of the gas (nitrogen in this case).

A similar plot for argon is shown in figure 2.50. In addition to the translational degree of freedom, a monatomic gas has two other degrees of freedom, namely, electronic excitation and ionization. Therefore, the transition lines for the imperfect argon diverge from their appropriate perfect gas transition lines due to the excitation of the ionizational degree of freedom. Each of the transition lines splits up into four lines (numbered 1 to 4). Each of these four lines corresponds to a different initial pressure, p_0. The lower is the initial pressure, the greater is the shift from the appropriate perfect gas transition line. Unlike the ionizational relaxation, which is pressure dependent, the electronic exitational relaxation does not depend on the pressure. Therefore, the transition lines for different initial pressures, but identical initial temperatures, converge to a single line.

Note that both figures 2.49 and 2.50 indicate that real gas effects cause the transition lines to shift away from the perfect gas transition lines to lower values of reflecting wedge angles. In addition, the smaller the initial pressure, the smaller is the incident shock wave Mach number at which this shift starts.

2.5.4 Viscous Effects

One of the assumptions upon which the foregoing analyses of the regular and Mach reflections were based was that the fluid is inviscid. This, however, is not true, for in reality all fluids have a finite viscosity. This viscosity will cause a momentum exchange between the fluid and the reflecting surface over which it flows in the case of RR and MR, and between the fluids on both sides of the slipstream in the case of MR. In the following, the viscous effects on RR and MR are treated separately.

2.5.4.1 Viscous Effects on RR

Consider figure 2.51a where an RR over a straight surface is schematically drawn. The figure is, in general, similar to figure 2.4.

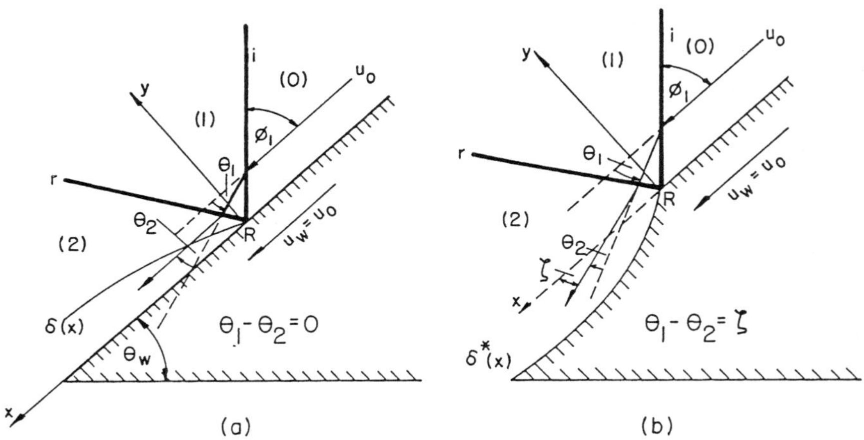

Figure 2.51 Schematic illustration of: a) viscous RR over the real reflecting surface; b) inviscid RR over the displaced reflecting surface.

However, unlike figure 2.4 the boundary layer $\delta(x)$, which develops in state (2) along the reflecting surface in a frame of reference attached to the reflection point, R, is added to figure 2.51a. This boundary layer, which starts to grow at the reflection point ($x = 0$), indicates the flow region in which viscous effects are dominant, and hence should not be neglected.

It is clear from figure 2.51a that if the boundary condition of a regular reflection, i.e., $\theta_1 - \theta_2 = 0$ (see equation 1.13) is to be retained, then the gas in state (2) cannot be treated as inviscid. Instead the full Navier-Stokes equations should be solved in state (2). This is not the case in states (0) and (1). In state (0) there is no friction between the gas and the reflecting surface as both of them move with identical velocity and state (1) is isolated from the reflecting surface. Therefore, the fluid in states (0) and (1) can be assumed to be inviscid, and a full solution of the Navier-Stokes equations is not required there. Unfortunately, solution of an analytical model with the full Navier-Stokes equations in state (2) is too difficult a task to undertake.

There is, however, a simple way, known as the boundary layer displacement technique, to overcome the above difficulty. By applying the boundary layer displacement technique, it is possible to change the geometry of the reflecting wedge surface in such a way that the flow over it can be treated as inviscid. The idea to use this

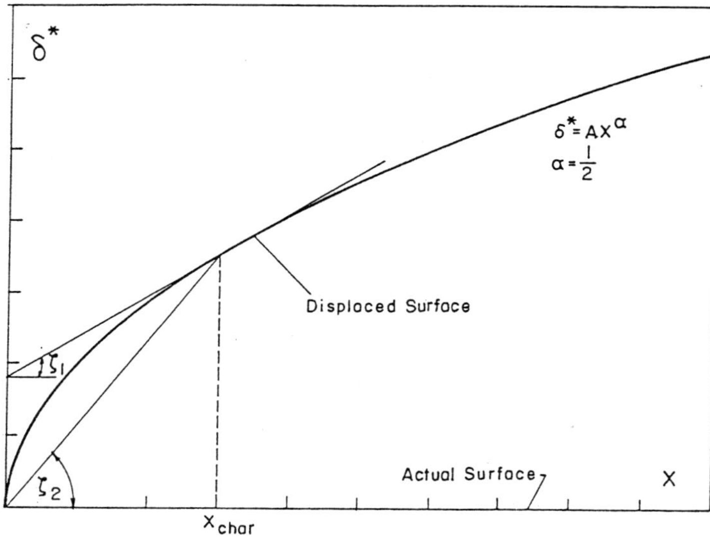

Figure 2.52 Schematic illustration of the displacement thickness profile over a straight surface and definition of various displacement angles, for a laminar boundary layer.

technique for shock wave reflections was first proposed by Hornung & Taylor (1982) who tried to resolve the so-called von Neumann paradox concerning the persistence of RR beyond its theoretical limit. Consider figure 2.51b where the reflecting wedge surface behind the reflection point, R, is displaced by $\delta^*(x)$ which can be simply calculated from $\delta(x)$, if $\delta(x)$ is known, in the way described by Shames (1982). Figure 2.51b indicates that due to the displacement of the reflecting surface $\theta_1 - \theta_2 \neq 0$, i.e., the flow behind the reflected shock wave is not parallel to the real reflecting wedge surface. Instead, it moves towards it with an angle $-\zeta$ which, in principle, could be obtained from δ^*. Unfortunately, however, a straightforward procedure by which ζ could be deduced from δ^* does not exist. Consider figure 2.52 where $\frac{d\delta^*}{dx}$ is plotted as a function of x which is measured from the reflection point, R. Since $\delta^*(x) = Ax^\alpha$ where $\alpha < 1$, $\frac{d\delta^*}{dx} \to \infty$ at $x = 0$. Hence, ζ cannot be equal to $\frac{d\delta^*}{dx}$ at $x = 0$ since no flow can smoothly negotiate a sudden

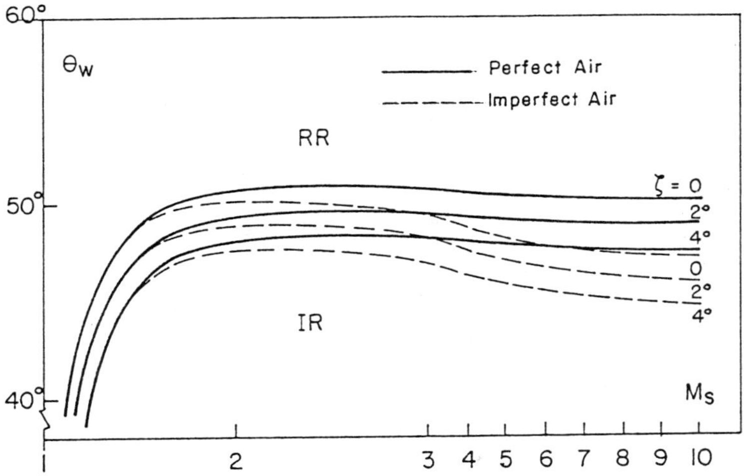

Figure 2.53 The dependence of the RR⇔MR transition line on the boundary layer displacement thickness angle. Solid lines are for frozen air with $\gamma = 1.4$, dashed lines are for air in equilibrium.

90° turn. Instead, Shirouzu & Glass (1982) proposed using either of the two following values for ζ:

1. the slope of the displaced boundary at a characteristic distance,

$$\zeta_1 = \tan^{-1} \frac{d\delta^*}{dx} \Big|_{x_{char}} \tag{2.55a}$$

2. the average slope of the displaced boundary at a characteristic distance

$$\zeta_2 = \tan^{-1} \frac{\delta^*}{x} \Big|_{x_{char}} \tag{2.55b}$$

Both of these suggestions require a choice of a characteristic length in order to define ζ. Shirouzu & Glass (1982) suggested using the average slope method and using x_{char} = 1mm which according to them is most realistic since it is the smallest practical resolved length on photographs in experiments. Unfortunately, neither they nor others have followed this procedure, i.e., evaluated the boundary layer δ to finally obtain ζ, for this case of RR. Instead, equations (1.5) to (1.12) together with the following boundary condition which replaces equation (1.13)

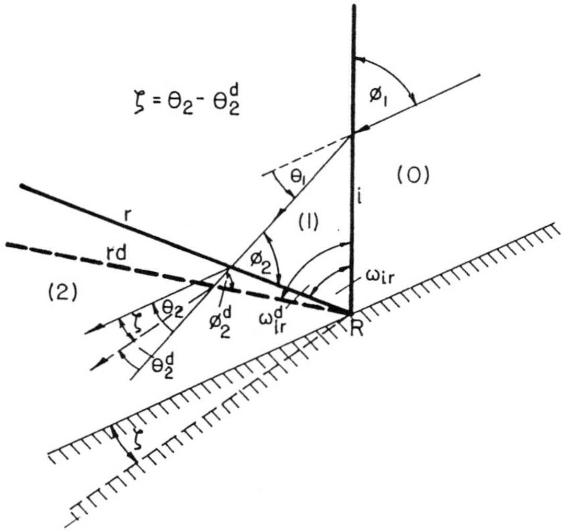

Figure 2.54 Schematic illustration of the displaced reflected shock wave of an RR due to the neglect of viscous effects.

$$\theta_1 - \theta_2 = \zeta \tag{2.56}$$

were solved by assigning different values to ζ and thereby forcing the RR⇔IR transition line to shift downwards. Figure 2.53 illustrates the results of this procedure for air. It can be clearly seen that the RR⇔IR transition line is shifted to lower values of transition wedge angles by assigning non-zero values to ζ. The larger is ζ, the greater is the shift of the RR⇔IR transition line. The shift is seen to be independent of whether or not real gas effects are accounted for. Thus, it is clear that, by using the boundary layer displacement concept, it is possible to explain the persistence of RR configurations below the RR⇔IR transition line as predicted by the inviscid two-shock theory.

The boundary layer displacement concept implies that if the viscous flow is to be solved, then the real reflecting surface is negotiated by the flow. However, if one wishes to use an inviscid approach, then the flow must be assumed to negotiate an imaginary reflecting surface. This imaginary surface has a change in its slope at the reflection point, R, as is shown in figure 2.54. In order to further prove the validity of the boundary layer displacement concept, Shirouzu & Glass (1982) illustrated that viscous effects are also responsible for the disagreement between the values predicted by the two-shock theory for the angle between the incident and the

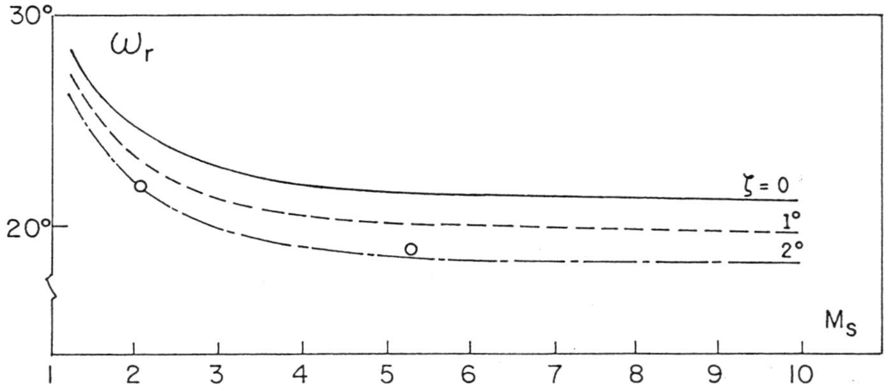

Figure 2.55 The dependence of the reflected shock wave angle on the boundary layer displacement angle for $\theta_w = 60°$.

reflected shock waves, ω_{ir}, and those measured in actual experiments. Consider figure 2.54 and note that if the flow negotiates an imaginary wedge, the reflected shock wave, r, (solid line) should assume a new orientation, r^d, (dashed line) because it is required to deflect the flow only by an angle θ_2^d which is smaller than θ_2 since $\theta_2 - \theta_2^d = \zeta$. This effect is clearly illustrated for Ar in figure 2.55, where the experimentally measured values of the angle between the reflected shock wave and the reflecting wedge surface, ω_r (note $\omega_r = 180° - \phi_1 - \omega_{ir}$) are compared with the predictions of the two-shock theory for $\zeta = 0$, 1° and 2°. If $\zeta = 0$, i.e., the value appropriate to the inviscid two-shock theory is used, then the predicted values of ω_r are higher than those measured experimentally, by about 2.6°. A value of $\zeta = 2°$, which could be attributed to viscous effects, results in excellent agreement with the experimentally measured values.

Although neither Shirouzu & Glass (1982) nor others tried to systematically obtain the correct value of ζ (2° for the case shown in figure 2.55) from the boundary layer, δ, which develops over the reflecting surface, the foregoing discussion leaves little doubt about the fact that viscous effects should be accounted for if an RR is to be accurately solved using the two-shock theory.

Wheeler (1986) has further illustrated that the persistence of RR below the RR⇔IR transition line as predicted by the inviscid two-shock theory is due to viscous effects. He showed experimentally

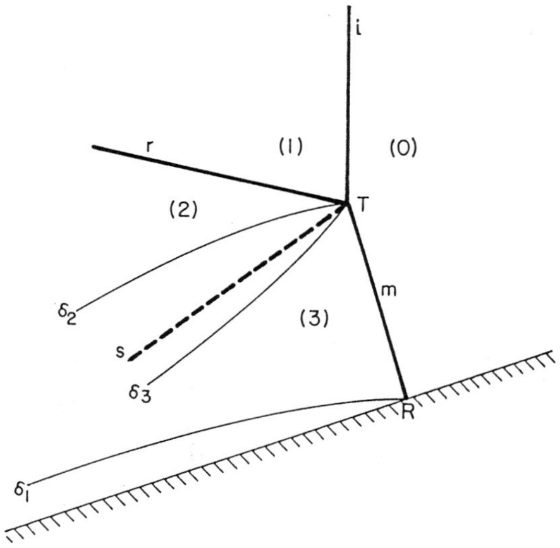

Figure 2.56 Schematic illustration of the various boundary layers associated with a Mach reflection.

that due to viscous effects the deviation of the actual RR⇔IR transition line from the one predicted using the inviscid two-shock theory increases with a drop in the initial pressure, p_0, in a manner consistent with the boundary layer theory.

2.5.4.2 Viscous Effects on MR

Figure 2.56 indicates that in the case of an MR viscous effects are dominant not only along the reflecting wedge surface but also on both sides of the slipstream.

The viscous effects along the reflecting wedge surface have little to do with the solution of the flow field in the vicinity of the triple point, T. They do have, however, some influence for they affect the flow field in the vicinity of reflection point, R, where the foot of the Mach stem touches the reflecting surface. Since the flow behind the Mach stem is subsonic, in a frame of reference attached to the triple point, T, the influence of the boundary layer, which develops along the reflecting surface, could be communicated to the triple point, and influence it too. The fact that the boundary layer affects the foot of the Mach stem was first reported by Dewey & McMillin (1985) who experimentally observed that the foot of the Mach stem is not exactly perpendicular to the reflecting wedge surface, as it should be if the flow is assumed to be inviscid. Their observation could again be explained by the earlier-mentioned argument that, if the flow is

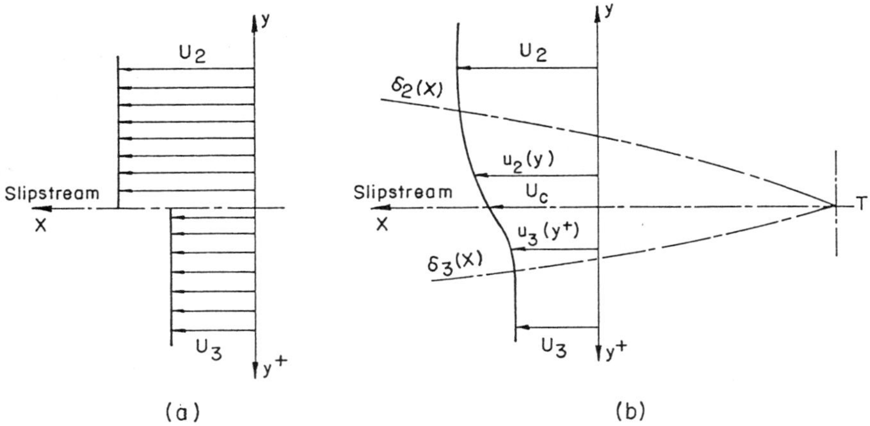

Figure 2.57 Illustration of the flow field in both sides of the slipstream of a Mach reflection: a) inviscid flow; b) viscous flow.

assumed to be inviscid, it should not be required to negotiate the real reflecting surface and hence the condition of zero flow deflection through the foot of the Mach stem is incorrect. Instead, if the flow is assumed to be inviscid, it should be required to negotiate an imaginary displaced reflecting surface for which an oblique Mach stem is needed. Unfortunately, no analytical work has been done so far to study the non-perpendicular nature of the foot of the Mach stem and its influence on the flow field in the vicinity of the triple point.

Viscous effects on both sides of the slipstream probably have significant influence on the flow field in the vicinity of the triple point, upon which the predictions of the SMR⇔TMR and TMR⇔DMR transition criteria, as well as the angles between the four discontinuities, i, r, m and s, using the three-shock theory, are based.

Figure 2.57a illustrates the situations as assumed by the inviscid three-shock theory. The flows above and below the slipstream are uniform with velocities U_2 and U_3, respectively. At the slipstream itself there is a discontinuity in the velocity field. However, since all fluids have a finite viscosity, the situation shown in figure 2.57a is aphysical. Instead, the fluid above and below the slipstream exchanges momentum resulting in a smooth transition from U_2 to U_3. The result of momentum exchange is shown in figure 2.57b. It is evident from figure 2.57b that the inviscid three-shock theory

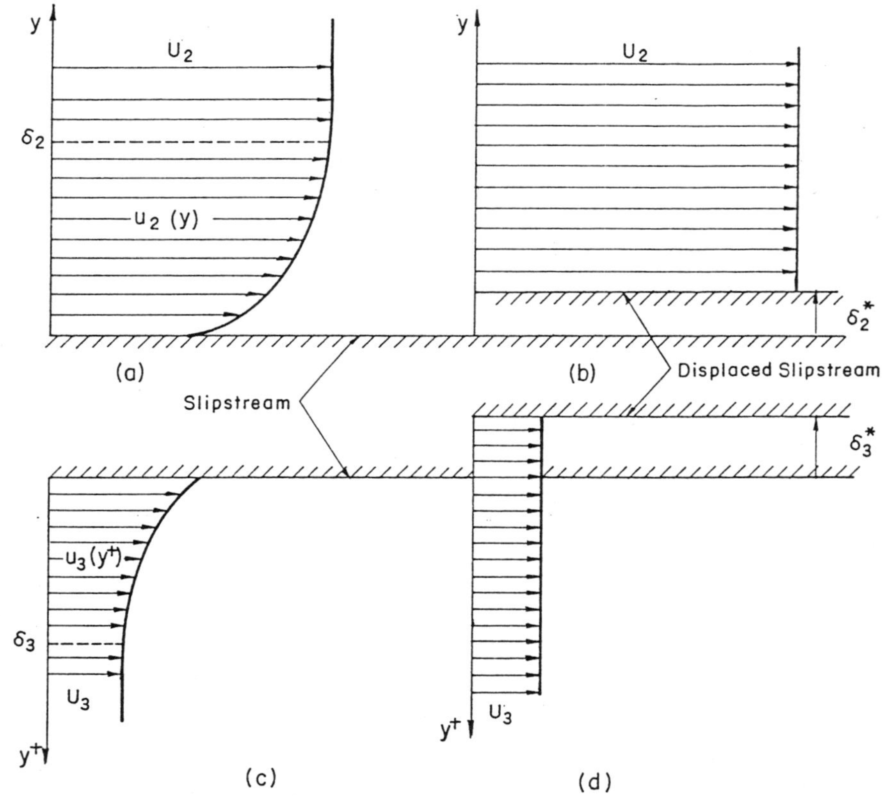

Figure 2.58 Definition of the displacement thicknesses on both sides of the slipstream.

cannot be used to treat this case. Instead, the full Navier-Stokes equations must be solved in states (2) and (3).

If, however, one still wishes to use the model given by the inviscid three-shock theory then the displacement thickness technique, which was described in section 2.5.4.1, could again be applied. This was done, quite successfully, by Ben-Dor (1987). Consider figure 2.58 where the actual velocity profiles, shown previously in figure 2.57b, are replotted on the left side of this figure. Each of these two velocity profiles is plotted separately. If uniform flow profiles are to be used in states (2) and (3), then the slipstream should be displaced. Owing to the shape of the velocity profiles, the displacement thickness of the flow field above the slipstream is positive while that of the flow field below the slipstream is negative. However, since both flows have a common surface along which they flow, i.e., the slipstream, the displacement

Table 1: Comparison between the experimental results and the predictions of the wave angles by the various modified models of the three-shock theory

Type of Model	ω_{ir}	ω_{im}	ω_{rs}
Experimental results	118°±1°	132°±1°	32°±1°
Three-shock theory for an inviscid perfect gas	123.19° (+5.19°)	132.74° (+0.74°)	27.16° (-4.84°)
Three-shock theory for a perfect gas with viscous effects along the slipstream	118.02° (+0.02°)	131.70° (-0.30°)	32.32° (+0.32°)
Three-shock theory for an inviscid real gas in rotational-vibrational equilibrium	117.32° (-0.68°)	131.95° (-0.05°)	29.81° (-2.19°)
Three-shock theory for a real gas in rotational-vibrational equilibrium with viscous effects along the slipstream.	120.04° (+2.04°)	130.70° (-1.30°)	27.09° (-4.91°)

$M_s = 2.71$, $\phi_1 = 39.9°$, $T_0 = 296$ K, $p_0 = 760$ torr

thicknesses above and below the slipstreams must satisfy the condition $\delta_2^* = \delta_3^*$.

In a detailed study, Ben-Dor (1987) compared the angles between the various discontinuities at the triple point, namely, ω_{ir}, ω_{im} and ω_{rs}, as predicted by the inviscid three-shock theory with actual experiments and found a discrepancy too large to be explained by experimental uncertainties. Then the boundary layers that develop on both sides of the slipstream, δ_2 and δ_3, were calculated. These boundary layers were used to obtain the displacement thicknesses δ_2^* and δ_3^*. By imposing the condition $\delta_2^* = \delta_3^*$ the orientation of the imaginary slipstream was obtained. The orientation of the imaginary slipstream was then used to define the imaginary flow streamlines in

Figure 2.59 Enlarged photograph of a double-Mach reflection illustrating the structure of the slipstream with $M_s = 2.71$ and $\theta_w = 47^o$ in air at $T_0 = 296K$ and $p_0 \approx 760$ torr (courtesy of Professor K. Takayama, Shock Wave Research Center, Institute of Fluid Science, Tohoku University, Sendai, Japan).

states (2) and (3) which were assumed to be parallel to it. Then the reflected and Mach stem shock waves were displaced in such a way, that they resulted in the required imaginary flow deflections. The obtained results agreed perfectly with those measured experimentally. Note that in order to obtain the angular displacement of the slipstream, a displacement angle had to be obtained from the analysis. This angle is analogous to the angle ζ mentioned earlier for the case of RR (see section 2.5.4.1). In order to obtain this angle a characteristic length was needed. It was found by Ben-Dor (1987) that the most appropriate characteristic length was of the order of the thickness of the incident shock wave.

Table 1 summarizes the results of Ben-Dor (1987) for the case of an MR with $M_S = 2.71$, $\phi_1 = 39.9^o$, $T_0 = 296K$ and $p_0 = 760$ torr in nitrogen. The results indicate that the predictions of the inviscid perfect three-shock theory for ω_{ir} and ω_{rs} are about 5° greater and smaller than the respective measured values from the experimental record which is shown in figure 2.59. When the foregoing boundary layer dispacement technique (viscous) model is used, the agreement between the theory and the experimental results is excellent. Table 1 also indicates that when the three-shock theory is used without viscous effects but assumes that the air is in rotational-vibrational equilibrium in the vicinity of the triple point, it is capable of predicting better values for ω_{ir}, ω_{im} and ω_{rs} than those predicted by the inviscid perfect gas theory. Unfortunately, however, as indicated in figure 2.48 the vibrational relaxation length of N_2 at $M_S = 2.71$ is more than 1m. This implies that the flow in the vicinity of the triple point must be considered as frozen with $\gamma = 1.4$. Thus the better agreement due to the inclusion of real gas effects must be considered as accidental. It is also evident from table 1, that when both viscous and real gas effects are integrated into the three-shock theory the predicted values of ω_{ir}, ω_{im} and ω_{rs} are poorer than the results which are obtained when these effects are separately integrated into the inviscid three-shock theory model.

The foregoing comparisons imply that viscous effects should be accounted for if the flow field in the vicinity of the triple point is to be calculated. Unfortunately, to the best of the author's knowledge, the SMR⇔TMR and the TMR⇔DMR transition lines and the additional requirement for the SMR⇔TMR transition have not been evaluated using the modified viscous three-shock theory model described in this section.

It should also be mentioned that the above method of accounting for viscous effects in the solution of the flow field in the vicinity of the triple point must also be adopted when the second triple point is treated, for it also has a slipstream across which momentum exchange takes place.

Wheeler (1986) found experimentally that the initial pressure influences the height of the Mach stem of an MR. Lower initial pressures, which result in greater viscous effects because both δ and δ^* depend on $p_0^{-1/2}$, reduced the height of the Mach stem to values which are smaller than those predicted by the inviscid three-shock theory. Note that the incident shock wave thickness, which was found by Ben-Dor (1987) to be the most appropriate characteristic length for evaluating the value of the displacement angle of the slipstream, ζ, is also p_0 dependent, since the mean free path λ is proportional to p_0^{-1}. Thus the experimental results of Wheeler (1986) might serve as a verification for the choice made by Ben-Dor (1987) for x_{char}.

2.5.4.3 Special Reflecting Surface Conditions

The recognition of the role of the kinematic boundary layer in influencing the RR⇔IR transition line, and thereby successfully explaining the persistence of RR-experiments below the theoretical RR⇔IR transition line as predicted by the inviscid two-shock theory, has in recent years been the impetus for various investigations of the reflection of shock waves over wedges, where the surface conditions were such that they could either enhance or reduce the rate of growth of the boundary layers. Some of these investigations are still in their early stages and hence they are only briefly discussed.

2.5.4.3.1 Rough Reflecting Surface

By imposing a roughness on the reflecting wedge surface the rate of the boundary layer growth could be strongly enhanced. Figures 2.60a and 2.60b are shadowgraphs of an RR and a TMR over rough reflecting straight surfaces. A detailed experimental and analytical study on the influence of surface roughness on the RR⇔IR transition was carried out by Ben-Dor, Mazor, Takayama & Igra (1987). The results of their study are summarized in figure 2.61 for nitrogen. The experimental results indicate that for a given incident shock wave Mach number the surface roughness results in a reduction in the wedge angle, at which the RR⇔IR transition occurs. The greater the roughness size, the larger is the reduction in the transition wedge angle. This behavior is consistent with the earlier discussion on the effect of viscosity on the RR wave configuration since the roughness results in an increase in the rate of growth of the boundary layer which, in turn, results in an increase

a)

Figure 2.60a Shadowgraph of a regular reflection over a rough reflecting surface with M_s = 2.74 and $\theta_w = 43^o$ in air at T_0 = 283.5K and $p_0 \approx$ 760 torr (courtesy of Professor K. Takayama, Shock Wave Research Center, Institute of Fluid Science, Tohoku University, Sendai, Japan).

b)

Figure 2.60b Shadowgraph of a transitional-Mach reflection over a rough reflecting surface with M_s = 2.30 and θ_w = 39° in air at T_0 = 283.4K and $p_0 \approx$ 760 torr (courtesy of Professor K. Takayama, Shock Wave Research Center, Institute of Fluid Science, Tohoku University, Sendai, Japan).

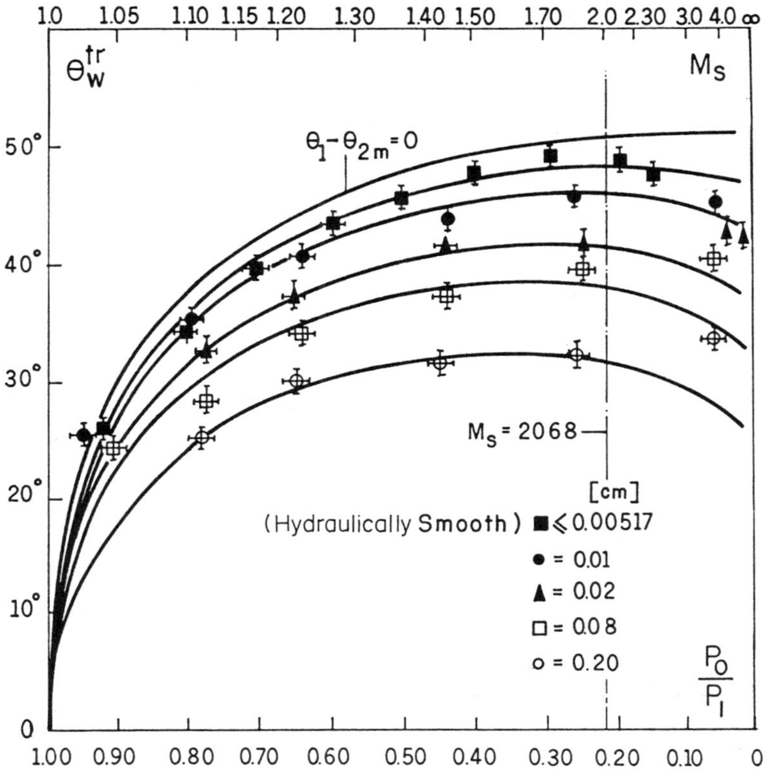

Figure 2.61 The dependence of the RR⇔MR transition line on the reflecting surface roughness and comparison with experiments.

in the rate of growth of the displaced boundary, i.e., larger values of ζ (see section 2.5.4.1) at the required characteristic length.

In their study Ben-Dor, Mazor, Takayama & Igra (1987) developed an analytical model by which the boundary layer growth for a supersonic compressible flow over a flat rough surface could be predicted. Then the surface was displaced in a way similar to that mentioned in section 2.5.4.1. By again using a characteristic length of the order of the thickness of the incident shock wave, a value the angle of ζ was obtained. This value was then used in the two-shock theory with the modified boundary condition [equation (2.56) instead of (1.13)]. The analytical results using this procedure are shown in figure 2.61. Excellent agreement between the experiments and the analytical predictions is evident in the range $1 \leq M_S < 2$. In the range $M_S > 2$ the predicted results failed to reproduce the experiments. Ben-Dor, Mazor, Takayama & Igra (1987)

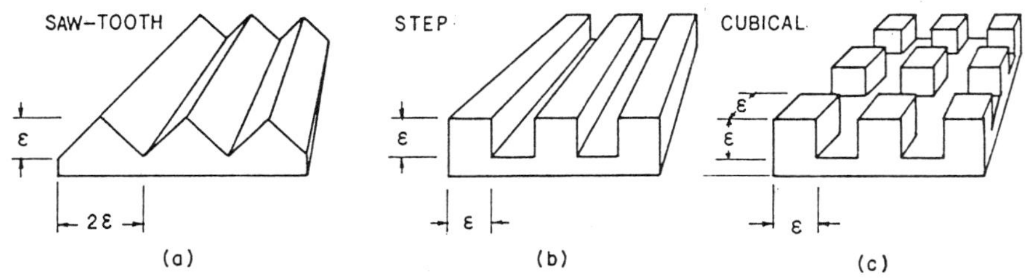

Figure 2.62 Definition and illustration of various types of surface roughness.

hypothesized that this might be due to the fact that the shock induced flow behind the incident shock wave is subsonic for $M_S < 2.068$, while it is supersonic for $M_S > 2.068$. Thus it is possible that this fact has an effect, unfortunately not yet understood, on the reflection process. The persistence of RR-experiments below the theoretical limit as predicted by the inviscid two-shock theory, is clearly seen in figure 2.61 where the experimentally recorded transition wedge angles, for a smooth wedge, are seen to lie below the RR⇔IR transition line as calculated using the inviscid perfect gas two-shock theory. Based on these findings, and on the fact that no surface is perfectly smooth, it must be concluded that the surface roughness, regardless of how small it is, does have some effect on the rate of growth of the boundary layer developing over the reflecting surface. This boundary layer affects the actual wave configuration in the manner described earlier.

Two more points regarding the surface roughness effect on the shock reflection phenomenon in pseudo-steady flows should be mentioned. The first is that the most appropriate characteristic length to be used in equation (2.55b) was found to be of the order of the thickness of the incident shock wave. This is surprisingly identical to the findings mentioned earlier in section 2.5.4.2, where the most appropriate characteristic length was also found to be of the same order of magnitude when the viscous effects were integrated into the inviscid three-shock theory. This is in spite of the fact that while integrating viscous effects into the three-shock theory the boundary layer was assumed to be laminar, and in the case of a boundary layer over a rough reflecting surface it was assumed to be turbulent. These findings might lead to the conclusion that, in shock wave reflection phenomena, the thickness of the incident shock wave is

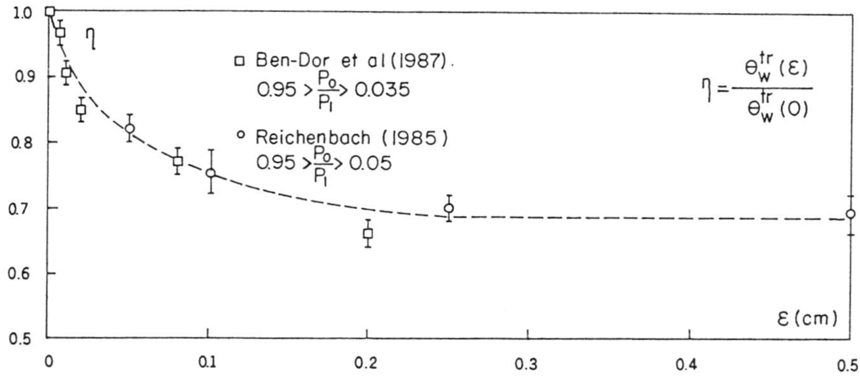

Figure 2.63 The dependence of the RR⇔MR transition angle ratio on the roughness height.

the characteristic length which should be used whenever viscous effects are accounted for by means of the boundary layer displacement technique.

The second striking point is that while Ben-Dor, Mazor, Takayama & Igra (1987) used in their experiments a saw-tooth shaped roughness, Reichenbach (1985) who used in his experiments different shapes of roughness reported the same shifts in the transition wedge angle for identical heights of the roughness. The shape of the roughness used by Ben-Dor, Mazor, Takayama & Igra (1987) (saw-tooth) and the shapes of the roughness used by Reichenbach (step and cubical) are shown in figure 2.62, with the definition of the roughness height-ε. Reichenbach (1985) defined the ratio between the transition wedge angle over a rough reflecting surface and the transition wedge angle over a smooth reflecting surface as - η, and presented his own and Ben-Dor, Mazor, Takayama & Igra's (1987) experimental results in the (η, ε)-plane as shown in figure 2.63. Figure 2.63 clearly indicates that the roughness height, ε, rather than its shape, is the dominant factor in determining the transition wedge angle. In his experimental study, Reichenbach (1985) also measured the height of the Mach stem, L_m, as a function of the distance of its foot from the leading edge of the reflecting wedge, x. The results for a step shape roughness are shown in figure 2.64. The results clearly indicate that the triple point is not formed at the leading edge of the reflecting wedge, i.e., at x = 0, but rather at some distance up the reflecting wedge surface. The

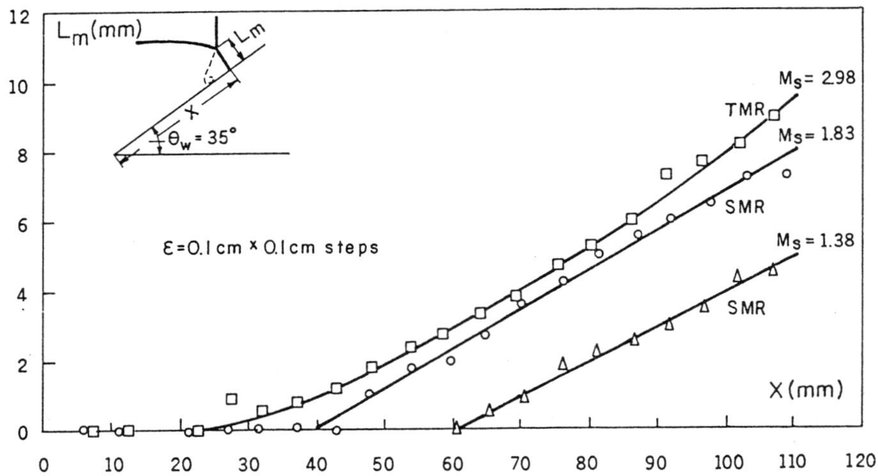

Figure 2.64 Experimentally measured height of the Mach stem of a Mach reflection as function of distance from the leading edge of the reflecting wedge for a rough surface.

distance from the leading edge of the reflecting wedge to the point along the reflecting surface, where the triple point is formed, was found by Reichenbach (1985) to increase with decreasing incident shock wave Mach numbers.

2.5.4.3.2 Perforated Reflecting Surface

Onodera (1986) reinitiated an experimental investigation of the reflection of a planar shock wave over a perforated plate which was originally initiated by Friend (1958). The effect of a perforated plate should be similar to that of a rough surface discussed earlier, because in the case of a rough surface, the boundary layer displacement technique results in a situation in which the flow behind the reflection point assumes a direction parallel to an imaginary reflecting wedge surface which extends below the real reflecting wedge surface. Thus fluid is being drawn away from the flow field since it is allowed to flow into the real reflecting wedge surface. A similar effect occurs when the plate is perforated since the fluid is now also drawn away from the flow field as it penetrates the perforated plate and flows across it. A typical experimental record from Onodera's (1986) study is shown in figure 2.65 for $M_S = 2.93$, $\theta_W = 33^o$ and a perforation ratio of 0.355 (i.e., the ratio between the perforated area and the overall surface area). The very complicated structure of the wavelets which are generated along the reflecting surface is clearly seen. The structure of these wavelets

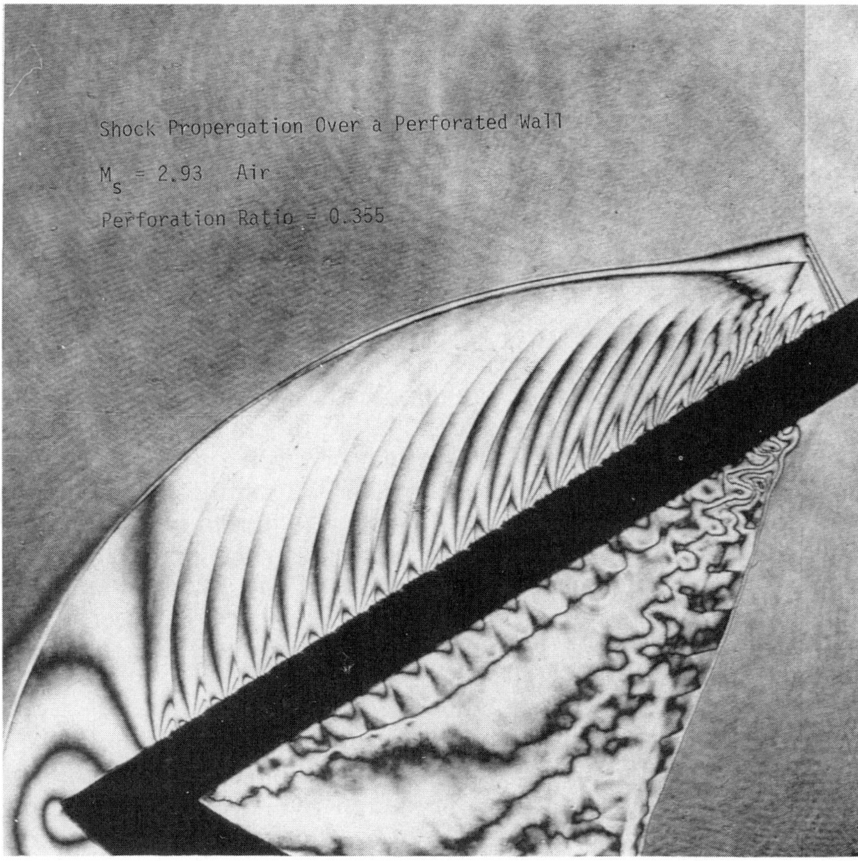

Figure 2.65 An interferogram of a Mach reflection in air over a perforated flat plate with a perforation ratio of 0.355 and $M_s = 2.93$ and $\theta_w = 33^o$ (courtesy of Professor K. Takayama, Shock Wave Research Center, Institute of Fluid Science, Tohoku University, Sendai, Japan).

show a similarity to those in figures 2.60a and 2,60b, where an RR and a TMR over rough surfaces are recorded.

Since there are not enough experimental data regarding the reflection phenomenon of planar shock waves over perforated plates, not much can be said, at this stage, about it. Perhaps the only encouraging conclusion which can be drawn from figure 2.65 is that the MR does seem to respond in the expected way to the fact that the reflecting surface is perforated. Consider figure 2.26, and note that for $M_S = 2.93$ and $\theta_W = 33^o$ the experimentally measured first triple point trajectory angle, χ, should be about 8°. In the case of a reflection over a perforated plate with identical initial conditions (figure 2.65) the value of χ is about 6°. This means that the triple point is lower and hence a transition to RR should be expected to occur at lower reflecting wedge angles. If this is the case, then the sucking effect in the case of a reflection over a perforated plate influences the RR⇔IR transition in a way similar to that of a reflection over a rough surface. Finally, it is important to note that Wheeler (1986) reported that an increase in the boundary layer height causes a decrease in the height of the Mach stem. This observation further supports the fact that the effect of a perforated reflecting surface on the shock wave reflection phenomenon is similar to that of viscous effects.

2.5.4.3.3 Slitted Reflecting Surface

Onodera & Takayama (1990) investigated both experimentally and analytically the reflection phenomenon over a slitted reflecting surface. Their experimental study was conducted using three different models. In their models A and B the slits were open (as in a perforated plate), whereas in their model C the slits were closed. Models A, B and C had 58, 36, and 36 slits, respectively. The perforation ratios, i.e., the ratio between the perforated area and the overall surface area, were 0.34, 0.40 and 0.40 for models A, B and C, respectively.

A schematic drawing of the wave configuration and the various flow velocity vectors, for the case of a reflection over a slitted surface, is shown in figure 2.66. In a laboratory frame of reference, the flow velocity along the slitted reflecting surface has two components, namely, V_2^L parallel to and V_4^L perpendicular to the reflecting surface.

Similar to the above-mentioned change in the boundary condition of an RR, from equation (1.13) to equation (2.56), when viscous effects are accounted for by means of the boundary layer displacement technique, Onodera & Takayama (1990) suggested

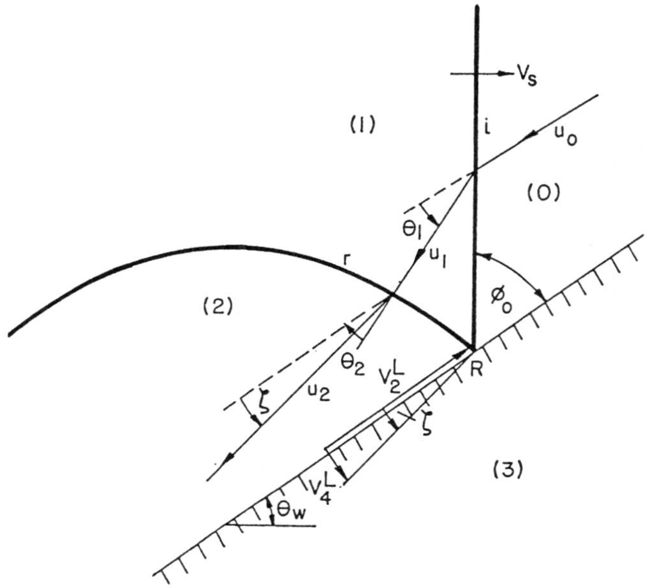

Figure 2.66 Schematic illustration of a regular reflection over a slitted surface and definition of various velocity vectors.

that, for the case of a slitted surface, the boundary condition of an RR should be

$$\theta_1 - \theta_2 = \zeta \tag{2.57}$$

Furthermore, they argued that ζ could be related to V_2^L and V_4^L in the following way

$$\tan \zeta = \xi \frac{V_4^L}{V_2^L} \tag{2.58}$$

where ξ is the discharge coefficient through the perforated area.

Unfortunately, they did not pursue this idea to calculate the RR⇔IR transition for a slitted surface. Instead, they proposed the following empirical approach based on their experimental observation. Consider figure 2.67 where a holographic interferogram of a shock wave having a Mach number $M_S = 2.96$ is seen to reflect as an RR over an inclined slitted surface (model A) having $\theta_W = 44°$. Note that an oblique shock wave is driven by a

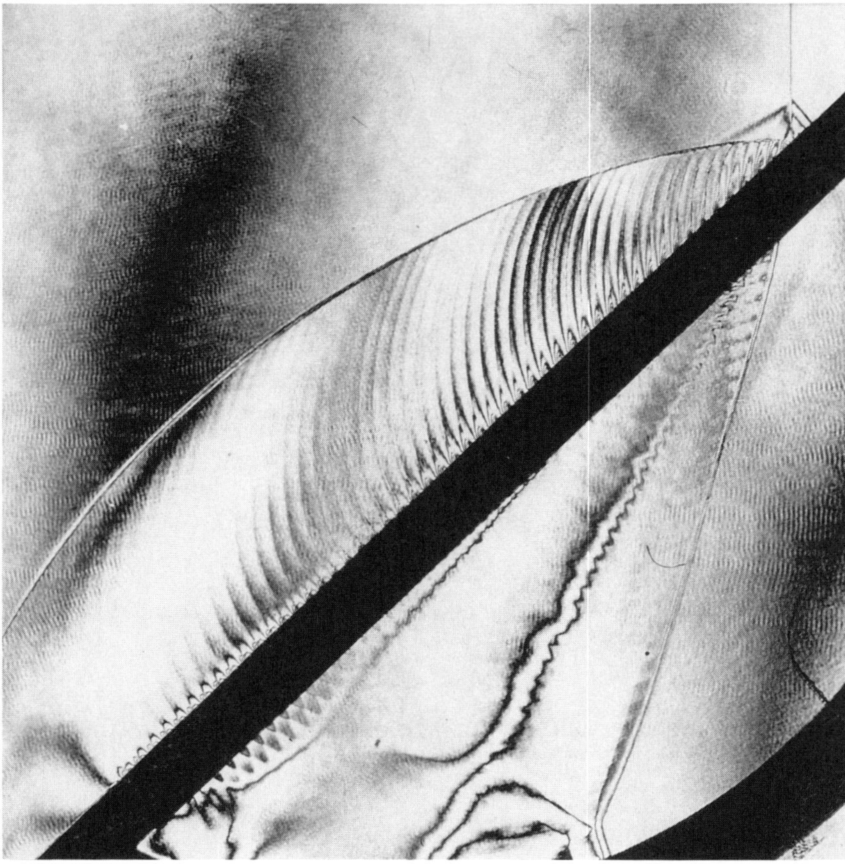

Figure 2.67 A holographic interferogram of a regular reflection over an inclined slitted surface with $M_s = 2.96$ and $\theta_w = 44^o$.

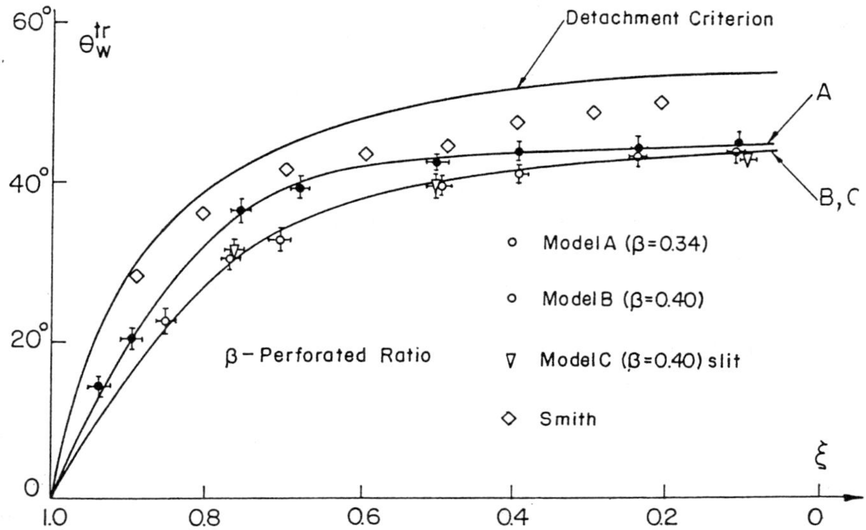

Figure 2.68 The empirical RR⇔MR transition lines over a slitted surface with experimental results (courtesy of Professor K. Takayama).

contact discontinuity in region (3) below the slitted surface. If the angle between the reflecting surface and the contact discontinuity is defined as ζ' then according to Onodera & Takayama (1990)

$$\zeta = \xi'\zeta' \tag{2.59}$$

where ξ' is a matching coefficient. This matching coefficient was determined empirically by Onodera & Takayama (1990) to be equal to 0.61 and 0.55 for models A and B, respectively.

Their experimental and empirical results for the RR⇔IR transition over a slitted surface are shown in figure 2.68. The predicted transition lines for models A, B and C are seen to agree excellently with the corresponding experimental results. Furthermore, the transition wedge angles for models A and B approach each other at strong incident shock waves, while at moderate incident shock waves, they differ by about 5°. In addition, the results indicate that the transition wedge angles for models B and C are almost the same. Hence, it can be concluded that the transition wedge angle does not depend on whether the slits are open or closed.

Figure 2.68 also implies that the transition wedge angle depends on the perforation ratio. The larger the perforation ratio, the greater is the reduction in the transition wedge angle. However, for incident shock wave Mach numbers with $M_S > 1.5$ the dependence of the transition wedge angle on the perforation ratio seems to be negligible, as the results indicate that the transition wedge angle attains a value of about 43.5° for models A, B and C.

2.5.4.3.4 Porous Reflecting Surface

Clarke (1984a, 1984b) investigated the case of a weak regular reflection over a porous surface both experimentally and analytically. The porous material was backed by an impermeable solid substance. A central feature of Clarke's model was that the porous material swallowed or regurgitated air until such time as the pressure within the porous material was uniform and equal to the external air pressure. Thus the reflection of a shock wave over a porous surface which is backed by a solid surface, should better resemble the situations of a reflection over a rough surface than a reflection over a perforated plate where the amount of fluid drawn away from the flow field is practically unlimited.

Although porous materials are often structurally weak and easily compressed or stretched, Clarke (1984a) mentioned experimental evidence that such bulk responses of the porous material played little or no part on the wave reflection process.

Unfortunately, Clarke (1984a) did not use his analysis to predict the RR⇔IR transition over a porous reflecting surface. However he concluded his study by saying that his theory seemed to provide an adequate description of the effects of porosity on what would be regular reflections of a plane shock wave from a porous surface, at least for urethane foam-like materials. Since his theory predicts significant reductions in the reflected shock wave strength over distances measured in tens of centimetres, the shock reflection phenomenon over porous surfaces should perhaps not be considered in this chapter of shock wave reflection in pseudo-steady flows. However, if the shock wave attenuation is not too strong, it is possible that along short reflecting surfaces the phenomenon can still be considered as close to being pseudo-steady.

2.5.4.3.5 Non-Solid Reflecting Surface

Using a non-solid reflecting surface, i.e., a liquid reflecting surface is another situation in which the boundary layer developing over the reflecting surface could be changed from that developing over a solid wedge. This is due to the following two reasons:

1. unlike a solid reflecting surface, a liquid reflecting surface, such

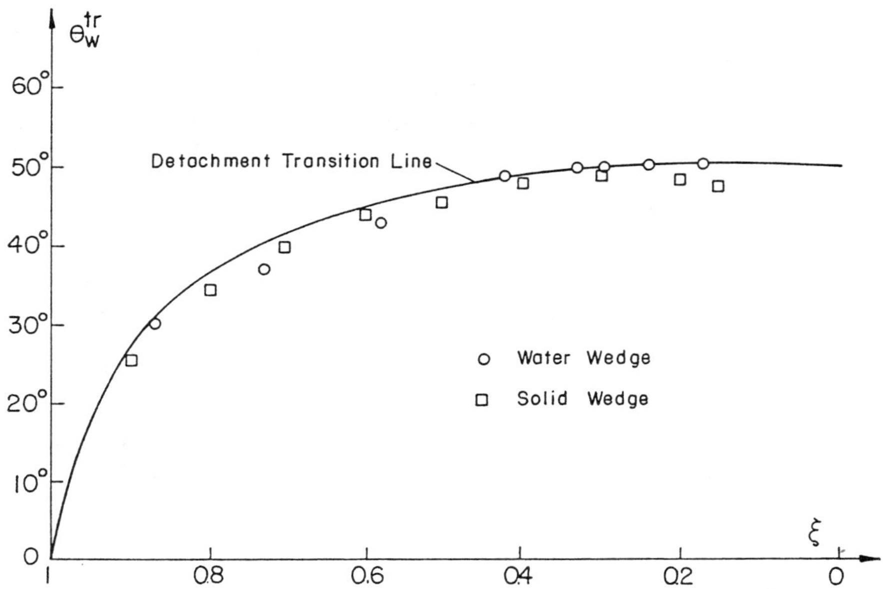

Figure 2.69 The RR⇔IR transition over a water and a solid surface and comparison with experimental results.

as water, is extremely smooth, and

2. the viscous interaction of the gas with the liquid surface is probably different from that with a solid surface.

An experimental investigation of the reflection of a planar shock wave over a water surface was conducted by Takayama, Miyoshi & Abe (1989), Takayama & Ben-Dor (1989) and Henderson, Ma, Sakurai & Takayama (1990). In order to reflect a planar shock wave over a water surface, a special shock tube was designed and constructed in such a way that it could be tilted in the vertical plane. Using this technique it was possible to adjust the shock tube inclination to obtain any desired water wedge angle. The experimentally measured transition wedge angles for a water and a solid surface are shown in figure 2.69 together with the RR⇔IR transition line arising from the detachment criterion. It is evident from figure 2.69 that for $1.47 \le M_S \le 2.25$ the actual transition wedge angles over the water wedge agree excellently with those predicted by the detachment criterion, and the well-known persistence of RR-configurations below the RR⇔IR transition line is not in evidence. However, for $1 < M_S \le 1.47$ an opposite behavior is evident. While the experimental results, both over the solid and the water reflecting surfaces, exhibit the well-known persistence of RR-configurations below their theoretical limit, the actual transition wedge angles over the water reflecting surface show worse agreement with the theoretical inviscid RR⇔IR transition line than the transition wedge angles over the solid reflecting surface.

The reason for this peculiar behavior might lie in the fact that for air (see figure 2.39) the SMR⇔TMR transition line intersects the RR⇔IR transition line at $M_S \approx 1.5$. Thus for $M_S < 1.5$ the termination of an RR results in an SMR while for $M_S > 1.5$ the termination of RR results in a TMR. In a TMR the flow behind the reflected shock wave in the vicinity of the triple point is supersonic with respect to the triple point, and hence the triple point is isolated from pressure signals generated along the reflecting surface. However, in the case of an SMR the flow behind the reflected shock wave is subsonic with respect to the triple point, and therefore pressure signals generated along the reflecting surface can reach the triple point and influence it. Perhaps this difference might explain the experimental results shown in figure 2.69.

2.5.5 Thermal Conduction Effects

The possibility that thermal conduction might also have a non-negligible effect on the solutions of the flow fields in the vicinities of the reflection point of an RR and the triple point of an MR have already been mentioned as a possible explanation for the persistence of TDMR-experiments beyond the DMR⇔RR transition line (see sections 2.4.3 and 2.4.4).

When the gas is assumed to be an ideal fluid, i.e., $\mu = 0$ and $k = 0$, the two- and three-shock theories inherently assume a discontinuity in the temperature field along the wedge which is usually at room temperature and along the slipstream. Consider figures 2.70a and 2.70c where the temperature profiles, as assumed by the three-shock theory, on both sides of the slipstream of an MR and behind the Mach stem over the reflecting wedge surface are drawn. In reality, however, due to thermal conductivity the temperature profiles must change smoothly as is shown in figures 2.70b and 2.70d. The change takes place inside the thermal boundary layer, δ_T. Outside the thermal boundary layer the temperatures are uniform and thermal conduction can be neglected. However, inside the thermal boundary layers, heat transfer due to thermal conduction must be accounted for, if one wishes to accurately solve the flow fields. Note that depending upon the Prandtl number, Pr, the thermal boundary layer δ_T can be either smaller or greater than the kinematic boundary layer, δ.

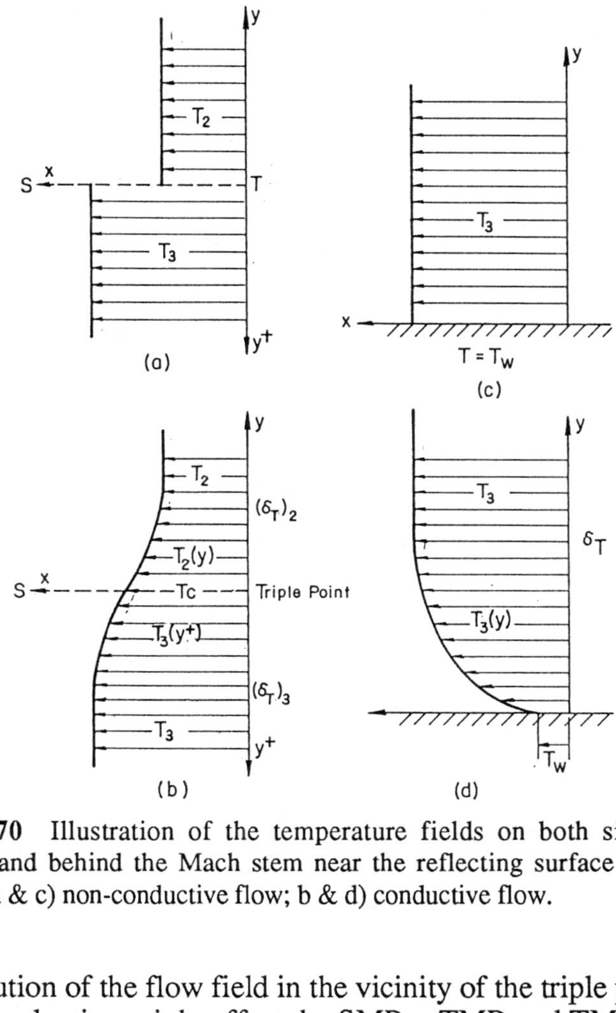

Figure 2.70 Illustration of the temperature fields on both sides of the slipstream and behind the Mach stem near the reflecting surface of a Mach reflection: a & c) non-conductive flow; b & d) conductive flow.

The solution of the flow field in the vicinity of the triple point with thermal conduction might affect the SMR⇔TMR and TMR⇔DMR transition lines, while the solution of the flow field at the foot of the Mach stem with thermal conduction might contribute to the earlier-mentioned experimental observation by Dewey & McMillin (1985) that the Mach stem is not perpendicular to the reflecting wedge surface. In the case of an RR (see figure 2.51) heat conduction might be important in state (2) behind the reflection point, R, where a high temperature gas flows over the cold reflecting wedge which is initially at room temperature.

The following two examples illustrate the magnitude of the heat fluxes due to the assumed temperature discontinuities in an RR and an MR. In the case of an RR with $M_S = 4.68$, $\phi_1 = 30^o$, $p_0 = 15.31$

torr and T_0 = 300K in nitrogen, the inviscid two-shock theory predicts a flow with temperature T_2 = 2907K over the reflecting wedge surface which has a temperature T_w = 300K. In a case of an MR with M_S = 2.71, ϕ_1 = 39.9°, p_0 = 760 torr and T_0 = 296K in nitrogen the inviscid three-shock theory predicts T_2 = 920.6K and T_3 = 1305K. The heat conductivity of N_2 is about 0.064 W/m/K at T = 920K and 0.084 W/m/K at T = 1305K. Thus assuming an average value of 0.074 W/m/K implies that the heat flux across the slipstream, which can be assumed to have a thickness similar to that of the incident shock wave, i.e., about 6.6×10^{-7}m (ten mean free paths), is of the order of $4.3 \times 10^7 \frac{W}{m^2}$. In the case of the high temperature flows over the reflecting wedges the heat flux might be even greater because the temperature differences are larger, e.g., ΔT = 1005K for the MR and 2607K for the RR in the above examples. In addition to the larger temperature differences, the heat conductivity which increases with increasing temperatures further increases the heat flux.

Unfortunately, an analysis of neither the two- nor the three-shock theories where thermal conduction is accounted for has ever been undertaken. Therefore, it is impossible to quantitatively assess its influence. However, owing to the fact that both viscous and real gas effects cause a shift in the RR⇔IR transition line to lower values of reflecting wedge angles, and owing to the fact that TDMR-experiments were recorded above the RR⇔IR transition line as calculated by the inviscid, non-conductive perfect two-shock theory, it is possible that the inclusion of heat transfer would shift the RR⇔IR transition line to higher values of reflecting wedge angles.

2.5.6 Non-Infinitely Thin Contact Discontinuity

Skews (1971/2) and Zaslavskii & Safarov (1973) suggested that the failure of the inviscid three-shock theory in correctly predicting the angles between the four discontinuities of a triple point might arise from inadequate choice of the boundary conditions across the contact discontinuity. As discussed in section 1.3.2 the contact discontinuity of the triple point was assumed, in the three-shock theory, to be infinitely thin, i.e., a slipstream. Hence, the flows on both sides of the slipstream were assumed to be parallel, i.e.,

$$\theta_1 - \theta_2 = \theta_3 \qquad (2.60)$$

Alternatively, Skews (1971/2) hypothesized that perhaps the contact discontuity is an angular contact zone rather than an infinitely

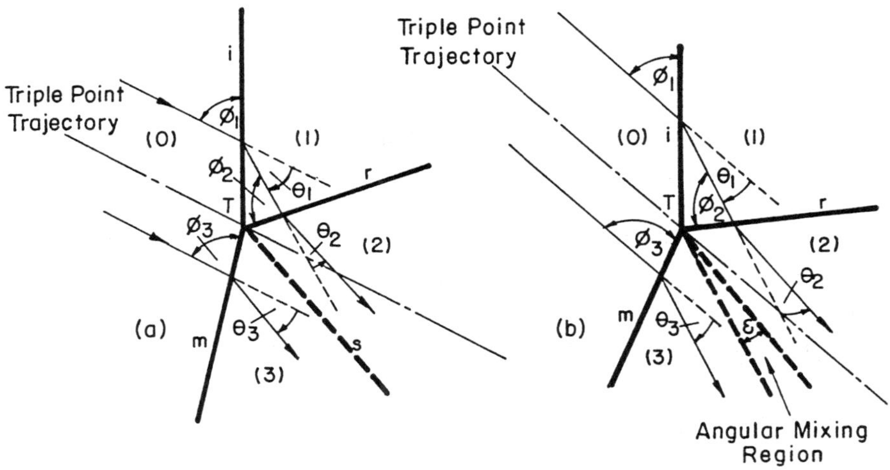

Figure 2.71 The wave configuration of a Mach reflection: a) with an infinitely thin slipstream; b) with an angular contact zone.

thin slipstream. The possibility of such a wave configuration was originally suggested by Courant & Friedrichs (1948).

An inspection of the enlarged photograph of the triple point of the DMR, shown in figure 2.59, indicates that the contact discontinuity indeed seems to be diverging.* If this is the case, then, according to Skews (1971/2), equation (2.60) should be replaced by

$$\theta_1 - \theta_2 = \theta_3 - \zeta \tag{2.61}$$

where ζ is the divergence angle of the angular contact zone, as shown in figure 2.71b.

Ben-Dor (1990) followed Skews' (1971/2) hypothesis and analyzed the photograph shown in figure 2.59 for which the measured value of ζ results in $\zeta = 4^\circ$. The results of Ben-Dor's (1990) study are summarized in table 2 for $M_S = 2.71$, $\phi_1 = 39.9^\circ$, $T_0 = 296$ K and $p_0 = 760$ torr in nitrogen. The results indicate that the predictions of the inviscid perfect three-shock theory with an infinitely thin contact discontinuity, i.e., a slipstream, for ω_{ir} and ω_{rs} are about 5° greater or smaller than the respective measured

---

* Many more photographs of MR wave configurations with clear diverging contact discontinuities can be found in Deschambault's (1984) comprehensive report.

Table 2: Comparison between the experimental results and the predictions of the wave angles by the various modified models of the three-shock theory

Type of Model	ω_{ir}	ω_{im}	ω_{rs}
Experimental results	118°±1°	132°±1°	32°±1°
Three-shock theory with a slipstream (inviscid perfect gas)	123.19° (+5.19°)	132.74° (+0.74°)	27.16° (-4.84°)
Three-shock theory with a 4°-divergence angular contact zone (inviscid perfect gas)	123.25° (+5.25°)	133.95° (+1.95°)	30.311° (-1.89°)
Three-shock theory with a 4°-divergence angular contact zone (inviscid gas in vibrational relaxation)	117.38° (-0.62°)	133.19° (+1.19°)	31.65° (-0.35°)

$M_S = 2.71$, $\phi_1 = 39.9°$, $T_0 = 296$ K, $p_0 = 760$ torr.

values from the experimental record which is shown in figure 2.59.

It is evident from table 2 that the analytical predictions using an angular diverging contact discontinuity are not much better than those obtained using a slipstream. The three-shock theory with a 4° divergence angular contact zone results in slightly worse values for ω_{ir} and ω_{im} and a much better value for ω_{rs} than the three-shock theory with a slipstream.

However, if the assumption that the gas behaves as a perfect gas is replaced, and the gas is allowed to be in vibrational equilibrium, the analytical predictions become fairly good. The discrepancies between the analytical predictions and the measured values for this model are within the experimental uncertainty.

The foregoing discussion indicates that only the inclusion of real gas effects was able to improve the analytical predictions of ω_{ir}. This might be due to the fact that ω_{ir} is influenced by the flow behind the reflected shock wave, which is double shocked and

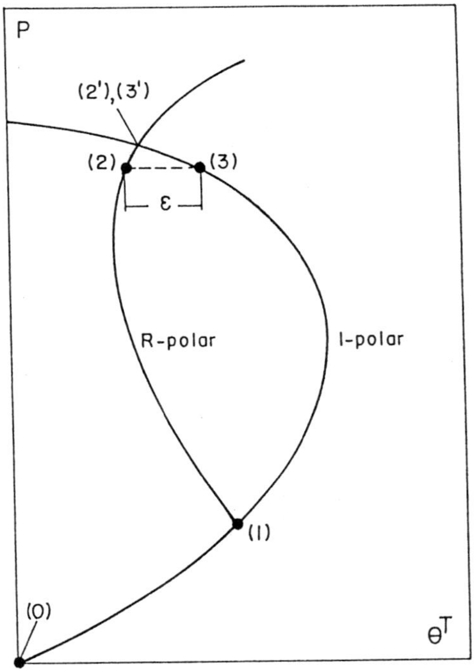

Figure 2.72 (p,θ)-solution of a Mach reflection with an angular contact zone.

therefore reaches temperatures probably high enough to excite the vibrational degree of freedom.

The above results also imply that the shock polar solution of an MR, such as the one shown in figure 1.12, is incorrect. The shock polar solution of an MR with $\zeta \neq 0$ is shown in figure 2.72. Here states (2) and (3) lie along a constant pressure line, but they are separated by the angle of divergence of the angular contact zone - ζ.

2.5.7 Non-Self-Similar Effects

Some comments suggesting that the flow field associated with shock wave reflections in pseudo-steady flows cannot be self-similar, were given earlier in section 2.5.1, where non-steady effects were considered.

The assumption of self-similarity was checked and verified experimentally by many investigators. However, in light of the foregoing discussions on viscous, heat transfer, and real gas effects one is left to wonder whether poor experimental resolution techniques mistakenly led to an incorrect conclusion of a self-similar behavior. The three flow zones arising from the above-mentioned

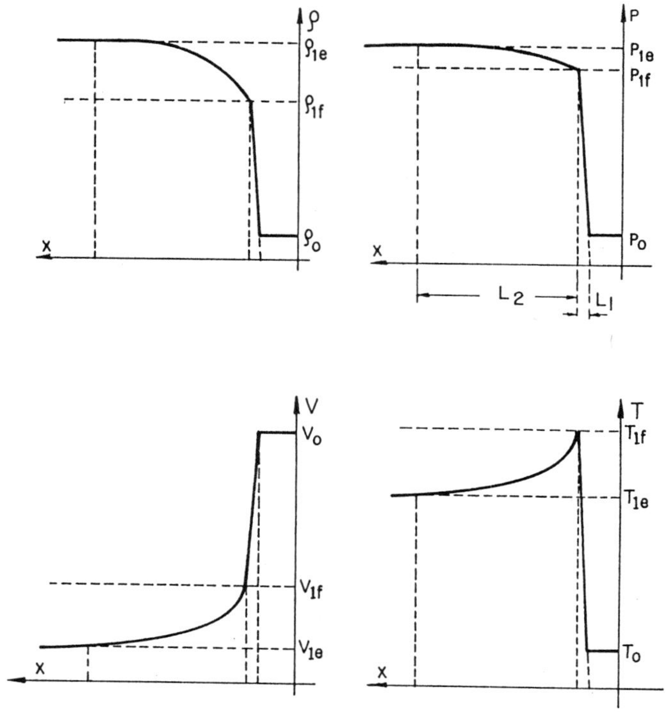

Figure 2.73 Illustration of the change in the flow properties behind normal stationary shock wave due to the relaxation of the internal degrees of freedom: a) density; b) pressure; c) flow velocity; d) tempertaure.

three effects, namely, the kinematic boundary layer, the thermal boundary layer and the relaxation zone, are all non-self-similar. The kinematic boundary layer thickness depends solely on the Reynolds number, the thermal boundary layer thickness depends on both the Reynolds and the Prandtl numbers, and the relaxation lengths depend solely on the shock wave Mach number. Thus, while the wave configurations seems to be growing linearly with time, these three lengths remain constant in a frame of reference attached to either the reflection point of an RR or the triple point of an MR. For these reasons it is unavoidable to conclude that actual shock wave reflections in pseudo-steady flows cannot be self-similar.

To further justify this conclusion consider figure 2.73 where the flow field in the relaxation zone behind a shock wave is schematically illustrated. All the flow conditions immediately behind the shock front (which has a thickness of L_1) can be obtained from the R-H equations. They are known as the frozen jump conditions. Once they are reached, the internal degrees of freedom of the gas are excited (provided the temperature T_{1f} is high enough) to result in a further increase in both the pressure and the density and a decrease in both the temperature and the flow velocity. The changes asymptotically approach the equilibrium values which are reached at the end of the relaxation zone, L_2. Thus it is clear that the properties of the flow field at any location depend on its distance from the shock wave front. Consider now figure 2.47 where a DMR at three different positions along a straight reflecting wedge is drawn. The second triple point of this DMR, which was initially adjacent to the first triple point, moves away from it as the DMR propagates up the reflecting wedge surface. Consequently, the properties of the gas in state (1), which flows through the second triple point, are changing continuously from their frozen values to their equilibrium values. Similarly, the second reflected shock wave, r', also faces continuously changing flow properties in region (2). Thus, it is clear that the second triple point encounters different flow properties as long as it is inside the relaxation zones behind the incident and reflected shock waves. This suggests that the flow in its vicinity cannot be self-similar, and the orientation of the four discontinuities must be continuously changing with time, to negotiate the continuous change in the flow properties.

Note that the above argument cannot be applied to the first triple point as it always faces frozen flow conditions. However, it is possible that while the wave configuration of the first triple point resembles self-similar behavior, the flow fields around it are changing with time in such a way that they are not self-similar. This behavior can be better understood if one refers to the study of Ben-Dor & Glass (1978) where the early attempts to numerically simulate RR and SMR were compared with actual reflections. In the different numerical approaches the predicted wave configurations were almost identical to those recorded experimentally. However, the shapes of the isopycnics (constant density contours) were different. Thus one must conclude that the fact that the wave configurations are self-similar does not necessarily imply that the flow fields associated with them are also self-similar. If this is true, then the location along the reflecting wedge where an actual experiment is recorded might play some role in determining the details of the wave structure of the reflection configuration.

Furthermore, the recent experimental results about the formation of the triple point, i.e., the fact that it does not form at the leading edge of the reflecting wedge (see section 2.2 for details), might also suggest that even the wave configuration of an MR is not self-similar.

Unfortunately, relaxing the assumption of self-similarity can be accomplished only by numerically solving the complete Navier-Stokes equations for an imperfect gas, which is too difficult a task. Due to the fact that such solutions do not exist as yet, it is difficult to assess if and how the various transition lines would shift in the (M_S, θ_w^C) or (M_S, θ_w)-planes. However, in view of the earlier remarks on the need for a mechanism to shift the RR⇔IR transition to higher wedge angles in order to account for the TDMR-experiments which lie above the RR⇔IR transition line, it is possible that the inclusion of non-self-similar effects would indeed cause a shift in the required direction. Note also, that the TDMR-experiments are obtained in the case of heavy gases at high Mach numbers where real gas effects undoubtedly play a significant role. Thus, the fact that the second triple point encounters different flow properties as it moves along the relaxation zone might suggest that this is indeed the mechanism causing the TMR⇔DMR transition line ($M_2^T = 1$) and the DMR$^-$⇔TDMR transition line ($\chi' = 0$) to show poor agreement with the experiments.

2.6 Additional Considerations

When a planar incident shock wave interacts with a straight wedge to result in one of the previously mentioned shock wave reflection configurations, there is an additional process, namely the deflection of the incident shock induced flow around the compressible corner of the reflecting wedge. Thus, the interaction of a planar incident shock wave with a straight reflecting wedge in pseudo-steady flows can be divided into two processes:

1. the shock wave reflection over the reflecting wedge surface,
2. the shock induced flow deflection around the compressive corner of the reflecting wedge.

The combined phenomenon is usually referred to as shock wave diffraction.

The shock wave reflection process in pseudo-steady flows has been dealt with, in detail, in the foregoing sections. In the following, a brief description of the flow deflection process, followed by the combined process, namely, the shock diffraction process, is given.

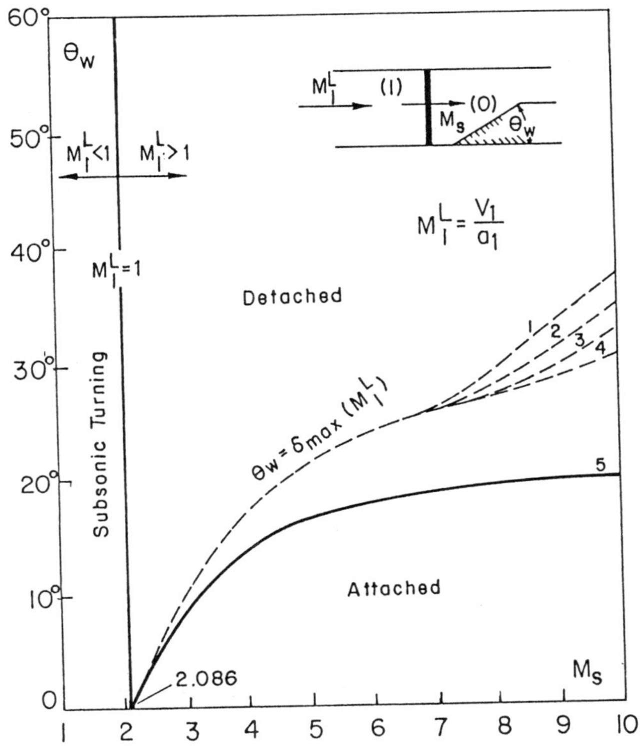

Figure 2.74 Domains and transition boundaries of different types of flow deflection processes in the (M_s,θ_w)-plane. Solid lines are for perfect nitrogen with $\gamma = 1.4$. Dashed lines are for nitrogen in dissociational equilibrium with $T_0 = 300K$ and $p_0 = 1, 10, 100, 1000$ torr for lines 1 to 4, respectively.

2.6.1 Flow Deflection Process

When a planar shock wave propagates along the shock tube towards a reflecting wedge it induces a flow behind it. Depending on the incident shock wave Mach number, this flow can be either subsonic or supersonic. The incident shock wave Mach number for which the induced flow is exactly sonic can be obtained from

$$M_1^L = V_{10}/A_{10} = 1 \tag{2.62}$$

where V_{10} and A_{10}, which depend solely on the incident shock wave Mach number, M_s, are defined by equations (2.39) and (2.40). In the case of N_2 and Ar, for example, the limiting incident shock wave Mach numbers are 2.068 and 2.758, respectively.

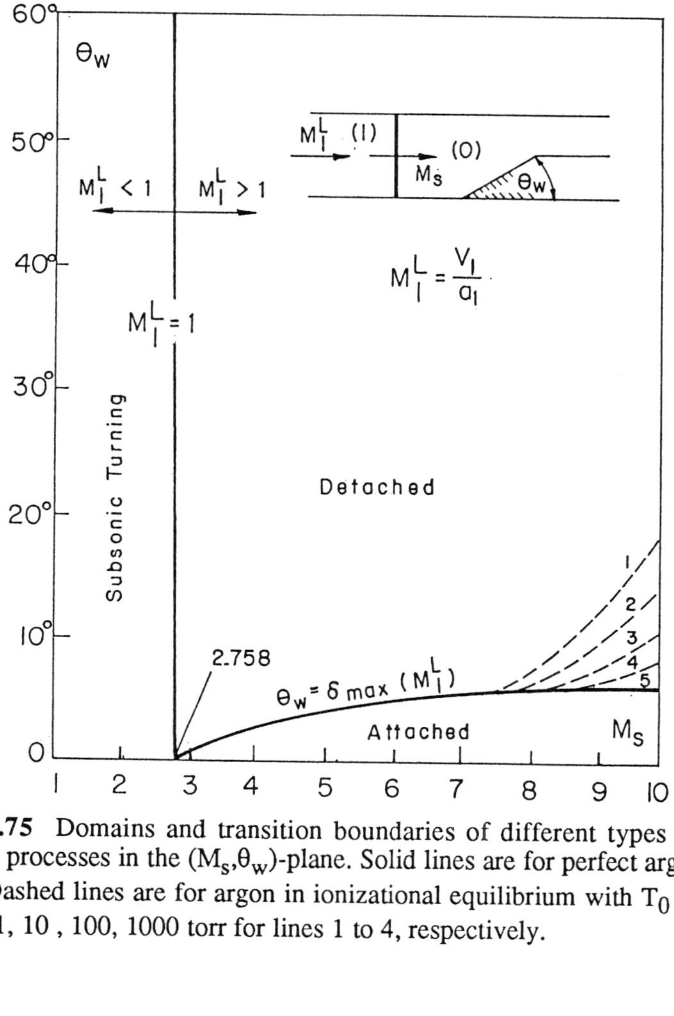

Figure 2.75 Domains and transition boundaries of different types of flow deflection processes in the (M_s,θ_w)-plane. Solid lines are for perfect argon with $\gamma = 5/3$. Dashed lines are for argon in ionizational equilibrium with $T_0 = 300$K and $p_0 = 1, 10, 100, 1000$ torr for lines 1 to 4, respectively.

As discussed earlier (see section 1.2 and figure 1.2), the shock induced flow can negotiate the wedge through a subsonic turning if it is subsonic or with the aid of a shock wave if it is supersonic. In the latter, the shock wave can be either attached or detached (see figures 1.2b and 1.2c, respectively) depending upon the shock induced flow Mach number and the reflecting wedge angle.

Figures 2.74 and 2.75 illustrate the various domains of flow deflection in the (M_S, θ_W)-plane for N_2 and Ar, respectively. The solid lines are for a perfect gas and the dashed lines are for dissociational equilibrium in the case of N_2 and ionizational equilibrium in the case of Ar.

Consider first the region for which the shock induced flow is

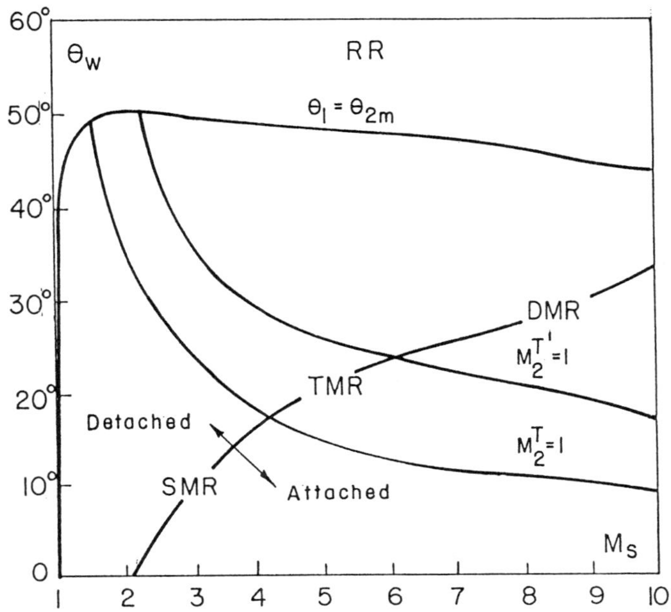

Figure 2.76 Domains and transition boundaries of different types of shock wave diffraction processes in the (M_s,θ_w)-plane. Nitrogen in dissociational equilibrium with $T_0 = 300K$ and $p_0 = 15$ torr.

supersonic in a laboratory frame of reference, i.e., $M_1^L > 1$. This region is divided by the line $\theta_w = \delta_{max}(M_1^L)$ where $\delta_{max}(M_1^L)$ is the maximum deflection that a flow having a Mach number $M_1^L > 1$ could experience by passing through an oblique shock wave. Based on the discussion in section 1.2, one can conclude that if $\theta_w < \delta_{max}(M_1^L)$ then the deflecting shock wave is attached to the leading edge of the reflecting wedge. If, however, $\theta_w > \delta_{max}(M_1^L)$ then the deflecting shock wave is detached. In the latter case it continuously moves towards the oncoming flow since the entire wave configuration grows with time.

The region for which $M_1^L < 1$, where a subsonic turning (see figure 1.2a) should be expected, does not materialize in reality. Instead a detached shock wave is obtained in this region too, and hence it degenerates to become a part of the region marked in figures 2.74 and 2.75 as "detached."

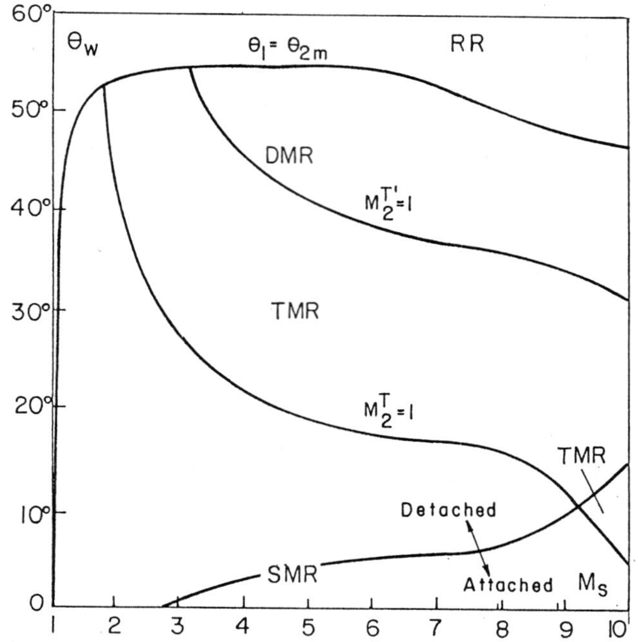

Figure 2.77 Domains and transition boundaries of different types of shock wave diffraction processes in the (M_s,θ_w)-plane. Argon in ionizational equilibrium with $T_0 = 300K$ and $p_0 = 15$ torr.

2.6.2 Shock Wave Diffraction Process

A superposition of figures of the type shown in figure 2.40, but for N_2 and Ar with figures 2.74 and 2.75, respectively, results in figures 2.76 and 2.77 where the domains and transition boundaries of the shock wave diffraction processes are shown in the (M_s, θ_w)-plane. Figures 2.76 and 2.77 clearly indicate that the four general types of reflections, i.e., RR, SMR, TMR and DMR, can all have either an attached or a detached shock wave at the leading edge of the reflecting wedge, although sometimes a reflection configuration with an attached shock wave requires a very high incident shock wave Mach number. In general the domain of an attached shock wave at the leading edge of the reflecting wedge is larger for N_2 than for Ar. Therefore, it should be expected that reflections having attached shock waves would be easier to obtain with gases having smaller values of γ, where γ is the specific heat capacities ratio.

Furthermore, real gas effects shift the transition lines between the domains of various types of reflections to smaller wedge angles while they shift the attached/detached transition line to larger wedge

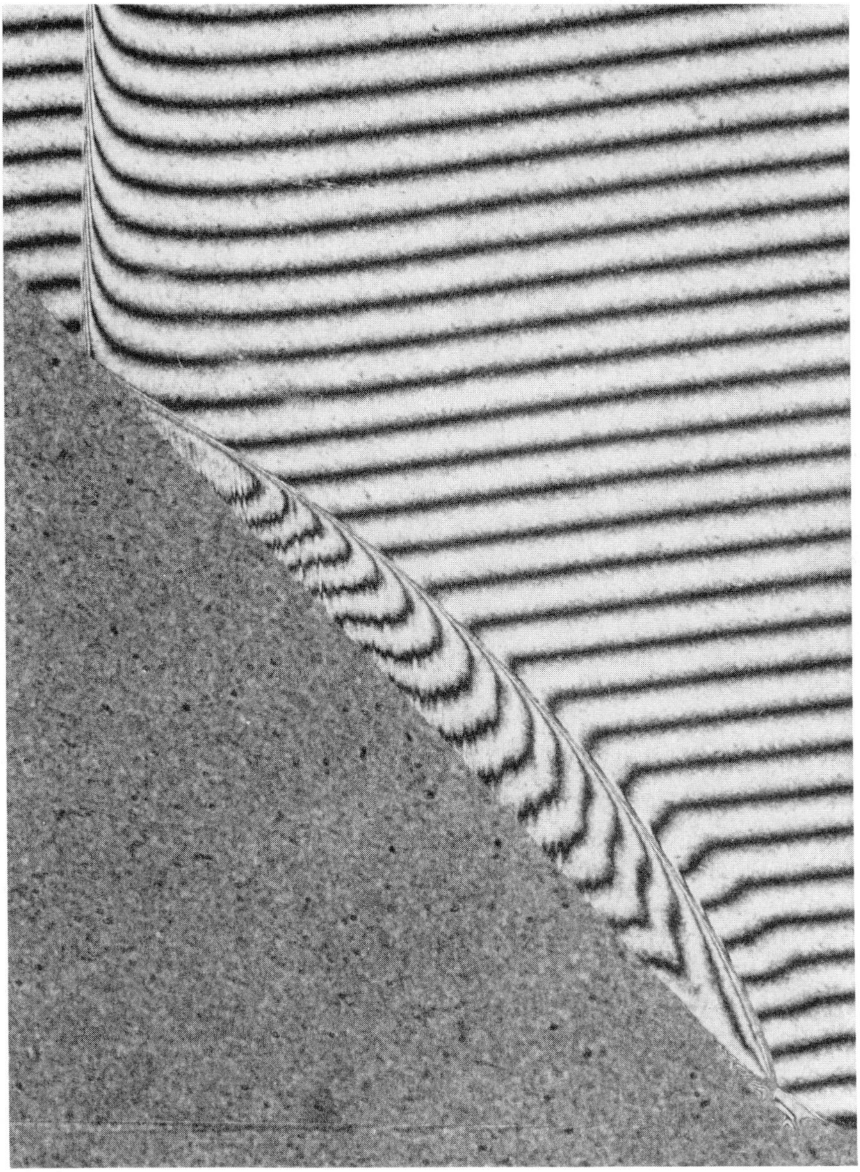

Figure 2.78 An interferogram showing a regular reflection with a reflected shock wave attached to the leading edge of the reflecting wedge with $M_s = 9.9$ and $\theta_w = 47^o$ in air at $T_0 = 296.4K$ and $p_0 = 15$ torr (courtesy of Professor I.I. Glass, Institute for Aerospace Studies, University of Toronto, Toronto, Ontario, Canada).

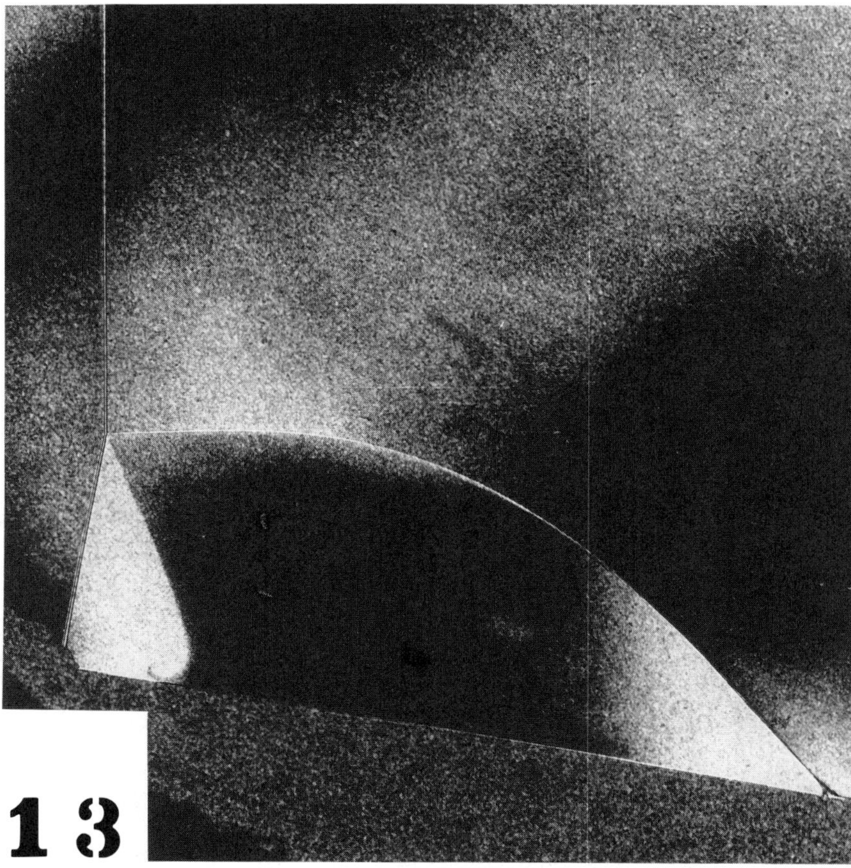

Figure 2.79 An interferogram showing a single-Mach reflection with a reflected shock wave attached to the leading edge of the reflecting wedge with $M_s = 4.73$ and $\theta_w = 10^o$ in air at $T_0 = 296.6K$ and $p_0 = 15$ torr (courtesy of Professor I.I. Glass, Institute for Aerospace Studies, University of Toronto, Toronto, Ontario, Canada).

angles. Hence it is easier to obtain reflections having attached shock waves when the internal degrees of freedom of the gas are excited. Recall that the practical effect of the excitation of the internal degrees of freedom is a reduction in γ.

An interferogram of an RR and a shadowgraph of an SMR having attached reflected shock waves are shown in figures 2.78 and 2.79, respectively. For comparison with an RR and an SMR having detached reflected shock waves, the reader is referred to figures 2.3 and 2.6, respectively.

As a final remark, it should be mentioned that in the case of an SMR the flow field in states (2) and (3) is subsonic with respect to the triple point, i.e., $M_2^T < 1$ and $M_3^T < 1$. Thus, in the case of an SMR the flow deflection process at the leading edge of the reflecting wedge might have an influence on the flow in the vicinity of the triple point, by means of the subsonic path connecting these to points.

2.7 References

Bazhenova, T.V., Fokeev, V.P. & Gvozdeva, L.G., "Regions of Various Forms of Mach Reflection and Its Transition to Regular Reflection", Acta Astro., Vol. 3, pp. 131-140, 1976.

Ben-Dor, G., "Regions and Transitions on Nonstationary Oblique Shock-Wave Diffractions in Perfect and Imperfect Gases", UTIAS Rep. 232, Inst. Aero. Studies, Univ. Toronto, Toronto, Ont., Canada, 1978.

Ben-Dor, G., "Analytical Solution of Double-Mach Reflection", AIAA J., Vol. 18, pp. 1036-1043, 1980.

Ben-Dor, G., "Relation Between First and Second Triple Point Trajectory Angles in Double Mach Reflection", AIAA J., Vol. 19, pp. 531-533, 1981.

Ben-Dor, G., "A Reconsideration of the Three-Shock Theory for a Pseudo-Steady Mach Reflection", J. Fluid Mech., Vol. 181, pp. 467-484, 1987.

Ben-Dor, G., "Structure of the Contact Discontinuity of Nonstationary Mach Reflections", AIAA J., Vol. 28, pp. 1314-1316, 1990.

Ben-Dor, G. & Glass, I.I., "Nonstationary Oblique Shock Wave Reflections: Actual Isopycnics and Numerical Experiments", AIAA J., Vol. 16, pp. 1146-1153, 1978.

Ben-Dor, G., Mazor, G., Takayama, K. & Igra, O., "The Influence of Surface Roughness on the Transition from Regular to Mach Reflection in a Pseudo-Steady Flow", J. Fluid Mech., Vol. 176, pp. 336-356, 1987.

Ben-Dor, G., Takayama, K. & Dewey, J.M., "Further Analytical Considerations of Weak Planar Shock Wave Reflections over Concave Wedges", Fluid Dyn. Res., Vol. 2, pp. 75-85, 1987.

Ben-Dor, G. & Whitten, B.T., "Interferometric Technique and Data Evaluation Methods for the UTIAS 10cm×18cm Hypervelocity Shock Tube", UTIAS Tech. Note 208, Inst. Aero. Studies, Univ. Toronto, Toronto, Ont., Canada, 1979.

Birkhoff, G., *Hydrodynamics, A Study in Logic, Fact and Similitude*, Princeton Univ. Press, N.J., U.S.A., 1950.

Clarke, J.F., "Regular Reflection of a Weak Shock Wave from a Rigid Porous Wall", Quar. J. Mech. & Math., Vol. 37, pp. 87-111, 1984a.

Clarke, J.F., "The Reflection of Weak Shock Waves from Absorbent Surfaces", Proc. Roy. Soc. Lond., Ser. A., Vol. 396, pp. 365-382, 1984b.

Colella, P. & Henderson, L.F., "The von Neumann Paradox for the Diffraction of Weak Shock Waves", J. Fluid Mech., Vol. 213, pp. 71-94, 1990.

Courant, R. & Friedrichs, K.O., *Hypersonic Flow and Shock Waves*, Wiley Interscience, New York, 1948.

Deschambault, R.L., "Nonstationary Oblique-Shock-Wave Reflections in Air", UTIAS Rep. 270, Inst. Aero. Studies, Univ. Toronto, Toronto, Ont., Canada, 1984.

Dewey, J.M. & McMillin, D.J., "Observation and Analysis of the Mach Reflection of Weak Uniform Plane Shock Waves. Part 1. Observation", J. Fluid Mech., Vol. 152, pp. 49-66, 1985.

Friend, W.H., "The Interaction of Plane Shock Wave with an Inclined Perforated Plate", UTIAS Tech. Note 25, Inst. Aero. Studies, Univ. Toronto, Toronto, Ont., Canada, 1958.

Glaz, H.M., Colella, P., Collins, J.P. & Ferguson, E., "Nonequilibrium Effects in Oblique Shock-Wave Reflection", AIAA J., Vol. 26, pp. 698-705, 1988.

Henderson, L.F. & Gray, P.M., "Experiments on the Diffraction of Strong Blast Waves", Proc. Roy. Soc. Lond., Ser. A., Vol. 377, pp. 363-378, 1981.

Henderson, L.F. & Lozzi, A., "Experiments on Transition of Mach Reflection", J. Fluid Mech., Vol. 68, pp. 139-155, 1975.

Henderson, L.F., Ma, J.H., Sakurai, A. & Takayama, K., "Refraction of a Shock Wave at an Air-Water Interface", Fluid Dyn. Res., Vol. 5, pp. 337-350, 1990.

Hornung, H.G. & Taylor, J.R., "Transition from Regular to Mach Reflection of Shock Waves. Part 1. The Effect of Viscosity on the Pseudo-Steady Case", J. Fluid Mech., Vol. 123, pp 143-153, 1982.

Ikui, T., Matsuo, K., Aoki, T. & Kondoh, N., "Mach Reflection of a Shock Wave from an Inclined Wall", Memoirs Faculty Eng., Vol. 41, pp. 361-380, Kyushu Univ., Fukuoka, Japan, 1981.

Jones, D.M., Martin, P.M. & Thornhill, C.K., "A Note on the Pseudo-Stationary Flow Behind a Strong Shock Diffracted or Reflected at a Corner", Proc. Roy. Soc. Lond., Ser. A., Vol. 209, pp. 238-248, 1951.

Law, C.K. & Glass, I.I., "Diffraction of Strong Shock Waves by a Sharp Compressive Corner", CASI Trans., Vol. 4, pp. 2-12, 1971.

Lee, J. -H. & Glass, I.I., "Pseudo-Stationary Oblique-Shock-Wave Reflections in Frozen and Equilibrium Air", Prog. Aerospace Sci., Vol. 21, pp. 33-80, 1984.

Mach, E., "Über den Verlauf von Funkenwellen in der Ebeme und im Raume", Sitzungsbr. Akad. Wiss. Wien, Vol. 78, pp. 819-838, 1878.

Mirels, H., "Mach Reflection Flowfields Associated with Strong Shocks", AIAA J., Vol. 23, pp. 522-529, 1985.

Onodera, H., "Shock Propagation over Perforated Wedges", M.Sc. Thesis, Inst. High Speed Mech., Tohoku Univ., Sendai, Japan, 1986.

Onodera, H. & Takayama, K., "Shock Wave Propagation over Slitted Wedges", Inst. Fluid Sci. Rep., Vol. 1, pp. 45-66, Tohoku Univ., Sendai, Japan, 1990.

Reichenbach, H., "Roughness and Heated Layer Effects on Shock-Wave Propagation and Reflection - Experimental Results", Ernst Mach Inst., Rep. E24/85, Freiburg, West Germany, 1985.

Schmidt, B., "Structure of Incipient Triple Point at the Transition from Regular Reflection to Mach Reflection", in *Rarefied Gas Dynamics: Theoretical and Computational Techniques*, Eds. E.P. Muntz, D.P. Weaver & D.H. Campbell, Progress in Astronautics and Aeronautics, Vol. 118, pp. 597-607, 1989.

Shames, I.H., *Mechanics of Fluids* , McGraw Hill, 2nd Ed., 1982.

Shirouzu, M. & Glass, I.I., "An Assessment of Recent Results on Pseudo-Steady Oblique Shock-Wave Reflection", UTIAS Rep. 264, Inst. Aero. Studies, Univ. Toronto, Toronto, Ont., Canada, 1982.

Shirouzu, M. & Glass, I.I., "Evaluation of Assumptions and Criteria in Pseudo-Stationary Oblique Shock-Wave Reflections", Proc. Roy. Soc. Lond., Ser. A., Vol. 406, pp. 75-92, 1986.

Skews, B.W., "The Flow in the Vicinity of a Three-Shock Intersection", CASI Trans., Vol. 4, pp. 99-107, 1971.

Skews, B.W., "The Effect of an Angular Slipstream on Mach Reflection", Departmental Note, McMaster Univ., Hamilton, Ont., Canada, 1971/2.

Smith, L.G., "Photographic Investigation of the Reflection of Plane Shocks in Air", OSRD Rep. 6271, Off. Sci. Res. Dev., Washington, DC., U.S.A., or NORC Rep. A-350, 1945.

Takayama, K. & Ben-Dor, G., "Pseudo-Steady Oblique Shock-Wave Reflections over Water Wedges", Exp. in Fluids, Vol. 8, pp. 129-136, 1989.

Takayama, K., Miyoshi, H. & Abe, A., "Shock Wave Reflection Over Gas/Liquid Interface", Inst. High Speed Mech. Rep., Vol. 57, pp. 1-25, Tohoku Univ., Sendai, Japan, 1989.

Wheeler, J., "An Interferometric Investigation of the Regular to Mach Reflection Transition Boundary in Pseudo-Stationary Flow in Air", UTIAS Tech. Note 256, Inst. Aero. Studies, Univ. Toronto, Toronto, Ont., Canada, 1986.

White, D.R., "An Experimental Survey of the Mach Reflection of Shock Waves", Princeton Univ., Dept. Phys., Tech. Rep. II-10, Princeton, N.J., U.S.A., 1951.

Zaslavskii, B.I. & Safarov, R.A., "Mach Reflection of Weak Shock Waves From a Rigid Wall", Zh. Prik. Mek. Tek. Fiz., Vol. 5, pp. 26-33, 1973.

3

Shock Wave Reflections in Steady Flows

List of Symbols

Latin Letters

k	-	thermal conductivity
L_m	-	length of the Mach stem
L_w	-	length of the reflecting wedge
M_i	-	flow Mach number in state (i)
M_s	-	incident shock wave Mach number
p_i	-	static pressure in state (i)
u_i	-	flow velocity in state (i)

Greek Letters

γ	-	specific heat capacities ratio
$\delta_{max}(M)$	-	maximum deflection angle for a flow having Mach number M through an oblique shock wave
θ_w	-	reflecting wedge angle
θ_w^C	-	complementary wedge angle
μ	-	dynamic viscosity
μ_i	-	Mach angle of the flow having a Mach number M_i
ρ_i	-	flow density in state (i)

ϕ_i - angle of incidence between the flow and the oblique shock wave across which the flow enters state (i)

ϕ^* - limiting angle of incidence

Subscripts

0 - flow state ahead of the incident shock wave, i.
1 - flow state behind the incident shock wave, i.
2 - flow state behind the reflected shock wave, r.
3 - flow state behind the Mach stem, m.

The analytical investigation of the reflection phenomenon of shock waves in steady flows is much simpler than that in pseudo-steady flows for the following reasons:

1) Unlike the reflection process in pseudo-steady flows, where it is coupled with an additional process, namely, the flow deflection process (see section 2.6.1), in steady flows the reflection process is independent.
2) Unlike the analysis of the reflection process in pseudo-steady flows, where there is a need to analytically predict the first triple point trajectory angle, χ, in order to transform the results from the (M_S, θ_w^C)-plane to the more physically meaningful and more applicable (M_S, θ_w)-plane, in steady flows the results are presented in the (M_0, ϕ_1)-plane. The presentation of the results in this plane can be done by solving equations (1.14) to (1.27) without the need for an additional expression for χ [equation (2.33)] as discussed in section 2.2.1.
3) Of the ten different shock wave reflection configurations mentioned in section 1.1 only two, namely, RR and SMR, are possible in steady flows.

Unfortunately, in spite of these obvious advantages, not too many experimental studies on the reflection of shock waves in steady flows have been reported thus far. Furthermore, most of the available experimental data were obtained more than two decades ago, with experimental equipment less accurate than that existing presently. In addition, many of the problems of interest to the scientific community today were not known at the time when the experimental investigations of the shock wave reflection phenomenon in steady flows were carried out.

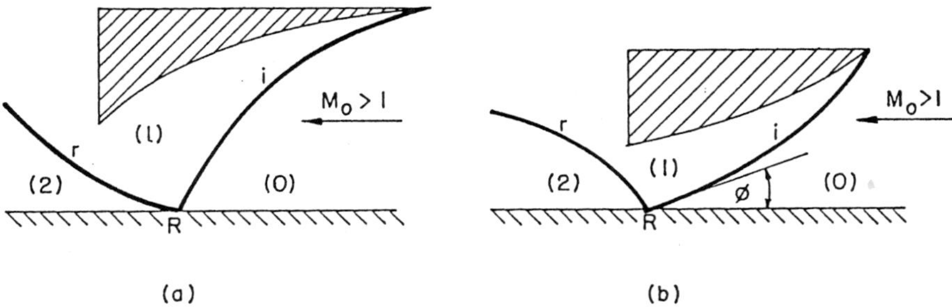

Figure 3.1 Schematic illustration of the wave configurations of a curved incident shock wave reflection over a straight surface: a) concave incident shock wave; b) convex incident shock wave.

3.1 Categories of Steady Reflection Phenomena

The shock wave reflection phenomenon in steady flows can, in general, be divided into four different categories:

1) the reflection of a curved incident shock wave from a straight surface,
2) the reflection of a straight incident shock wave from a curved surface,
3) the reflection of a curved incident shock wave from a curved surface, and
4) the reflection of a straight incident shock wave from a straight surface.

3.1.1 Curved Incident Shock Wave Reflections over Straight Reflecting Surfaces

If a supersonic flow, $M_0 > 1$, encounters a concave or a convex reflecting wedge, then the shock wave which results in order to enable the flow to negotiate the wedge, is also concave or convex. The reflection of the incident shock wave for these two possibilities are shown schematically in figures 3.1a and 3.1b, respectively. The intermediate case of a straight reflecting wedge, of course, results in a straight attached shock wave provided that the reflecting wedge angle is smaller than the maximum flow deflection appropriate to M_0, as shown in figure 1.2b [i.e., $\theta_w < \delta_{max}(M_0)$]. If, however, the reflecting wedge angle is greater than the maximum flow

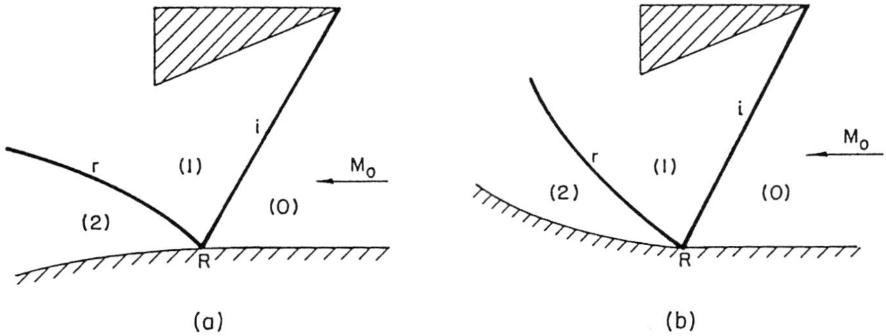

Figure 3.2 Schematic illustration of the wave configurations of a straight incident shock wave reflection over a curved surface: a) convex surface; b) concave surface.

deflection, as shown in figure 1.2c, then the straight reflecting wedge results in a detached bow shock wave which results in a situation similar to that shown in figure 3.1b.

Pant (1971), who analytically studied the reflection of steady curved shock waves, showed that for weak incident shock waves there is a wave angle, ϕ, (see figure 3.1b) for which the reflected shock wave is straight. This specific wave angle, ϕ^*, which was found to be independent of the incident flow Mach number, M_0, can be obtained from:

$$\phi^* = \cos^{-1}\left[\frac{(\gamma + 1)^{1/2}}{2}\right] \tag{3.1}$$

Thus, in the regular reflection of weak shock waves of all strengths for $\phi < \phi^*$ the incident and the reflected shock waves have curvatures of opposite sign. As the wave angle in the vicinity of the reflection point approaches ϕ^* the reflected shock wave straightens until it becomes straight at $\phi = \phi^*$. For $\phi > \phi^*$ the curvatures of the incident and the reflected shock waves have the same sign.

Mölder (1971) numerically investigated this type of steady flow reflection. In the case of an RR a zero downstream curvature on the streamline behind the reflected shock wave in the vicinity of the reflection point - R was imposed, and in the case of an MR the pressure gradients and curvatures of the streamlines along the slipstream, in the vicinity of the triple point - T, were matched. Mölder's (1971) results showed many possible combinations of

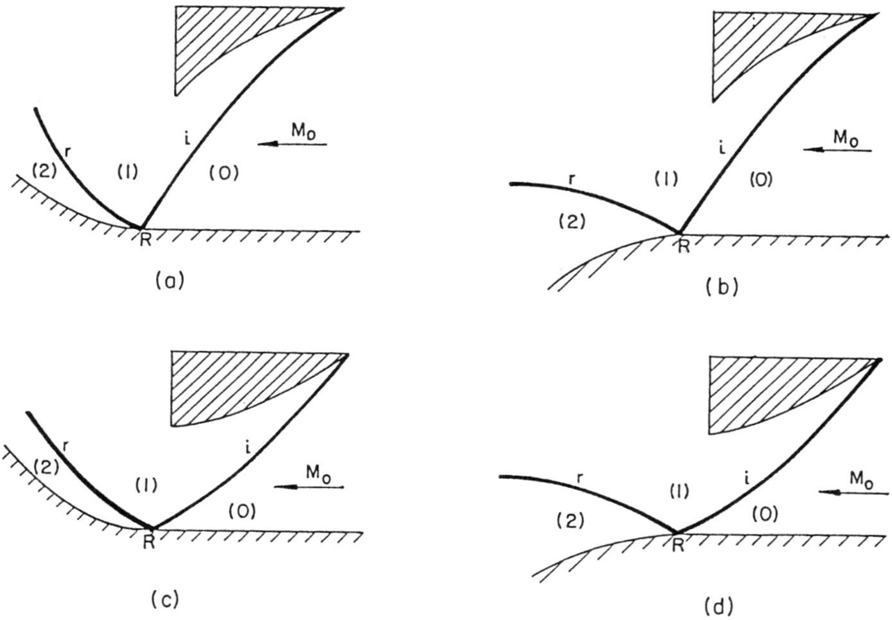

Figure 3.3 Schematic illustration of the wave configurations of a curved incident shock wave reflection over a curved surface: a) concave incident shock wave over a convex surface; b) concave incident shock wave over a concave surface; c) convex incident shock wave over a convex surface; d) convex incident shock wave over a concave surface.

reflected shock curvature, streamline curvature and pressure gradients.

In addition Mölder (1971) presented both theoretical arguments and experimental evidence that the RR⇔MR transition occurs when the Mach stem is normal to the incident flow, i.e., at the point predicted by the length scale criterion, namely

$$\theta_1 - \theta_2 = \theta_3 = 0$$

Although only RR-configurations are shown in figures 3.1a and 3.1b, MR-configurations are also possible in this steady flow reflection category.

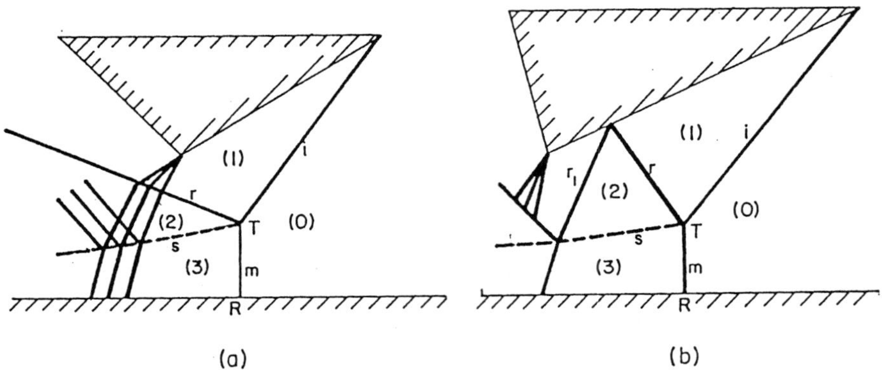

Figure 3.4 Schematic illustration of the reasons for not obtaining TMR and DMR in steady flows.

3.1.2 Straight Incident Shock Wave Reflections over Curved Reflecting Surfaces

Two general cases which belong to this category of shock wave reflections in steady flows are shown schematically in figures 3.2a and 3.2b. For these two cases the incident shock waves are straight and the reflecting surfaces are straight up to the reflection point, R, after which they are either concave or convex. Depending upon the curvature of the reflecting surface beyond the reflection point, R, a concave or a convex reflected shock wave can be obtained. It is hypothesized here, that the curvature of the reflected shock wave will be of the same sign as the curvature of the reflecting surface as shown schematically in figures 3.2a and 3.2b.

The remark in the previous section about the possibility of obtaining MR-configurations in this steady flow reflection category is correct also in this case.

3.1.3 Curved Incident Shock Wave Reflections over Curved Reflecting Surfaces

Four general cases which belong to this category of shock wave reflections in steady flows are shown schematically in figures 3.3a to 3.3d. The incident shock wave in each of these cases is curved and the reflecting surface is straight up to the reflection point, R, beyond which it is either concave or convex. It is again hypothesized here that the reflected shock waves assume a curvature with the same sign as the curvature of the reflecting surface, as is shown schematically in figures 3.3a to 3.3d.

It is quite obvious that there should be conditions in this steady-

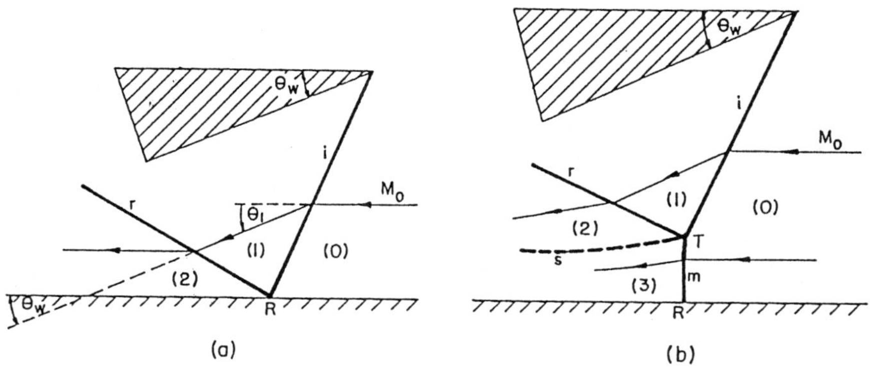

Figure 3.5 Definition of the various flow parameters in steady shock wave reflections: a) RR; b) MR.

flow reflection category, for which MR-configurations are obtained rather than RR-configurations.

3.1.4 Straight Incident Shock Wave Reflections over Straight Reflecting Surfaces

This category of shock wave reflections in steady flows is undoubtedly the easiest one to treat analytically as both the incident shock wave, the reflected shock wave, and the reflecting surface are straight. In addition, most of the reported analytical and experimental studies on the reflection of shock waves in steady flows fall into this steady flow reflection category.

3.1.4.1 Types of Reflections

Experimental evidence indicates that in this case of steady flows only two reflection configurations are possible, namely, regular reflection - RR and single-Mach reflection - SMR. The reason for this might be as follows:

Consider figure 3.4 where two different reflecting wedges are used to generate the incident shock wave, i. The one shown in figure 3.4a is shorter than the one shown in figure 3.4b. In the case of the shorter wedge the flow in state (2), behind the reflected shock wave, r, interacts with the expansion wave generated by the shoulder of the reflecting wedge. By making the reflecting wedge longer (figure 3.4b), this expansion wave can be pushed backwards. However, now the reflected shock wave, r, reflects from the surface of the extended reflecting wedge (either as an RR or as an SMR) to result in a situation in which the flow in state (2) behind the reflected shock wave, r, interacts with the additional reflected shock wave, r_1. Thus, in both cases, the flow in state (2) is interrupted by an additional wave (either an expansion or a shock). This additional interaction probably prevents a situation in which the supersonic flow in state (2) can find itself propagating towards the bottom surface, thereby eliminating the possibility of obtaining more complex reflection configurations which would be required to turn the flow in state (2) away from the bottom surface.

A schematic illustration of the wave configurations of an RR and an SMR in steady flows is shown in figures 3.5a and 3.5b, respectively. The interaction between the supersonic flow, M_0, and the reflecting wedge, generates a straight oblique attached shock wave, the incident shock wave, i, through which the flow is deflected by an angle of θ_1 to become parallel to the reflecting wedge surface, i.e., $\theta_1 = \theta_W$. The flow behind the incident shock wave, i, is supersonic as the incident shock wave, i, arises from the weak shock solution (for details see section 1.4). The deflected flow obliquely approaches the bottom surface with an angle θ_W (see figure 3.5a). If θ_W is smaller than the maximum deflection angle appropriate to the flow in state (1), i.e., if $\theta_W < \delta_{max}$ (M_1), then an RR as shown in figure 3.5a is obtained. If, however, θ_W is greater than the maximum deflection angle of the flow in state (1), i.e., if $\theta_W > \delta_{max}$ (M_1), then an RR is impossible, and the reflection obtained is an SMR as shown in figure 3.5b.

3.1.4.2 RR⇔MR Transition Criterion

Hornung & Robinson (1982) showed that the RR⇔MR transition line in steady flows is that arising from the length scale criterion (see section 1.5.4). In the case of steady flows, the length scale criterion results in two different transition formulae depending upon whether the incident flow Mach number, M_0, is smaller or greater than a certain critical value, M_{0C}. This critical value which was discussed earlier (see section 1.5.1) is the point where the transition lines arising from the mechanical equilibrium and the detachment criteria intersect (see figure 1.20b). The I-R shock polar combination at this critical value of M_{0C} is shown in figure 1.14c.

For values of $M_0 > M_{0C}$ the length scale criterion results in a transition at

$$\theta_1 - \theta_2 = \theta_3 = 0 \qquad (3.2)$$

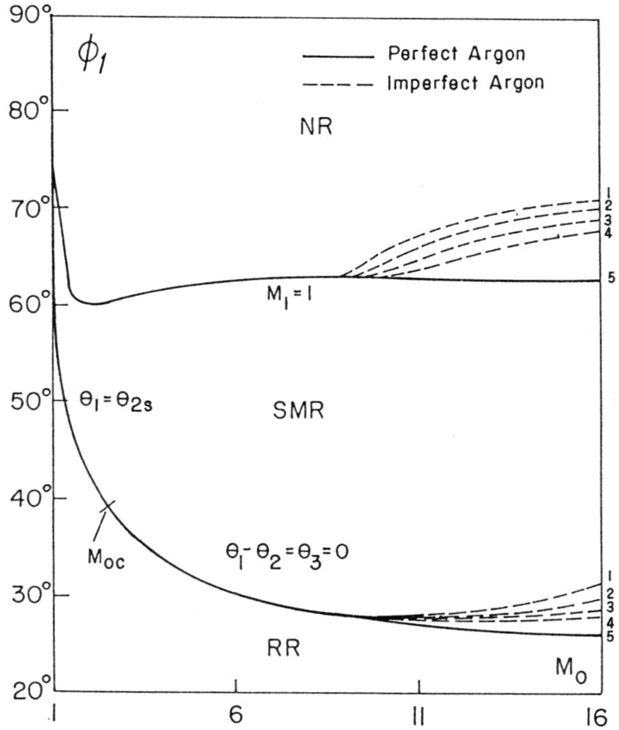

Figure 3.6a Domains and transition boundaries of various types of shock wave reflections in the (M_0, ϕ_1)-plane. Solid lines are for perfect argon with $\gamma = 5/3$, dashed lines are for argon in ionizational equilibrium with T_0 = 300K and p_0 = 1, 10, 100 and 1000 torr for lines 1 to 4, respectively.

which, incidently, is identical to that predicted by the mechanical equilibrium criterion (see section 1.5.2). For values of $M_0 < M_{0C}$, for which equation (3.2) cannot be satisfied, the length scale criterion predicts transition at the point where the flow behind the reflected shock wave is sonic, i.e.,

$$M_2 = 1 \tag{3.3}$$

Equation (3.3) which is also known as the sonic criterion (see section 1.5.3) can be rewritten as

$$\theta_1 = \theta_{2S} \tag{3.4}$$

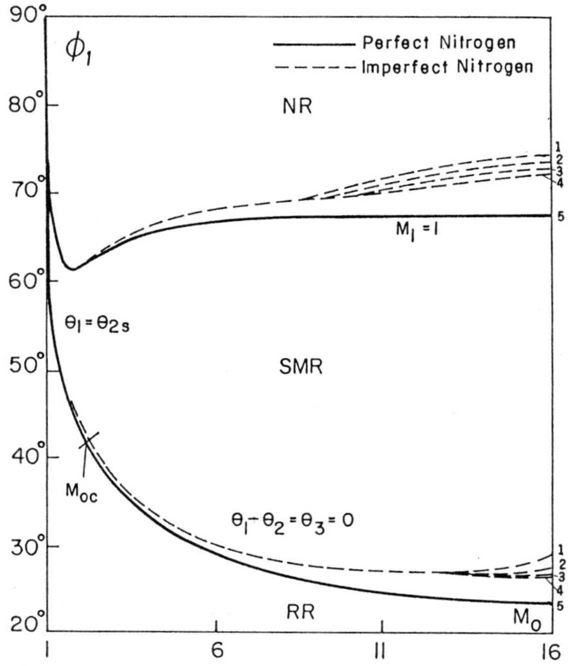

Figure 3.6b Domains and transition boundaries of various types of shock wave reflections in the (M_0, ϕ_1)-plane. Solid lines are for perfect nitrogen with $\gamma = 1.4$, dashed lines are for nitrogen in dissociational equilibrium with $T_0 = 300K$ and $p_0 = 1, 10, 100$ and 1000 torr for lines 1 to 4, respectively.

The transition lines which result from equations (3.2) and (3.3) or (3.4) join at the point $M_0 = M_{0C}$. The exact values of M_{0C} for a diatomic ($\gamma = 7/5$) and a monatomic ($\gamma = 5/3$) perfect gas were calculated by Mölder (1979) to be $M_{0C} = 2.202$ and $M_{0C} = 2.470$, respectively.

3.1.4.3 Domains of Different Types of Reflections in the (M_0, ϕ_1)-Plane

The domains of different types of reflections in steady flows in the (M_0, ϕ_1)-plane are shown in figures 3.6a and 3.6b for N_2 and Ar, respectively. The dashed boundary lines are for an imperfect gas (dissociational equilibrium for N_2 and ionizational equilibrium for Ar) and the solid lines are for a perfect gas. Lines (1) to (4) correspond to $p_0 = 1, 10, 100$ and 1000 torr, respectively. The NR (no reflection) domains correspond to the unobtainable strong shock

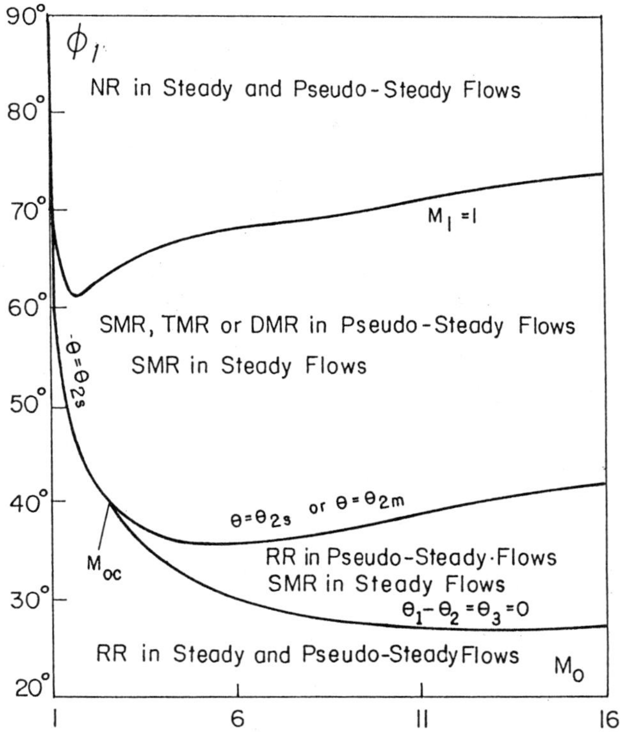

Figure 3.7 Comparison between the domains of RR and MR in steady and pseudo-steady flows in the (M_0,ϕ_1)-plane. Nitrogen in dissociational equilibrium at $T_0 = 300K$ and $p_0 = 15$ torr.

solution, i.e., $M_1 < 1$. Note that the RR⇔SMR transition line is calculated differently depending upon whether M_0 is smaller or greater than M_{0c} as discussed in section 3.1.4.2. The significance of real gas effects (when their inclusion is justified) on shifting the RR⇔SMR and the SMR⇔NR transition lines is self-explanatory.

3.1.4.3.1 Comparison Between Steady and Pseudo-Steady Reflection Domains

It was mentioned earlier (see section 1.5.5) that the RR⇔IR transition line differs in pseudo-steady flows from that in steady flows. The difference between these two transition lines for N_2 in dissociational equilibrium is shown in figure 3.7 in the (M_0, ϕ_1)-plane. In the case of pseudo-steady flows M_0 and ϕ_1 are obtained using the following relations:

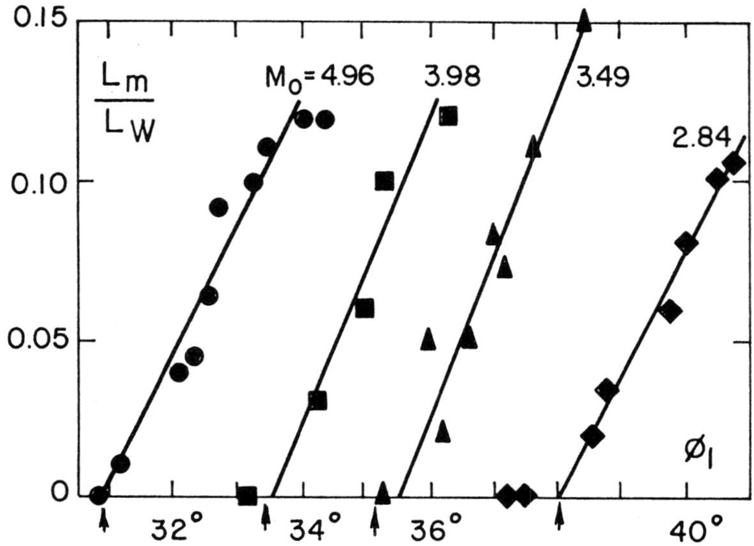

Figure 3.8 Experimental measurements of the height of the Mach stem as a function of the angle of incidence for various incident flow Mach numbers and the determination of the transition angle.

$$M_0 = M_S/\cos \theta_w^C$$

and

$$\phi_1 = 90° - \theta_w^C$$

Thus the transition lines in figure 3.7 for the pseudo-steady case are those shown in the (M_S, θ_w^C)-plane (e.g., see figure 2.49) transformed by the above relations.

Figure 3.7 indicates that there is a large area in which the type of the reflection, i.e., RR or MR, depends on the type of the flow under consideration, i.e., steady or pseudo-steady.

In pseudo-steady flows the RR-domain is larger than the corresponding domain in steady flows. The SMR-domain, however, is smaller in pseudo-steady flows than that in steady flows as the MR-domain is further divided in pseudo-steady flows into SMR-, TMR- and DMR-domains.

3.1.4.4 Comparison of Analysis with Experiments

The RR⇔MR transition lines as given by equations (3.2) and (3.4) were verified experimentally by Henderson & Lozzi (1975,

1979), Hornung, Oertel & Sandeman (1979) and Hornung & Robinson (1982).

The results of the experimental investigation of Hornung & Robinson (1982) are shown in figure 3.8, where the normalized height of the Mach stem, L_m/L_w, as a function of the angle of incidence, ϕ_1, for different flow Mach numbers, M_0, is shown (L_w is defined in figure 1.19b). By extrapolating their experimental results to $L_m = 0$, they showed that the actual RR⇔MR transition occurs at a value of ϕ_1 appropriate to that obtained from equation (3.2). The analytical values for the given incident flow Mach numbers are shown in figure 3.8 by arrowheads.

In spite of the excellent agreement between the experiments and the theory regarding the RR⇔MR transition, one must recall that the two-shock theory as applied to obtain the transition lines shown in figures 3.6 and 3.7 assumes that the fluid is an ideal one, i.e., $\mu = 0$ and $k = 0$ where μ is the dynamic viscosity and k is the thermal conductivity. These, of course, are only simplifying assumptions as a real fluid always has a finite viscosity and thermal conductivity. Discussion on these effects are given in the following sections.

3.2 Modifications of the Perfect Inviscid Two- and Three-Shock Theories

The reflection phenomenon in steady flows might be affected by the non-ideal effects mentioned for the case of the reflection phenomenon in pseudo-steady flows (see section 2.5). In the following, each of the relevant effects is discussed separately.

3.2.1 Non-Straight Discontinuities

Similar to the MR-configuration in pseudo-steady flows, in steady flows not all the discontinuities at the triple point of an MR are straight. This implies that the use of the three-shock theory to calculate the flow field in the vicinity of the triple point introduces an inherent error into the predicted results. Note that while in the case of a pseudo-steady SMR, where only the incident shock wave is straight, in the case of a steady SMR both the incident and the reflected shock waves are straight. This might imply that predictions based on the three-shock theory should better agree with steady SMR-configurations than with pseudo-steady SMR-configurations.

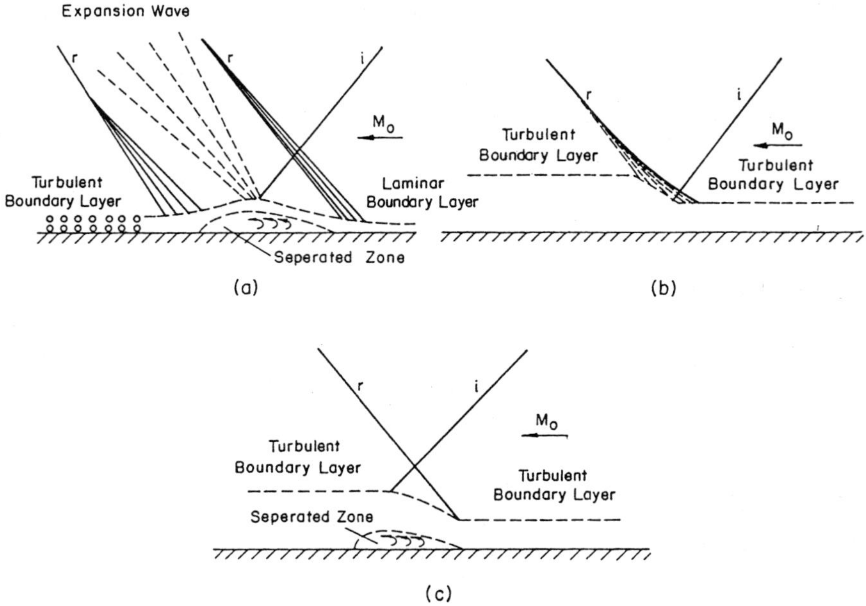

Figure 3.9 Schematic illustration of the way by which the boundary layer over the reflecting surface affects the wave structure of an RR near the reflection point.

3.2.2 Viscous Effects

The flow in state (0) develops a boundary layer along the bottom surface, and hence the incident shock wave, i, interacts with this boundary layer to result in a relatively complex structure near the reflection point at bottom surface. The interaction with the boundary layer depends on whether the boundary layer is laminar or turbulent as shown schematically in figure 3.9, where the structure of the incident shock wave interacting with the boundary layer in the vicinity of the reflection point, R, of an RR is shown schematically.

Figure 3.9 shows that if one is to accurately solve the flow field in the vicinity of the reflection point of an RR then a very complicated flow field must be dealt with.

Henderson (1967) analytically investigated the reflection of a shock wave from a rigid wall in the presence of a boundary layer by treating the problem not as a reflection but as a refraction process. He found that a Mach stem was always present and that the bottom of this wave was bifurcated (a lambda foot). The reflection was said to be regular (RR) if the Mach stem and the lambda foot were

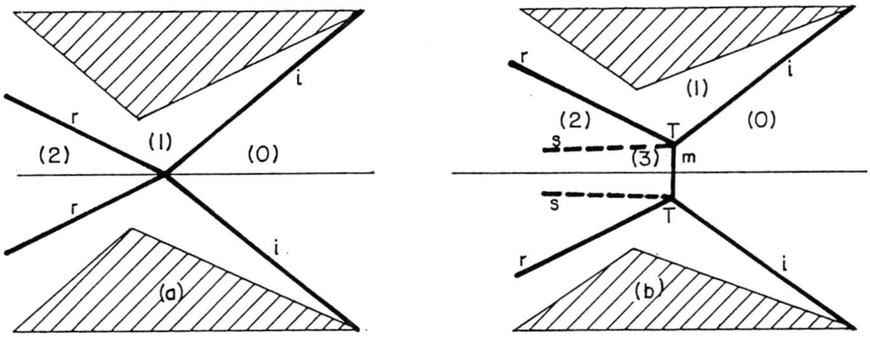

Figure 3.10 Schematic illustration of an experimental set-up for eliminating the boundary layer effect shown in figure 3.9: a) RR; b) MR.

confined to the boundary layer and irregular (IR) if either extends into the main stream. Two types of regular reflection were found, one that had a reflected compression wave and the other that had both reflected compression waves and expansion waves. Henderson (1967) presented initial conditions that enable one to decide which type of reflection would appear. Henderson (1967) reported furthermore, that they were two types of IR, one that had a Mach stem present in the main stream and the other that was characterized by a four-wave configuration. There were also two processes by which RR became IR. One was due to the formation of a downstream shock wave that subsequently swept upstream to establish the irregular system and the other was due to boundary layer separation which forced the lambda foot into the main stream.

There is, however, a possibility by which the above-mentioned viscous effects can be eliminated in steady flow reflections. By using a relatively simple experimental set-up the above illustrated interaction with the boundary layer developing over the bottom surface can be avoided. This is shown in figure 3.10 where two identical reflecting wedges are placed in such a way that they produce two symmetrical regular (figure 10a) and single-Mach (figure 10b) reflections. In this case, the line of symmetry replaces the bottom wedge, thus completely eliminating viscous effects along it. Hence, it is possible to generate inviscid RR-configurations in steady flows. Viscous effects along the slipstream of an SMR would have, in the case of a steady SMR, similar influence as in the pseudo-steady MR (see section 2.5.4.2). Thus, the modification of the three-shock theory presented briefly in section 2.5.4.2 is probably applicable also to a steady SMR if one is to accurately predict the angles between the various discontinuities in the vicinity of the triple point. Using an experimental set-up similar to that mentioned previously, the boundary layer with which the foot of the Mach stem interacts can also be eliminated. This is shown in figure 3.10b.

Some excellent photographs illustrating the interactions of both the incident shock wave of an RR and the Mach stem of an MR with the boundary layer can be found in section 28.3 of the book titled "The Dynamics and Thermodynamics of Compressible Fluid Flow", by A. H. Shapiro (1953).

3.2.3 Thermal Conduction Effects

The fact that a real gas has a finite thermal conductivity, introduces an additional mechanism, heat transfer, which might affect the flow fields in the vicinity of the reflection point of an RR and the triple point of an SMR in a way similar to that presented in section 2.5.5 for the case of pseudo-steady reflections. Here again, not much can be added, as neither experimental nor analytical studies of this effect are available.

The foregoing remark on the elimination of viscous effects by using a double wedge experimental set-up is also valid for the heat transfer effects which can also be eliminated along the line of symmetry. Along the slipstream, however, heat transfer effects might still play an important role.

3.2.4 Real Gas Effects

The question whether real gas effects are important and therefore must be accounted for depends again on the ratio between the relaxation length of a certain degree of freedom and a physical characteristic length of the reflection phenomenon.

When the flow fields in the vicinity of the reflection point, R, of an RR or the triple point, T, of an MR are concerned, it is quite appropriate to assume that the flow is frozen with its pre-shock thermodynamic state. However, as the flow moves away from the incident shock front, its internal degrees of freedom are excited (provided the temperatures are sufficiently high) and the flow fields ahead of and behind the reflected shock wave are no longer uniform.

Consider figure 3.11 where two reflecting wedges with identical wedge angles, θ_w, are used to produce two regular reflections, which are different in their size since the reflecting surface of the same incident shock wave, i, is located at different levels. Note that the two reflecting wedges of figure 3.11 overlap. The corresponding two reflection points are labelled R_1 and R_2, respectively. Although the flow fields associated with these two regular reflections seem to

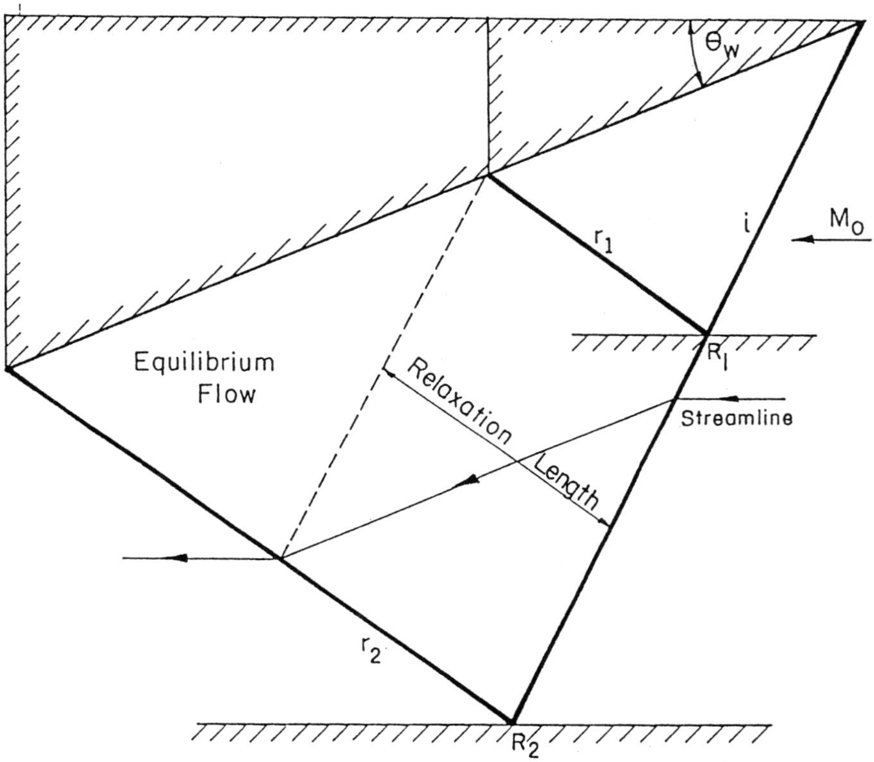

Figure 3.11 A schematic illustration of the way by which the reflection in steady flows is affected by the excitation of the internal degrees of freedom.

be identical (one is an enlargement of the other) this is not the case, because the relaxation length behind the incident shock wave is the same for both cases. Thus, while the reflected shock wave of the upper RR-configuration, r_1, faces a non-equilibrium flow along its entire length, the reflected shock wave of the lower RR-configuration, r_2, is seen to face an equilibrium flow along its portion which extends above the drawn streamline. This, of course, suggests that the flow fields which develop behind the shock waves r_1 and r_2 are different although they were generated using identical initial conditions. Thus, when the shock reflection phenomenon is investigated in steady flows, the influence of real gas effects should be carefully considered prior to their neglect or inclusion.

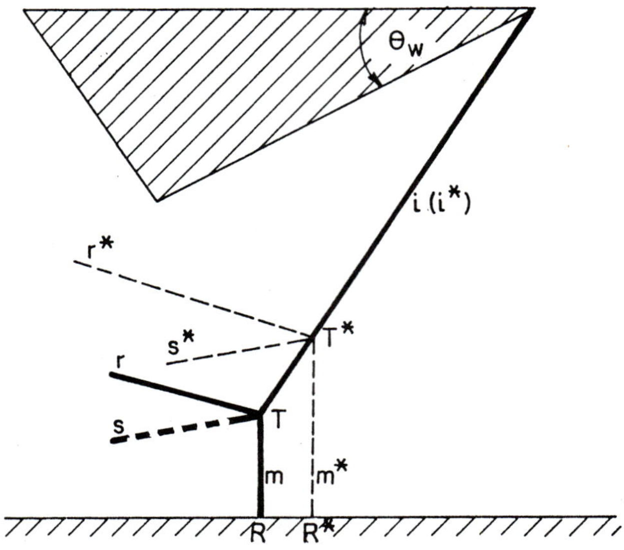

Figure 3.12 A schematic illustration of two theoretically possible MR wave configurations for identical initial conditions.

3.3 Prediction of the Mach Stem Height

There is, however, one experimental observation which has not been completely explained so far and which is very important as it suggests that the analysis of an MR in steady flows overlooks a significant parameter.

Consider figure 3.12 where the solid lines describe the four discontinuities of a steady SMR-configuration with triple point, T, shock waves i, r and m, and slipstream, s. If one selects any point along the incident shock wave, i, and draws three lines from it parallel respectively to the reflected shock wave, r, to the Mach stem, m, and to the slipstream, s, then one has a new triple point T*, with its four discontinuities. The two triple points, T and T*, as well as all the other triple points which could be obtained by choosing a different location for T* along the incident shock wave, completely satisfy the conservation equations given by the three-shock theory, i.e., equations (1.14) to (1.27). However, if an experiment with the same initial conditions (i.e., M_0 and θ_w) is repeated, then only one reflection configuration out of the infinity of possible ones is always obtained. Thus, the three-shock theory is

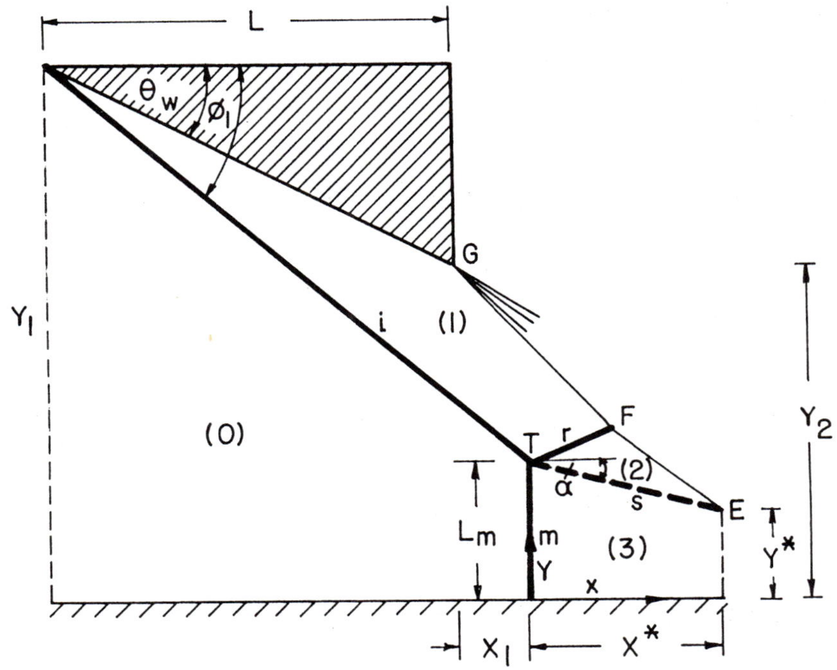

Figure 3.13 A schematic illustration of Azevedo's (1989) model and definition of the flow parameters.

incapable of predicting the actual size of the SMR since it is, inherently, independent of any physical length scale.

Azevedo (1989) suggested an ingenious physical model for predicting the height of the Mach stem. Consider figure 3.13, where a schematic drawing of a reflecting wedge, which generates an SMR, is shown.

Azevedo (1989) assumed that:

1. the Mach stem, the slipstream and the bottom surface form a one-dimensional converging nozzle,
2. the throat of this converging nozzle is at the point where the leading characteristic of the expansion wave, generated by the shoulder of the reflecting wedge, intersects the slipstream, (point E in figure 3.13),
3. the flow in region (3) is isentropic, and reaches sonic conditions at the throat, and
4. the gas is an ideal fluid, i.e., $\mu = 0$ and $k = 0$.

By applying the conservation laws of mass and momentum on the control volume shown in figure 3.13, Azevedo (1989) developed the following three equations:

- conservation of mass

$$\rho_0 u_0 \frac{Y_1}{L} = \rho_1 u_1 \sin\mu_1 \frac{\overline{GF}}{L} + \rho_2 u_2 \sin\mu_2 \frac{\overline{FE}}{L} + \rho * u * \frac{Y *}{L} \tag{3.5}$$

- conservation of linear momentum in the x-direction

$$p_0 \frac{Y_1}{L} - p_1 \frac{L \tan\theta}{L} - p_1' \sin(\mu_1 + \theta_w) \frac{\overline{GF}}{L} - p_2' \sin(\mu_2 + \alpha) \frac{\overline{FE}}{L}$$

$$-p * \frac{Y *}{L} = (\rho_1 u_1 \sin\mu_1) u_1 \cos\theta_w \frac{\overline{GF}}{L} + (\rho_2 u_2 \sin\mu_2) u_2 \cos\alpha \frac{\overline{FE}}{L}$$

$$+\rho * (u *)^2 \frac{Y *}{L} - \rho_0 u_0^2 \frac{Y_1}{L} \tag{3.6}$$

- conservation of linear momentum in the y-direction

$$p_0 \frac{L + X_1}{L} + p_3 \frac{L *}{L} - p_1 \frac{L}{L} - p_1' \cos(\mu_1 + \theta_w) \frac{\overline{GF}}{L} -$$

$$p_2' \cos(\mu_2 + \alpha) \frac{\overline{FE}}{L} = -(\rho_1 u_1 \sin\mu_1) u_1 \sin\theta_w \frac{\overline{GF}}{L}$$

$$-(\rho_2 u_2 \sin\mu_2) u_2 \sin\alpha \frac{\overline{FE}}{L} \tag{3.7}$$

The primed quantities indicate that the flow properties behind the leading characteristic of the expansion wave are slightly different from those ahead of it. The starred quantities denote conditions at the sonic throat. Azevedo (1989) simplified the foregoing model by further assuming that;

1. the pressures on either side of the leading characteristic of the expansion wave are close enough, that it can sensibly be posited that

$$p_1 = p_1' \text{ and } p_2 = p_2'$$

2. the gas obeys the equation of state of a perfect gas, i.e., $p = \rho RT$, and
3. the conditions at the throat are those resulting from an isentropic one-dimensional flow acceleration to sonic conditions.

In addition he used the following relations:

$$\sin\mu_1 = \frac{1}{M_1} \quad \text{and} \quad \sin\mu_2 = \frac{1}{M_2} \tag{3.8}$$

$$p_3 = p_{4av} = \frac{\int_0^{y_{max}} p_4 dy}{y_{max}}$$

where y_{max} represents the height above the bottom surface at which the slipstream is encountered at a given x-location.

Combining these additional assumptions, the above relations as well as the well-known isentropic relations together with equations (3.5) to (3.7), results in:

$$\frac{Y_1}{L} = \frac{\sin\phi_1}{M_1 \sin(\phi_1 - \theta_1)} \frac{\bar{GF}}{L} +$$

$$\frac{\sin\phi_1 \sin\phi_2}{M_2 \sin(\phi_1 - \theta_1) \sin(\phi_2 - \theta_2)} \cdot \frac{\bar{FE}}{L} + f_0 \frac{Y^*}{L} \tag{3.9}$$

$$\frac{Y_1}{L} - \frac{p_1}{p_0} \frac{\tan\theta_1}{1 + \gamma M_0^2} = \frac{p_1}{p_0} \frac{[\sin(\mu_1 + \theta_1) + \gamma M_1 \cos\theta_1]}{1 + \gamma M_0^2} \frac{\bar{GF}}{L} +$$

$$\frac{p_2}{p_0} \frac{[\sin(\mu_2 + \alpha) + \gamma M_2 \cos\alpha]}{1 + \gamma M_0^2} \frac{\bar{FE}}{L} + \frac{p^*}{p_0}\left(\frac{\gamma + 1}{1 + \gamma M_0^2}\right) \frac{Y^*}{L} \tag{3.10}$$

$$1 - \frac{p_1}{p_0} = \frac{p_1}{p_0} [\cos(\mu_1 + \theta_1) - \gamma M_1 \sin\theta_1] \frac{\bar{GF}}{L} +$$

$$\frac{p_2}{p_0}[\cos(\mu_2+\alpha) - \gamma M_2 \sin\alpha]\frac{\bar{FE}}{L} - \frac{p_3}{p_0}\frac{L^*}{L} - \frac{X_1}{L} \tag{3.11}$$

The above set of three equations contains five unknowns, namely,

Y^*/L, $\bar{GF}/L$, $\bar{FE}/L$, L^*/L and X_1/L, thus two additional independent equations are needed to make the set solvable. The two additional equations are the following two geometric relations:

$$\frac{L^*}{L} = \cos(\mu_1 + \theta_1)\frac{\bar{GF}}{L} + \cos(\mu_2+\alpha)\frac{\bar{FE}}{L} - \frac{X_1}{L} \tag{3.12}$$

$$\frac{Y^*}{L} = \frac{Y_2}{L} - \sin(\mu_1+\theta_1)\frac{\bar{GF}}{L} - \sin(\mu_2+\alpha)\frac{\bar{FE}}{L} \tag{3.13}$$

where $Y_2 = Y_1 - L\tan\theta_w$ and Y_1, the parameter through which the length scale is introduced into the solution, is considered as a known parameter.

The above set of five equations and five unknowns can be solved to obtain the value of X_1/L. Note that the set of equations is linear and, hence, it results in a unique solution. The length of the Mach stem - L_m can finally be obtained from

$$\frac{L_m}{L} = \frac{Y_1}{L} - \left(\frac{X_1}{L} + 1\right)\tan\phi_1 \tag{3.14}$$

The predicted values of the height of the Mach stem as obtained using Azevedo's model are shown in figures 3.14a and 3.14b, together with the experimental results of Hornung & Robinson (1982). The predicted results show a trend similar to that for the Mach stem height dependence on the angle of incidence in the experimental results. In addition, the RR⇔MR transition angles, as predicted by the analytical model, agree excellently with those predicted by the length scale criterion (compare value obtained for ϕ_1 at $L_m = 0$ and the corresponding arrowheads along the ϕ_1-axis which indicate the measured transition angles).

Although the agreement between the predicted values for the Mach stem height and the experimental ones is far from satisfactory, one must admit that, in general, the analytical predictions are

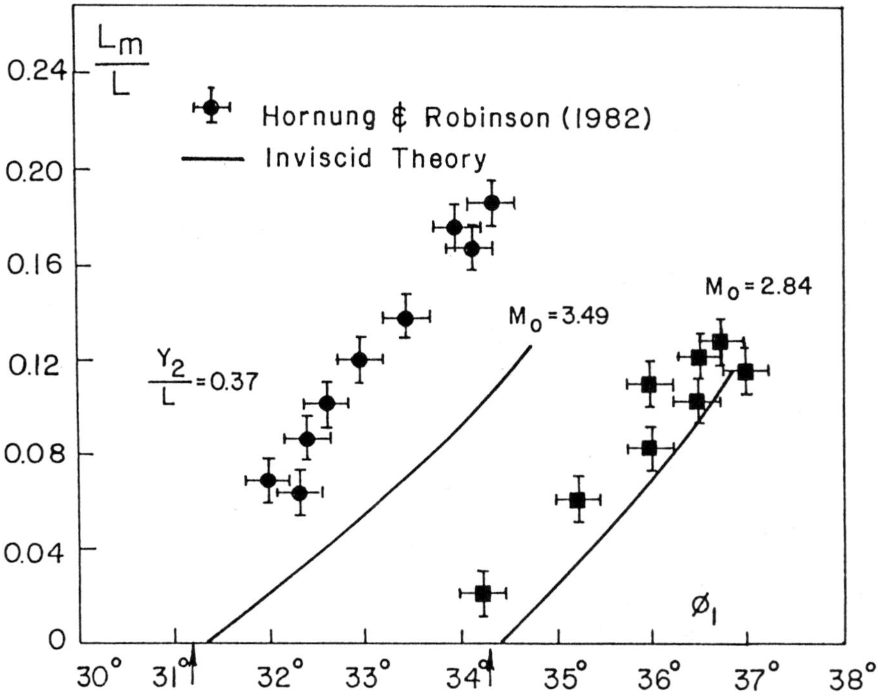

Figure 3.14a Predicted values of the height of the Mach stem for $M_0 = 2.84$ and 3.49 and comparison with experimental results.

surprisingly good in view of the over simplifying assumptions upon which the analytical model is based. Furthermore, it should be noted that no other model capable of predicting the height of the Mach stem exists!

As a final remark, it should be mentioned that Azevedo's model was developed for the case when the reflecting wedge, by which the incident shock wave is generated, is short and hence the reflected shock wave, r, interacts with the expansion wave, prior to any other interactions. Therefore, equations (3.9) to (3.14) cannot be used for the case of a long reflecting wedge, such as the one shown in figure 3.4b, where the reflected shock wave is reflected from the surface of the reflecting wedge prior to its interaction with the expansion wave. The foregoing discussion clearly indicates that a physical length associated with the reflecting wedge, which is communicated by the expansion wave to the triple point, determines the actual height of the Mach stem. Whether this occurs in the way suggested by Azevedo (1989) or in another way is yet to be clarified.

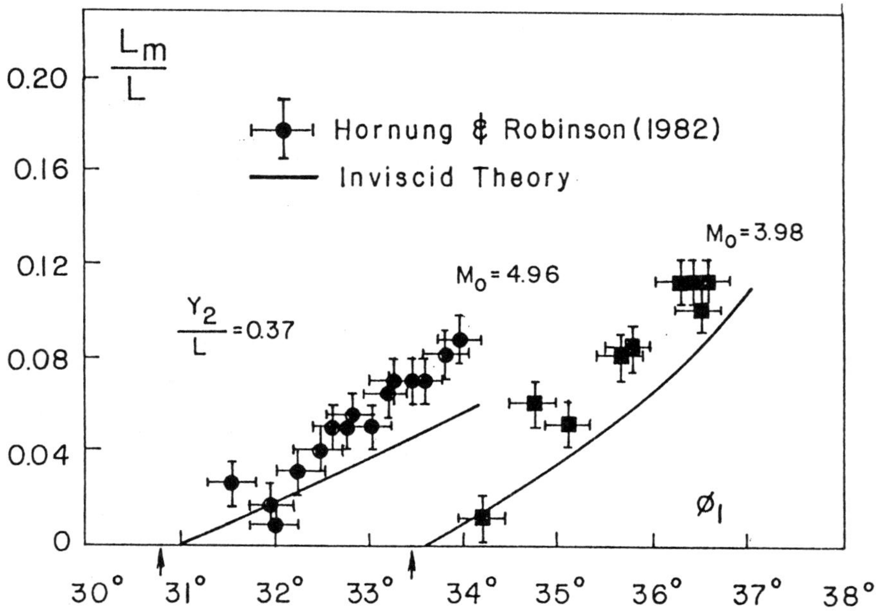

Figure 3.14b Predicted values of the height of the Mach stem for $M_0 = 3.98$ and 4.96 and comparison with experimental results.

3.4 References

Azevedo, D.J., "Analytic Prediction of Shock Patterns in a High-Speed, Wedge-Bounded Duct", Ph.D. Thesis, Dept. Mech. & Aero. Eng., State Univ. N.Y. Buffalo, N.Y., U.S.A., 1989.

Henderson, L.F., "The Reflection of a Shock Wave at a Rigid Wall in the Presence of a Boundary Layer", J. Fluid Mech., Vol. 30, pp. 699-722, 1967.

Henderson, L.F. & Lozzi, A., "Experiments of Transition to Mach Reflection", J. Fluid Mech., Vol. 68, pp. 139-155, 1975.

Henderson, L.F. & Lozzi, A., "Further Experiments of Transition to Mach Reflection", J. Fluid Mech., Vol. 94, pp. 541-560, 1979.

Hornung, H.G., Oertel, H. Jr. & Sandeman, R.J., "Transition to Mach Reflection of Shock Waves in Steady and Pseudo-Steady Flow with and without Relaxation", J. Fluid Mech., Vol. 90, pp. 541-560, 1979.

Hornung, H.G. & Robinson, M.L., "Transition from Regular to Mach Reflection of Shock Waves. Part 2. The Steady-Flow Criterion", J. Fluid Mech., Vol. 123, pp. 155-164, 1982.

Mölder, S., "Reflection of Curved Shock Waves in Steady Supersonic Flow", CASI Trans., Vol. 4, pp. 73-80, 1971.

Mölder, S., "Particular Conditions for the Termination of Regular Reflection of Shock Waves", CASI Trans., Vol. 25, pp. 44-49, 1979.

Pant, J.C., "Reflection of a Curved Shock from a Straight Rigid Boundary", Phys. Fluids, Vol. 14, pp. 534-538, 1971.

Shapiro, A.H., *The Dynamics and Thermodynamics of Compressible Fluid Flow,* Vol. II, The Ronald Press Co., New York, N.Y., U.S.A., 1953.

4

Shock Wave Reflections in Unsteady Flows

List of Symbols

Latin Letters

a_i - local speed of sound in state (i)
A_{ij} = a_i/a_j
L_m - length of the Mach stem in MR
L_r - length of the reflected shock wave in TRR
m_3 - flow mass in region (3) of a TRR
M_i - flow Mach number in state (i)
M_m - Mach stem Mach number
M_s - incident shock wave Mach number
R - radius of curvature of cylindrical wedges
s - coordinate along the cylindrical wedge surface
S - propagation distance of the corner-generated signals
t - time
u_i - flow velocity in state (i) with respect to R in RR and TRR and T in MR
V_i - flow velocity in state (i) in a laboratory frame of reference
V_{ij} = V_i/a_j
V_n - normal shock wave velocity of a TRR with respect to the reflection point R
x^{tr} - x coordinate of the MR→RR transition point

X_T - x coordinate of the triple point of an MR
y - distance from Q to P in a TRR
Y_T - y coordinate of the triple point of an MR
z - distance from R to Q in a TRR

Greek Letters

α - angle between the incident shock wave and the reflecting wedge surface in a TRR
β - angle between the slipstream and the reflecting wedge surface in a TRR
γ - specific heat capacities ratio
Δt - time interval
$\Delta\theta_w$ - change of the slope of the reflecting surface of a double wedge
θ - angular position of a flow particle
θ_T - angular position of the triple point
θ_w - reflecting wedge angle
θ_w^1 - wedge angle of the first surface of a double wedge
θ_w^2 - wedge angle of the second surface of a double wedge
$\theta_w^{initial}$ - initial wedge angle of a cylindrical concave or convex wedge
θ_w^{tr} - transition wedge angle
$\theta_w^{tr}|_M$ - transition wedge angle for shock wave Mach number M
$\theta_w[A \Leftrightarrow B]$ - transition wedge angle from reflection A to reflection B
ρ_i - flow density in state (i)
ϕ_i - angle of incidence between the flow and the oblique shock wave across which the flow enters state (i) with respect to R in RR and TRR and with respect to T in MR.
χ - triple point trajectory angle
χ_g - triple point trajectory angle at glancing incidence
χ^{tr} - triple point trajectory angle at transition

Subscripts

0 - flow state ahead of the incident shock wave, i
1 - flow state behind the incident shock wave, i
2 - flow state behind the reflected shock wave, r
3 - flow state behind the Mach stem, m, of an MR or the normal shock wave, n, of a TRR

Unlike the shock reflection phenomena in pseudo-steady flows (chapter 2) and in steady flows (chapter 3), where the flow fields basically depend on two independent variables, namely; x/t and y/t in the former and x and y in the latter, here the flow field depends on three parameters x, y and t. For this reason, the analytical consideration of the reflection phenomenon in unsteady flows is much more difficult and as a matter of fact very limited progress has been made.

In addition to this difficulty, all the difficulties which were mentioned for the case of shock wave reflections in pseudo-steady flows exist also in the case of shock wave reflections in unsteady flows.

Unsteady shock wave reflections can basically be obtained in one of the following ways:

1) reflecting a shock wave propagating with a constant velocity over a non-straight surface,
2) reflecting a shock wave propagating with a non-constant velocity over a straight surface, and
3) reflecting a shock wave propagating with a non-constant velocity over a non-straight surface.

In the following, each of these three cases is dealt with, separately.

4.1 Constant Velocity Shock Wave Reflections over Non-Straight Surfaces

Consider figure 1.4a and note that if the reflecting wedge surface is not straight but curved, either concavely or convexly, then point R at which the foot of the incident shock wave touches the reflecting wedge surface would have a constant velocity (the incident shock wave velocity) in the horizontal direction only. The velocity of the reflection point, point R, in the vertical direction would not be constant as in the case of a straight reflecting surface.

Thus, if a frame of reference is attached to point R, then the angle of incidence of the oncoming flow, ϕ_1, would change continuously as the shock wave propagates. Furthermore, since $M_0 = M_S / \sin \phi_1$, it is clear that the flow Mach number of the oncoming flow, M_0, would also change continuously.

4.1.1 Shock Wave Reflections over Cylindrical Concave Surfaces

When an incident planar shock wave encounters a cylindrical concave surface it reflects over it either as an RR or as an MR depending upon the initial wedge angle and the incident shock wave Mach number. The initial type of reflection can be determined using the pseudo-steady shock reflection theory presented in chapter 2.

Consider, for example, figure 2.40 and note that if one draws a constant M_s line (i.e., a vertical line for a given M_S) then the number of transition boundaries intersected by this line depends on the chosen value of M_S. In the case of nitrogen for example (figure 2.76) if $M_s < 1.57$ then only the SMR⇔RR transition line is intersected. If, however, $1.57 < M_s < 2.36$ then both the SMR⇔TMR and the TMR⇔RR transition lines are intersected. In the range $M_s > 2.36$ all the three transition lines, i.e., SMR⇔TMR, TMR⇔DMR and DMR⇔RR transition lines are intersected.

Let us consider the most general case of an incident shock wave having a Mach number, M_S, for which all the four types of reflection, i.e., SMR, TMR, DMR and RR are possible. Let us also define the corresponding three transition wedge angles as θ_W[SMR⇔TMR], θ_W[TMR⇔DMR] and θ_W[DMR⇔RR].

Consider figure 4.1 where four different cylindrical concave wedges are drawn. While they all have the same radius of curvature, R, their initial wedge angles (i.e., the slopes of the cylindrical wedges at their leading edges) are different. The slopes are zero for the wedge aa'e, θ_W[SMR⇔TMR] for the wedge bb'e, θ_W[TMR⇔DMR] for the wedge cc'e, and θ_W[DMR⇔RR] for the wedge dd'e. Thus for all the cylindrical wedges with an initial angle in the range $0 \le \theta_w^{\text{initial}} < \theta_W$[SMR⇔TMR] the initial reflection will be an SMR, for all the wedges with an initial angle in the range θ_W[SMR⇔TMR] $< \theta_w^{\text{initial}} < \theta_W$[TMR⇔DMR] the initial reflection will be a TMR, for all wedges with an initial angle in the range θ_W[TMR⇔DMR] $< \theta_w^{\text{initial}} < \theta_W$[DMR⇔RR] the initial reflection will be a DMR and for the wedges with an initial angle in the range $\theta_w^{\text{initial}} > \theta_W$[DMR⇔RR] the initial reflection will be an RR.

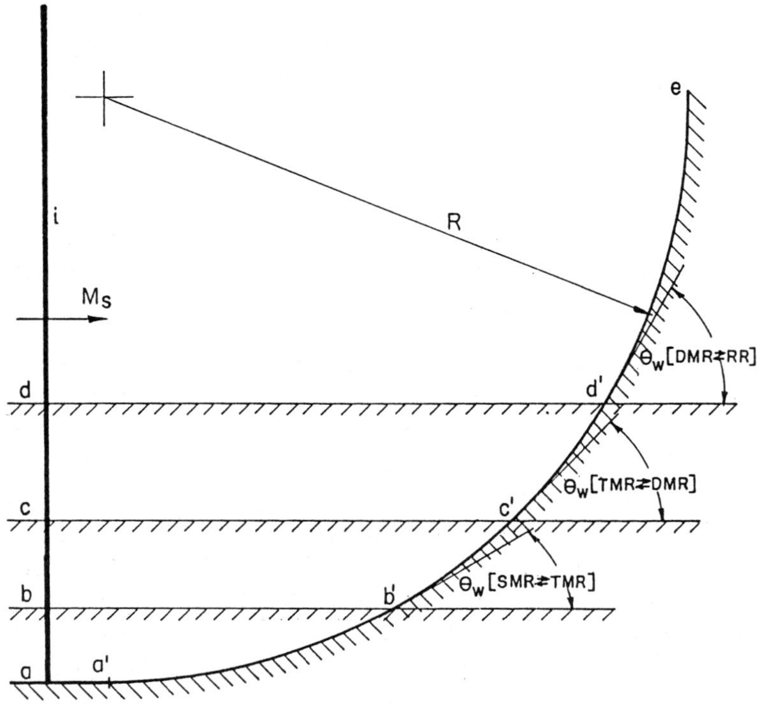

Figure 4.1 Schematic illustration of four different cylindrical concave surfaces over which different reflection processes are obtained.

If the initial reflection is an MR (i.e., $\theta_w^{\text{initial}} < \theta_w[\text{DMR}\Leftrightarrow\text{RR}]$) then the incident shock wave propagates along the cylindrical concave wedge, and encounters an ever increasing wedge angle, which will eventually force the MR to change into an RR. If, however, the initial reflection is an RR (i.e., $\theta_w^{\text{initial}} > \theta_w[\text{DMR}\Leftrightarrow\text{RR}]$) then the RR will persist.

The foregoing discussion implies that the initial reflection over a cylindrical concave wedge can be either an SMR, a TMR, a DMR or an RR depending upon the initial wedge angle. If it is an SMR, it will change first to a TMR, then to a DMR and finally to an RR; if the initial reflection is a TMR, it will first change to a DMR and then to an RR; if the initial reflection is a DMR it will change to an RR; and if the initial reflection is an RR it will remain an RR. Photographs of an SMR, a TMR, a DMR and an RR over a cylindrical conave surface are shown in figures 4.2a to 4.2d, respectively.

a)

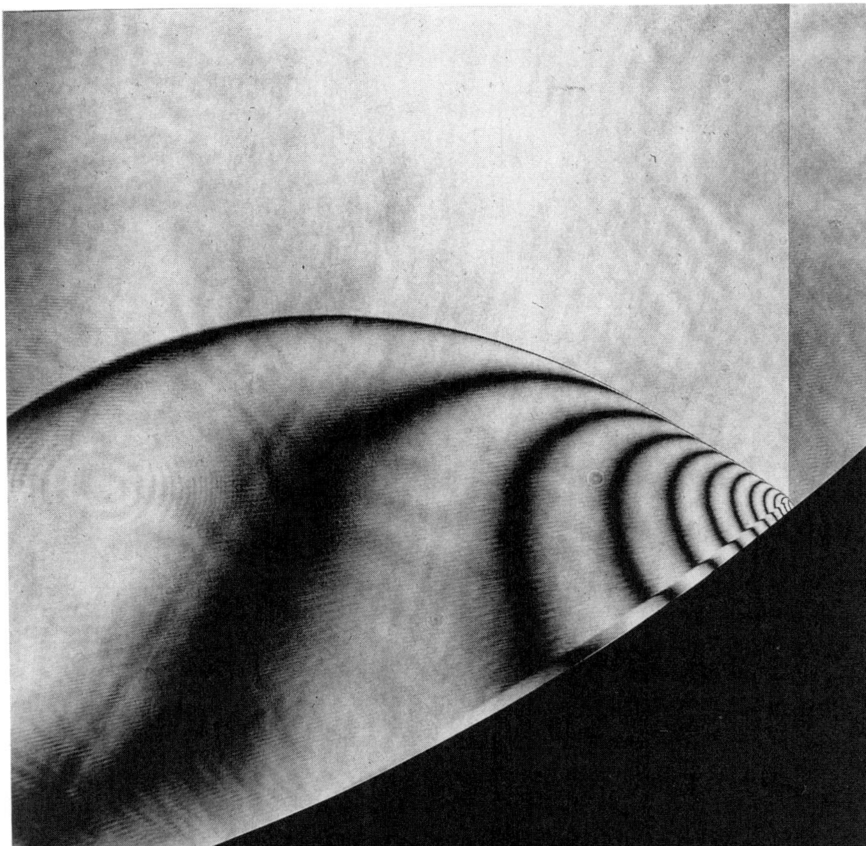

Figure 4.2a A holographic interferogram illustrating an actual SMR over a cylindrical concave surface with $M_s = 1.17$ in air at $T_0 \approx 290K$ and $p_0 \approx 760$ torr (courtesy of Professor K. Takayama, Shock Wave Research Center, Institute of Fluid Science, Tohoku University, Sendai, Japan.)

b)

Figure 4.2b A shadowgraph illustrating an actual TMR over a cylindrical concave surface with $M_s = 2.19$ in air at $T_0 = 291.3K$ and $p_0 \approx 760$ torr (courtesy of Professor K. Takayama, Shock Wave Research Center, Institute of Fluid Science, Tohoku University, Sendai, Japan.)

c)

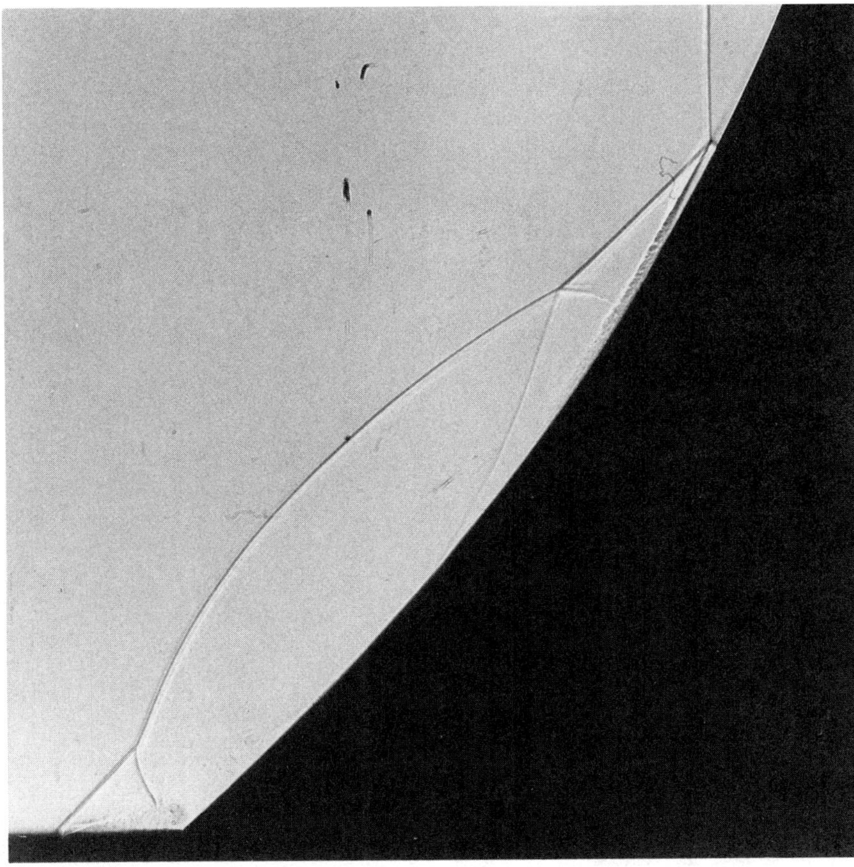

Figure 4.2c A shadowgraph illustrating an actual DMR over a cylindrical concave surface with M_s = 3.02 in air at T_0 = 292.4K and $p_0 \approx$ 760 torr (courtesy of Professor K. Takayama, Shock Wave Research Center, Institute of Fluid Science, Tohoku University, Sendai, Japan.)

d)

Figure 4.2d A shadowgraph illustrating an actual RR over a cylindrical concave surface with $M_s = 2.1$ in air at $T_0 \approx 290K$ and $p_0 \approx 760$ torr (courtesy of Professor K. Takayama, Shock Wave Research Center, Institute of Fluid Science, Tohoku University, Sendai, Japan.)

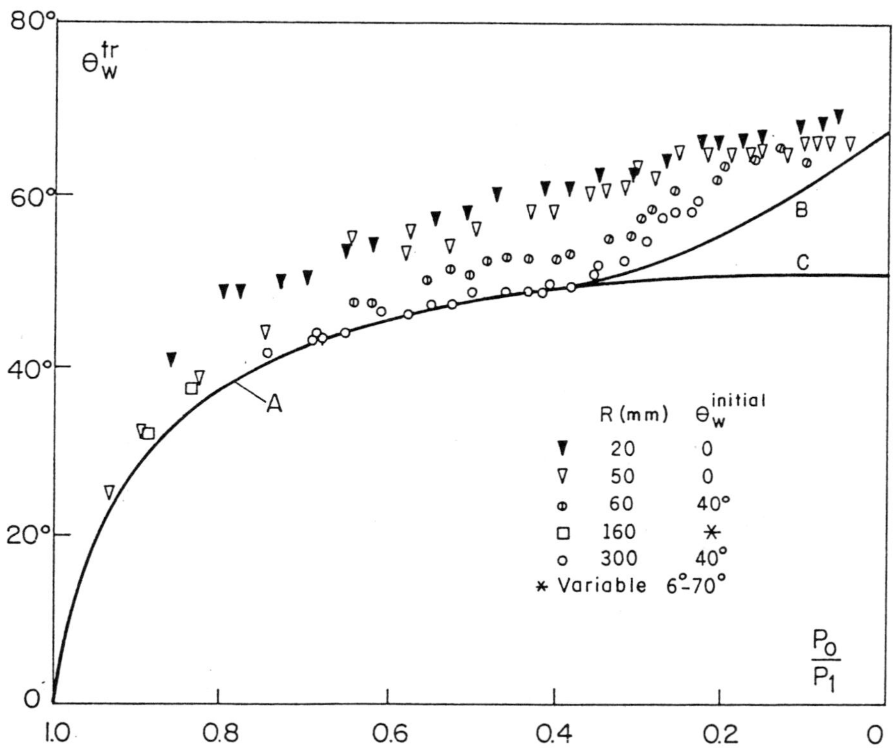

Figure 4.3 Experimental data with air illustrating the dependence of the transition wedge angle over a cylindrical concave surface on the radius of curvature and the initial wedge angle.

4.1.1.1 MR → RR Transition

As mentioned in section 4.1.1, if an incident shock wave reflects initially over a cylindrical concave wedge as an MR it will eventually change to an RR.

In an extensive experimental study Takayama & Sasaki (1983) showed that the MR → RR transition wedge angle, θ_w^{tr} depends in addition to the incident shock wave Mach number, M_S, on both the radius of curvature of the cylindrical wedge, R, and the initial wedge angle $\theta_w^{initial}$. The results of their experimental study are shown in figure 4.3 together with the MR⇔RR transition lines of steady (line AB) and pseudo-steady flows (line AC).

It is evident from figure 4.3 that all the experimentally recorded transition wedge angles lie above the MR⇔RR transition line of steady flows. Furthermore, as the radius of curvature increases the

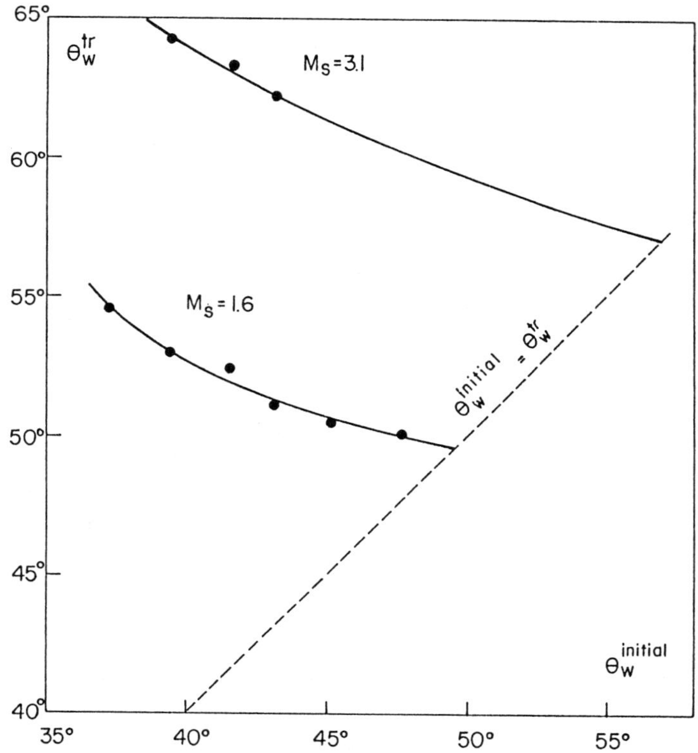

Figure 4.4 Experimental data with air illustrating the dependence of the transition wedge angle over a cylindrical concave surface on the initial wedge angle for $M_S = 1.6$ and 3.1.

transition wedge angles decrease and approach the steady flow MR⇔RR transition line. Similarly, the experimental results indicate that the transition wedge angle decreases with increasing initial wedge angles. The dependence of θ_w^{tr} on $\theta_w^{initial}$ is shown in a different way in figure 4.4 for two incident shock wave Mach numbers. The results for $M_S = 1.6$ reveal a continuous non linear decrease of θ_w^{tr} as $\theta_w^{initial}$ increases. The solid line is hand-drawn to illustrate the dependence of θ_w^{tr} on $\theta_w^{initial}$. Since θ_w^{tr} cannot be smaller than $\theta_w^{initial}$ this line must terminate at the point $\theta_w^{initial} = \theta_w^{tr}$. The initial wedge angle at this point results in an RR for $M_S = 1.6$. Similar behavior is evident for $M_S = 3.1$; however, due to the lack

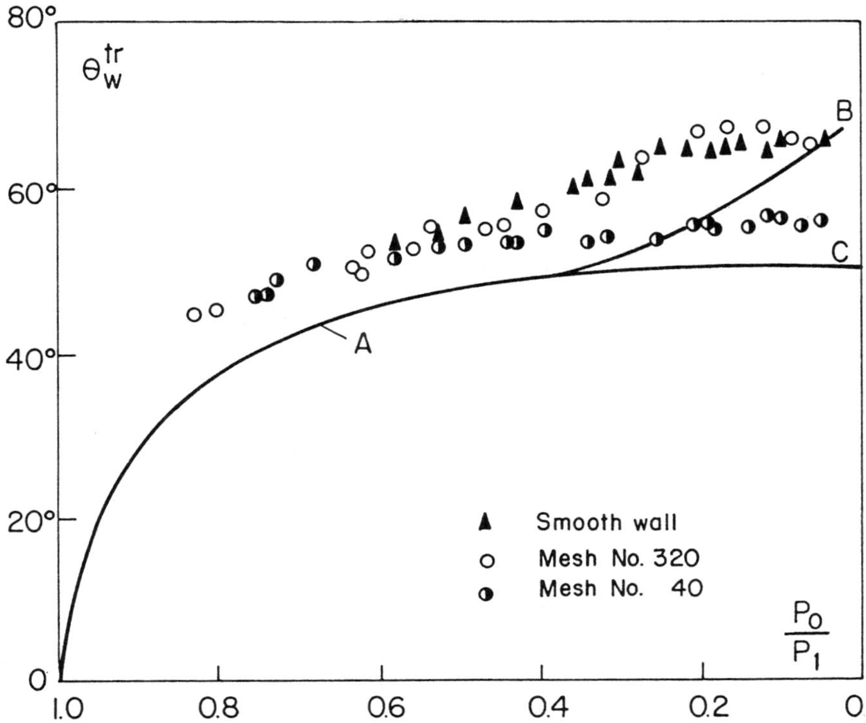

Figure 4.5 Experimental data with air illustrating the dependence of the transition wedge angle over a cylindrical concave surface on its roughness.

of enough experimental data points the hand-drawn line should not be trusted. It should also be noted that the experimental results suggest that as the radius of curvature approaches infinity, i.e., $R \rightarrow \infty$, the actual transition wedge angles approach the transition line appropriate to a steady flow.

4.1.1.1.1 Effect of Surface Roughness

The effect of surface roughness on the MR→RR transition was investigated experimentally by Takayama, Ben-Dor & Gotoh (1981). The roughness was imposed on the surface by pasting sand paper with different mesh sizes on the reflecting wedge surface. Their results for a cylindrical concave wedge with R = 50 mm, $\theta_w^{\text{initial}} = 0$ and three different roughnesses are shown in figure 4.5. Again, the MR⇔RR transition lines of steady (line AB) and pseudo-steady (line AC) flows are added.

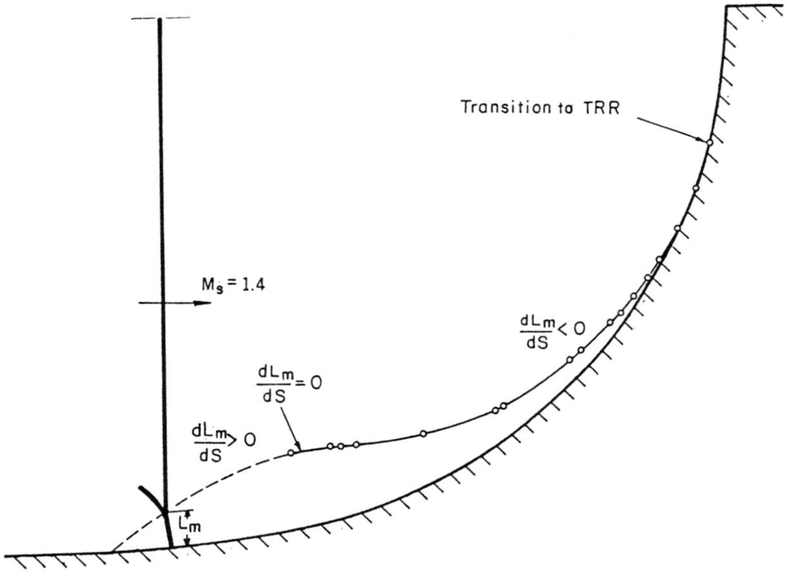

Figure 4.6 Measured triple point trajectory of an MR with M_s = 1.4 over a cylindrical concave surface.

It is evident from figure 4.5 that the transition wedge angle, θ_w^{tr}, decreases as the roughness of the reflecting surface increases. In the case of an extremely rough surface (mesh No. 40 sand paper) θ_w^{tr} is almost independent of the incident shock wave Mach number since it assumes a value of about 54.5°. Furthermore, while for the case of a smooth wedge, θ_w^{tr} was always greater than the appropriate value in steady flows (see figure 4.3), here there are cases where θ_w^{tr} is smaller than the appropriate value in steady flows. However, even with the extremely high roughnesses the recorded transition wedge angles were always greater than those appropriate to pseudo-steady flows.

4.1.1.2 Dynamics of the MR→RR Transition

Figure 4.6 illustrates the measured triple point trajectory of an MR with M_s = 1.4 over a cylindrical concave surface. The experimental results indicate that the length of the Mach stem, L_m, increases from L_m = 0 at the leading edge of the reflecting wedge, to a maximum after which it decreases until it vanishes at the point where the MR terminates and a TRR is formed. It should be noted that the first measured point is quite far away from the leading edge of the reflecting wedge; however, since L_m must start from zero a reasonable trajectory is drawn (dashed lines) in the region where experimental data are unavailable.

The triple point trajectory, as shown in figure 4.6, can be divided into two parts: a part along which $dL_m / ds > 0$, and a part along which $dL_m / ds < 0$ where s is a distance measured along the wedge surface.

Courant & Friedrichs (1948) indicated that, theoretically, three different types of MR-configuration are possible, depending on the direction of propagation of the triple point of the MR. If the triple point moves away from the reflecting surface, then the MR is called a direct-Mach reflection - D_iMR, if the triple point moves parallel to the reflecting surface, then it is called a stationary-Mach reflection - S_tMR, and if the triple point moves towards the reflecting surface then it is called an inverse-Mach reflection - I_nMR.

Following these definitions one can conclude that the experimental data shown in figure 4.6 suggest that the reflection of a planar shock wave over a cylindrical concave surface goes through the following sequence of events; a direct-Mach reflection, D_iMR, along the part where $dL_m / ds > 0$, a momentarily stationary-Mach reflection, S_tMR, at the point where $dL_m / ds = 0$, an inverse-Mach reflection, I_nMR, along the part where $dL_m / ds < 0$, and finally termination of the I_nMR and a formation of an RR at the point where $L_m = 0$.

Since the RR-configuration which is formed after the termination of an I_nMR has a special structure attached to it (see section 4.1.1.2.3), it is called transitioned regular reflection-TRR.

In summary, experimental investigations of the MR→RR transition over cylindrical concave surfaces suggest that the reflection process follows the following sequence of events: [for more details see Ben-Dor & Takayama (1986/7)]:

$$D_i\text{MR} \rightarrow S_t\text{MR} \rightarrow I_n\text{MR} \rightarrow \text{TRR}$$

Figure 4.7 illustrates this sequence of events in the (p,θ)-plane. The intersection of the I_1-R_1 shock polars at point 'a', which is on the right branch of the I_1-polar, results in a D_iMR, the intersection of the I_2-R_2 shock polars at point 'b', which lies also on the p-axis, results in an S_tMR, the intersection of the I_3-R_3 shock polars at point 'c', which is on the left branch of the I_3-polar, results in an I_nMR, and the intersection of the R_3-polar with the p-axis at point 'd' results in a TRR there. Note, that since the pressure in state (1) is determined solely by the incident shock wave Mach number, all

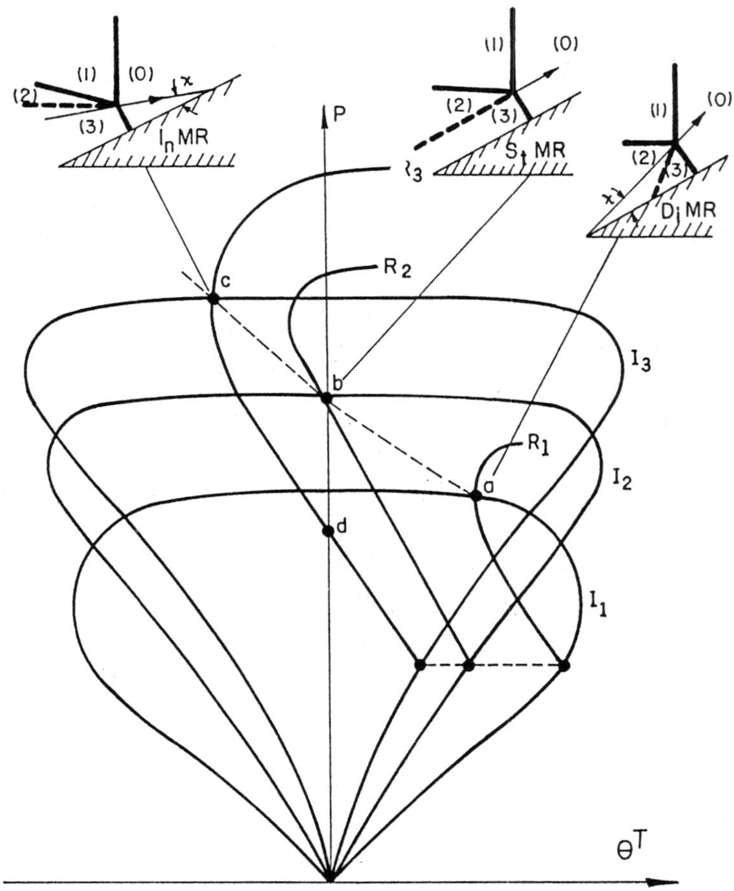

Figure 4.7 Schematic illustration in the (p,θ)-plane of the transition from D_iMR to TRR.

the R-polars emanate from their corresponding I-polars at the same pressure. The dashed line connecting points 'a', 'b' and 'c' in the locus of the Mach reflection configurations which are obtained from the moment a D_iMR was formed at point 'a' until the moment the I_nMR was terminated at point 'c' to form a TRR at point 'd'. D_iMR-configurations exist along the portion 'a-b', a momentary S_tMR-configuration exists at point 'b', and I_nMR-configurations exist along the portion 'b-c'. Figures 4.6 and 4.7 indicate that the part along which $dL_m/ds > 0$ corresponds to the section 'a-b', the point where $dL_m/ds = 0$ corresponds to point 'b', and the part along which $dL_m/ds < 0$ corresponds to the section 'b-c'.

Unlike the D_iMR, which is obtainable in pseudo-steady flows also, the S_tMR and the I_nMR can be obtained only in unsteady flows.

4.1.1.2.1 Stationary-Mach Reflection - S_tMR

The wave configuration of an S_tMR and the I-R shock polar combination of an S_tMR are shown in figure 4.7.

The wave configuration of an S_tMR is basically the same as that of a D_iMR. The only difference is that its slipstream is parallel to the reflecting surface. This is due to the fact that the triple point of an S_tMR moves parallel to the reflecting surface.

4.1.1.2.2 Inverse-Mach Reflection - I_nMR

The wave configuration of an I_nMR and the I-R shock polar combination of an I_nMR are shown in figure 4.7.

The wave configuration of an I_nMR is very similar to that of a D_iMR. The only difference is in the orientation of the slipstream. While in a D_iMR the slipstream extends from the triple point towards the reflecting surface (see figure 2.5), in the case of an I_nMR the slipstream emanating from the triple point extends away from the reflecting surface, because the triple point in the case of an I_nMR moves towards the reflecting surface. For this reason, the I_nMR is a temporary reflection. It will terminate as soon as its triple point collides with the reflecting surface. Upon the termination of an I_nMR, a new wave structure, transitioned regular reflection - TRR, is formed.

4.1.1.2.3 Transitioned Regular Reflection - TRR

Figure 4.8 is a schematic illustration showing the dynamics of the termination of an I_nMR and the formation of a TRR. Figure 4.8a shows an I_nMR moving towards the reflecting surface. Figure 4.8b illustrates the wave configuration at the moment the triple point, T, collides with the reflecting surface. The Mach stem, m, vanishes and the incident shock wave, i, and reflected shock wave, r, meet on the reflecting surface. The wave configuration which is obtained after the termination of an I_nMR is shown in figure 4.8c. The major part of the wave structure is an RR (i and r meet on the reflecting surface). In addition a new triple point, T^*, is formed on the reflected shock wave. The nature of this triple point is yet to be investigated. Figure 4.8c also indicates that an additional shock wave connects this new triple point with the reflecting surface. This additional shock wave, which is not necessarily straight, is perpendicular to the reflecting surface at its foot.

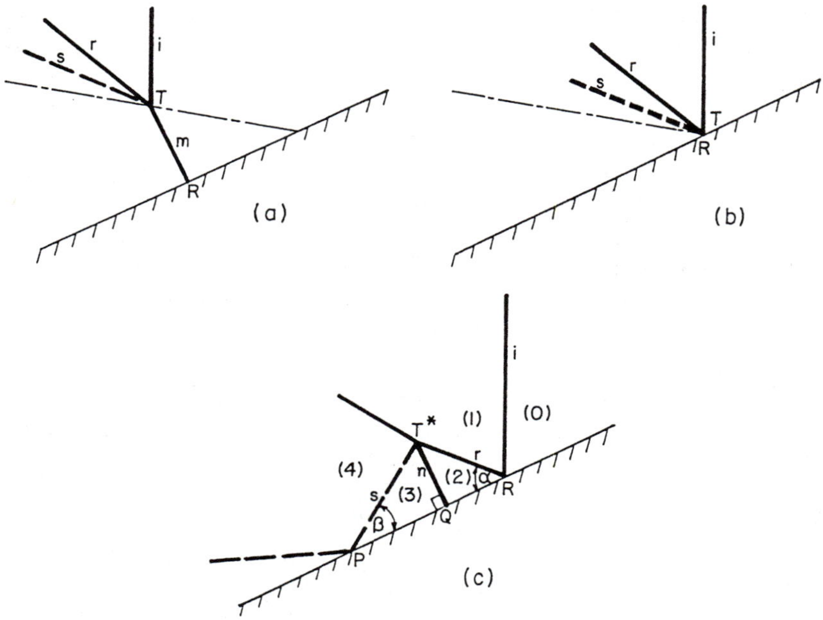

Figure 4.8 Schematic illustration showing the dynamics of the termination of an I_nMR and the formation of a TRR.

The physical reason for the formation of this addtional shock wave is the need to support a sudden pressure drop which is associated with the $I_nMR \to TRR$ transition. The need for this additional shock wave was first hypothesized by Henderson & Lozzi (1975). Consider figure 4.7 and note that the termination of an I_nMR at point 'c' and the formation of a TRR at point 'd' is associated with a sudden pressure drop from p_c to p_d. The reason for the existence of the additional shock wave is to support this sudden pressure drop. Hence, the pressure ratio across the additional shock wave should be p_c/p_d.

4.1.1.2.3.1 Analytical Solution of a TRR

The various flow regions of a TRR are shown in figure 4.8c. The reflection point of the RR is R, the foot of the additional shock wave, n, is at point Q, and the slipstream reflects from the reflecting surface at point P.

If one assumes that the wave configuration of a TRR is self-similar, then the entire structure shown in figure 4.8c grows linearly with time. Consider figure 4.9 where the wave configuration of a TRR is drawn at two different times, say t and t + Δt. Within the

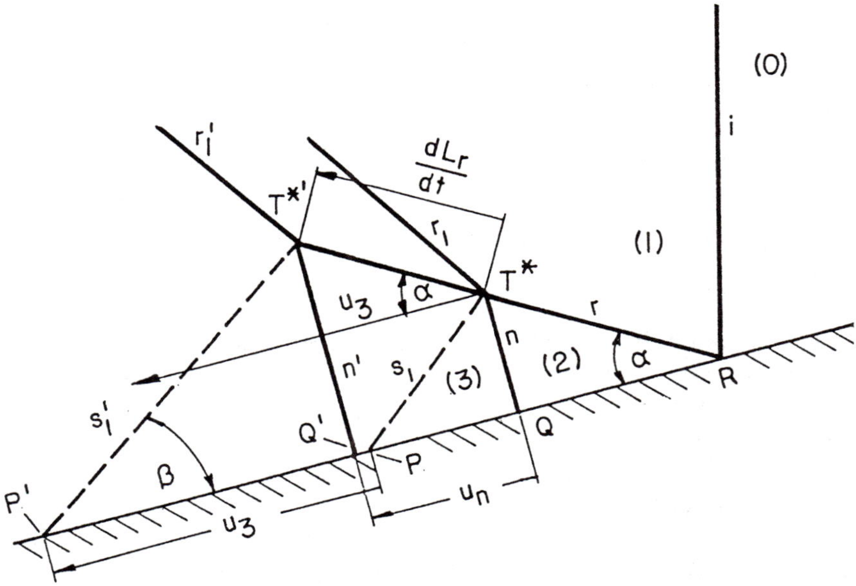

Figure 4.9 The wave configuration of a transitioned regular reflection at two different times.

time interval Δt the points Q, P and T* moved to new locations marked Q', P' and T*'.

Let us define the velocity of point Q with respect to the reflection point R as u_n, and the velocity of the triple point T* with respect to the reflection point R as $\frac{dL_r}{dt}$ where L_r is the length of the reflected shock wave, r, as shown in figure 4.9. In addition, it is important to note that since point P is on the slipstream, s_1, its velocity must be equal to u_3 where u_3 is the flow velocity in state (3) with respect to point R. For the reader's convenience, the velocity vectors u_n, u_3 and $\frac{dL_r}{dt}$ are added to figure 4.9.

As mentioned earlier, the additional shock wave, n, is not necessarily straight. However, experimental evidence indicates that its curvature is small. Thus, if one assumes that the additional shock wave, n, is straight, then it is normal to the flow passing through it from state (2) to state (3). Thus, the conservation equations across the normal shock wave for a perfect gas, are as follows:

- conservation of mass

$$\rho_2(u_2 - u_n) = \rho_3(u_3 - u_n) \tag{4.1}$$

- conservation of linear momentum

$$p_2 + \rho_2(u_2 - u_n)^2 = p_3 + \rho_3(u_3 - u_n)^2 \tag{4.2}$$

- conservation of energy

$$\frac{\gamma}{\gamma - 1}\frac{p_2}{\rho_2} + \frac{1}{2}(u_2 - u_n)^2 = \frac{\gamma}{\gamma - 1}\frac{p_3}{\rho_3} + \frac{1}{2}(u_3 - u_n)^2 \tag{4.3}$$

Note that since the normal shock wave is moving with a velocity u_n in the frame of reference under consideration, the relative velocities $(u_2 - u_n)$ and $(u_3 - u_n)$ are used in the above conservation equations. It is also important to note that all the flow properties in state (2) can be easily calculated by solving the RR of the incident shock wave over the reflecting surface.

The above set of conservation equations consists of three equations, (4.1) to (4.3), and four unknowns, namely, u_3, p_3, ρ_3 and u_n. Thus an additional equation is required to have a closed set which can, in principle, be solved.

Consider figure 4.8c and note that the gas which flows from region (2) into region (3) through the normal shock wave cannot leave region (3) because region (3) is bounded by a slipstream, s_1, and the reflecting surface. Thus using the following definitions:

L_r - the length of the reflected shock wave, r,
y - the distance from point Q to point P, and
z - the distance from point R to point Q,

it can easily be shown that the mass of the fluid in region (3) is

$$m_3 = \frac{1}{2}\rho_3 (L_r)^2 \frac{\sin^2\alpha}{\tan\beta} \tag{4.4}$$

where α and β are defined in figure 4.9. Differentiating equation (4.4) with respect to t results in the rate by which the fluid mass increases in region (3) i.e.,

$$\frac{dm_3}{dt} = \rho_3 L_r \frac{dL_r}{dt}\frac{\sin^2\alpha}{\tan\beta} \tag{4.5}$$

The rate by which the fluid mass increases in region (3) is equal to the rate of flow from region (2) to region (3) through the normal shock wave, n, i.e.,

$$\frac{dm_3}{dt} = \rho_2 (u_2 - u_n) L_r \sin\alpha \qquad (4.6)$$

Setting equation (4.5) equal to (4.6), results in

$$\frac{dL_r}{dt} = \frac{\rho_2}{\rho_3} (u_2 - u_n) \frac{\tan\beta}{\sin\alpha}$$

It could easily be shown that $\alpha = \phi_2 - \theta_1$; thus, the above expression can be rewritten as

$$\frac{dL_r}{dt} = \frac{\rho_2}{\rho_3} (u_2 - u_n) \frac{\tan\beta}{\sin(\phi_2 - \theta_1)} \qquad (4.7)$$

From simple geometrical considerations it is obvious that

$$u_n = \frac{dL_r}{dt} \cos\alpha = \frac{dL_r}{dt} \cos(\phi_2 - \theta_1) \qquad (4.8)$$

Inserting $\frac{dL_r}{dt}$ from equation (4.8) into equation (4.7) results in

$$u_n = \frac{\rho_2}{\rho_3} (u_2 - u_n) \frac{\tan\beta}{\tan(\phi_2 - \theta_1)} \qquad (4.9)$$

Equation (4.8) can be simplified by combining it with equation (4.1) to result in

$$u_n = \frac{u_3}{1 + \dfrac{\tan(\phi_2 - \theta_1)}{\tan\beta}} \qquad (4.10)$$

It is interesting to note that equation (4.10) can also be obtained using a different approach.

Since point P belongs to the slipstream, s_1, the velocity at which it moves away from the reflection point R (to which the frame of reference is attached) is u_3, Thus

$$\frac{d}{dt}(y + z) = u_3$$

By relating y and z to L_r, α and β, the above relation can be rewritten as

$$\frac{dL_r}{dt} = \frac{u_3}{\dfrac{\sin\alpha}{\tan\beta} + \cos\alpha}$$

which by the aid of equation (4.8) becomes identical to equation (4.10).

The information upon which equation (4.10) is based has not been used in equations (4.1) to (4.3), and hence it is an additional independent equation. However, it carries with it an additional unknown, β, and hence we now have a set of four equations [(4.1), (4.2), (4.3) and (4.10)] and five unknowns [u_3, p_3, ρ_3, u_n and β]. Thus there is still a need for an additional equation to have a solvable set of equations.

Since the wave configuration under consideration does not provide any information which can lead towards an additional independent equation let us assume that due to the similarity between the general wave configurations of a TRR and a DMR, (see comparison in figure 4.10) the triple point T* of a TRR has features similar to that of the second triple point T' of a DMR.

Once this assumption is made, we recall that while investigating the double-Mach reflection, Law & Glass (1971) assumed that the second triple point of a DMR, i.e., T', moves with a velocity identical to that induced by the incident shock wave (see section 2.1.3.1). This assumption which was later adopted by Ben-Dor (1980) who presented an analytical solution of a DMR was found to be quite good for a wide range of incident shock wave Mach numbers and reflecting wedge angles. Although a physical explanation for the validity of this assumption does not exist as yet, it is hypothesized here that it should be related to the communication of a geometrical length scale to the triple points T* or T' since both the DMR and the TRR wave configurations involve shock waves having finite lengths. Recall that Bazhenova, Fokeev & Gvozdeva (1976) showed that the "Law-Glass" assumption is fairly good for double Mach reflections in the range $\theta_w < 40°$. Consequently, the "Law-Glass" assumption might not apply in the entire range of wedge angles where a TRR is possible.

If the shock induced flow velocity in state (1) in a laboratory frame of reference is defined as V_1 then using the conservation of mass across the incident shock wave an expression equivalent to equation (2.4) can be developed.

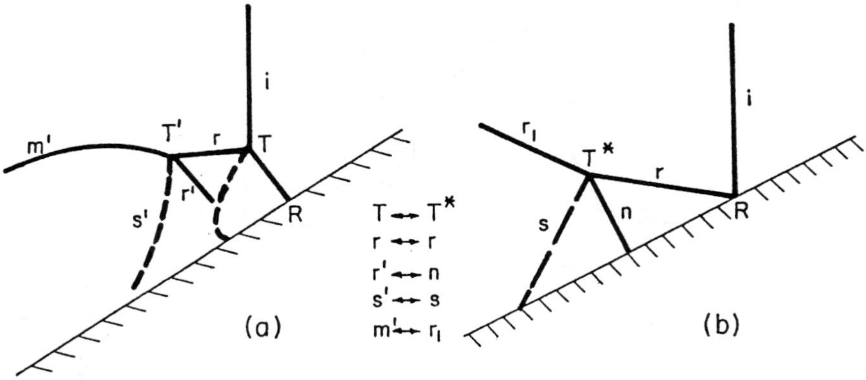

Figure 4.10 A comparison between the wave configurations around the triple point T' of a DMR and T* of a TRR.

$$V_1 = V_s\left(1 - \frac{\rho_0}{\rho_1}\right) \tag{4.11}$$

Thus while the reflection point R moves with the velocity V_s in the x-direction, the triple point T* moves, in the x-direction, with the velocity V_1. Using this fact it can easily be shown that the rate at which the length of the reflected shock wave increases is

$$\frac{dL_r}{dt} = \frac{V_s - V_1}{\cos(\alpha - \theta_w)}$$

Inserting equation (4.11) into the above relation yields

$$\frac{dL_r}{dt} = V_s\frac{\rho_0}{\rho_1}\frac{1}{\cos(\alpha - \theta_w)}$$

which together with equation (4.8) finally results in

$$u_n = V_s\frac{\rho_0}{\rho_1}\frac{\cos(\phi_2 - \theta_1)}{\cos(\phi_2 - \theta_1 - \theta_w)} \tag{4.12}$$

Equation (4.12) is the additional independent equation needed to have a closed set since it does not contain any new unknowns.

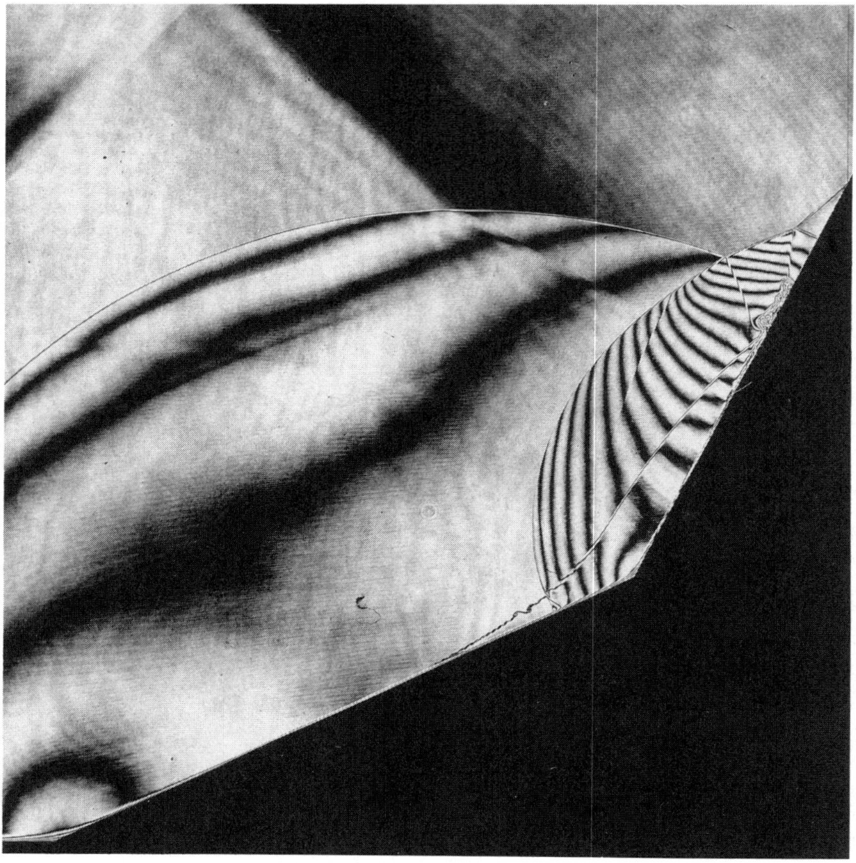

Figure 4.11 A holographic interferogram illustrating the wave configuration of a TRR as obtained from the reflection of a planar shock wave with $M_s = 1.52$ over a double wedge with $\theta_w^1 = 25^o$ and $\theta_w^2 = 60^o$ in air at $T_0 \approx 290K$ and $p_0 \approx 760$ torr (courtesy of Professor K. Takayama, Institute of Fluid Science, Tohoku University, Sendai, Japan).

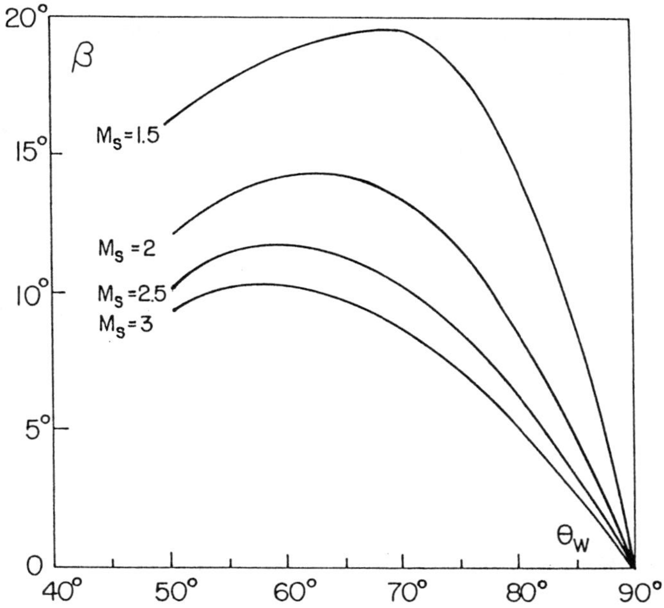

Figure 4.12 The dependence of β on the reflecting wedge angle θ_w for various incident shock wave Mach numbers.

The above set of five governing equations, (4.1), (4.2), (4.3), (4.10) and (4.12), contains five unknowns, namely, ρ_3, p_3, u_3, u_n and β. Thus, in principle, the set of equations is solvable.

There are only a few reported experiments illustrating the wave configuration of a TRR which is formed after an inverse-Mach reflection terminates at the reflecting wedge. Figure 31b in Ben-Dor, Dewey & Takayama (1987) illustrates such a wave configuration which was obtained by reflecting a planar incident shock wave having a Mach number $M_S = 1.3$ over a double wedge with $\theta_w^1 = 25°$ and $\theta_w^2 = 60°$. The value of the angle β as measured from the photograph is 21.5°. The predicted value of β using the above described analytical model is 21.7°.

The photograph shown in figure 4.11 again illustrates the wave configuration of a TRR for yet another reflection of a planar incident shock wave over a double wedge. The initial conditions for this case were: $M_S = 1.5$, $\theta_w^1 = 20°$ and $\theta_w^2 = 60°$. The value of β as measured from this photograph is about 19°. The value of β as predicted by the above described model is 18.6°.

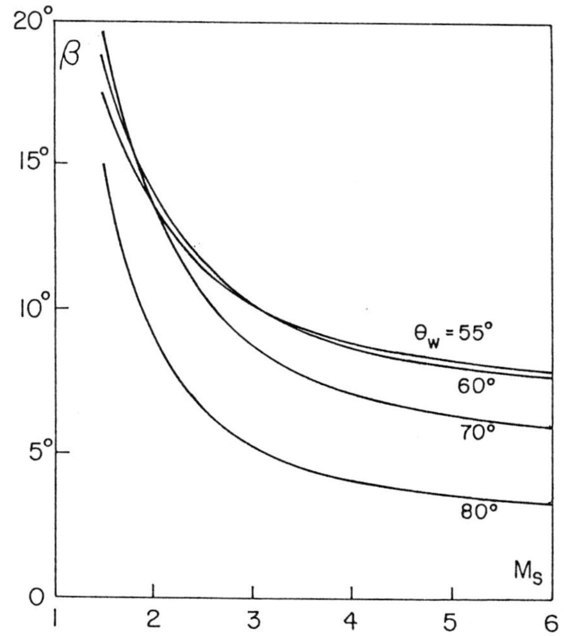

Figure 4.13 The dependence of β on the incident shock wave Mach number M_S for various reflecting wedge angles.

The foregoing discussion clearly indicates that the above developed model is capable of accurately predicting the wave structure of a TRR which results after an inverse-Mach reflection terminates.

Figure 4.12 illustrates the dependence of the angle β on the reflecting wedge angle, θ_W, for various incident shock wave Mach numbers in the range $1.5 \leq M_S \leq 3$. It is evident from figure 4.12 that for a given reflecting wedge angle, θ_W, the value of β increases as the incident shock wave Mach number decreases. As the reflecting wedge angle θ_W approaches 90°, the value of β approaches 0 regardless of the incident shock wave Mach number. In addition, it can be seen that for a given incident shock wave Mach number, M_S, the value of β increases as the reflecting wedge angle θ_W is increased, until it reaches a maximum value after which β decreases and approaches 0 as θ_W approaches 90°. It is also seen in figure 4.12 that the smaller is the incident shock wave Mach number, the larger is the value of the reflecting wedge angle at which β reaches its maximum. While for $M_S = 1.5$, β reaches its maximum at about $\theta_W = 68°$, it reaches its maximum at about $\theta_W = 59°$ for $M_S = 3$.

The dependence of the angle β on the incident shock wave Mach number, M_S, for various reflecting wedge angles, θ_W, is shown in figure 4.13. In general, it is seen that as the incident shock wave Mach number increases, the value of β decreases. It is also evident from figure 4.13 that as $M_S \rightarrow \infty$ the value of β asymptotically approaches a constant value. The smaller is the reflecting wedge angle, the larger is the asymptotic value approached by β.

4.1.1.3 Analytical Considerations

Although the analytical consideration of the reflection phenomenon over cylindrical concave wedges is extremely difficult for the reasons outlined in the introduction to this chapter, there exist some simplified analytical models predicting the MR→TRR transition, and some other aspects related to the reflection phenomenon. Unfortunately, however, these simplified analytical models are limited to relatively weak incident shock waves.

4.1.1.3.1 Analytical Prediction of the MR→TRR Transition

The first approach to analytically predicting the MR→TRR transition over cylindrical concave wedges was presented by Ben-Dor & Takayama (1985). Their analysis was based on the transition criterion of Hornung, Oertel & Sandeman (1979), namely, the length scale criterion (see section 1.5.4), which was found to be the correct criterion in both steady and pseudo-steady flows. The length scale criterion implies that an MR cannot exist unless a physical length scale can be communicated to its triple point.

Figure 4.14 illustrates an MR over a cylindrical concave wedge prior to its transition which would occur when its triple point T reaches point R. Since the transition occurs at a point beyond which an MR cannot exist, one must conclude that based on the length scale concept of Hornung, Oertel & Sandeman (1979), an MR with a triple point in the vicinity of point R (just before transition) represents the last time the corner-generated signals can catch-up with the triple point. Upon further propagation of the incident shock wave, i, an MR will become impossible. The actual reflection which is formed after the termination of the MR (shown in figure 4.2c) is a TRR as shown in figure 4.2d. The incident shock wave is clearly seen to reflect from the reflecting surface as an RR. A new triple point is formed along the reflected shock wave. A short shock wave, which emanates from this triple point, terminates

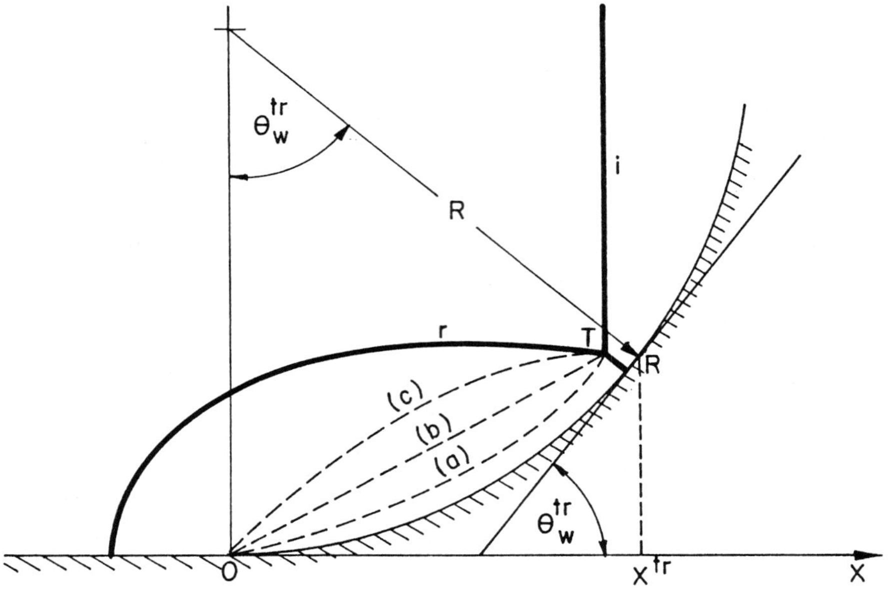

Figure 4.14 Illustration of an MR over a cylindrical concave wedge just before transition.

perpendicularly on the reflecting surface. The new triple point probably indicates the position reached by the corner-generated signals. The corner-generated signals cannot precede the additional shock wave. Thus, the reflected point, R, of the RR is isolated from the corner-generated signals by a supersonic flow region. Let us assume that the signals which are generated at the leading edge of the reflecting wedge (point 0) are moving with a velocity V + a where V is the flow velocity and a is the local speed of sound. In general, V + a changes inside the flow field. If Δt is the time required for the incident shock wave to travel from x = 0 to $x = x^{tr}$, where x^{tr} is the point along the x-axis where the corner-generated signals have caught-up with the triple point for the last time, then the corner generated signals have propagated during this time interval a distance of

$$S = \int_0^{\Delta t} (V + a)dt \tag{4.13}$$

Since the variation of (V + a) in the flow field is unknown, the integration on the right hand side of equation (4.13) cannot be carried out. However, inspecting a typical MR over a concave wedge (see figure 4.2a) indicates that, as the reflected shock wave approaches the surface, it becomes very weak. This, in turn, implies that the flow properties do not change significantly while passing through the reflected shock wave. Consequently, they can be assumed to retain their pre-reflected shock values (i.e., the values behind the incident shock wave). Therefore, it can be assumed that

$$V + a = V_1 + a_1 \tag{4.14}$$

where V_1 and a_1 are the flow velocity and the local speed of sound behind the incident shock wave, respectively. Using this assumption, the integration on the right hand side of equation (4.13) can be easily carried out to result in

$$S = (V_1 + a_1)\,\Delta t \tag{4.15}$$

Unfortunately, the exact path of the corner-generated signals is also unknown. Consequently, the value of S in equation (4.15) is unknown.

In the following, two possible propagation paths will be examined. The first path is along the reflecting surface and the second path is along a straight line connecting the leading edge of the reflecting wedge (point O in figure 4.14) and the triple point, T when it coincides with the reflection point R (i.e., exactly at transition). These two propagation paths are shown in figure 4.14 as lines (a) and (b), respectively. It should be noted again, that the actual propagation path need not be either of these two suggested propagation paths. It can be any path connecting the leading edge of the reflecting wedge and the triple point, e.g., line (c) in figure 4.14.

If the propagation path is assumed to be along the reflecting surface, then

$$S = R\theta_w^{tr} \tag{4.16}$$

where θ_w^{tr} is the reflecting wedge angle at the point where the MR→TRR transition takes place, and R is the radius of curvature of the cylindrical concave wedge.

Combining equations (4.15) and (4.16) results in

$$R\theta_w^{tr} = (V_1 + a_1)\Delta t \tag{4.17}$$

However, $\Delta t = x^{tr}/V_s$ where V_s is the incident shock wave velocity. Thus equation (4.17) could be rewritten as

$$R\theta_w^{tr} = (V_1 + a_1)\frac{x^{tr}}{V_s} \tag{4.18}$$

From figure 4.14 it is clear that

$$x^{tr} = R \sin \theta_w^{tr} \tag{4.19}$$

Combining equations (4.18) and (4.19) results in

$$\frac{\sin \theta_w^{tr}}{\theta_w^{tr}} = \frac{V_s}{V_1 + a_1} \tag{4.20}$$

Dividing both the numerator and the denominator on the right-hand side of equation (4.20) by the local speed of sound ahead of the incident shock wave, a_0, finally results in

$$\frac{\sin \theta_w^{tr}}{\theta_w^{tr}} = \frac{M_s}{V_{10} + A_{10}} \tag{4.21}$$

where M_s is the incident shock wave Mach number, $V_{10} = V_1/a_0$ and $A_{10} = a_1/a_0$. For a perfect gas both V_{10} and A_{10} depend solely on M_s through equations (2.39) and (2.40).

If the propagation path of the corner-generated signals is assumed to be the shortest path connecting the leading edge of the reflecting wedge and the triple point, i.e., a straight line, and if the height of the Mach stem, L_m, is assumed to be much shorter than the radius of curvature of the reflecting wedge, i.e., $L_m << R$, then it is evident from figure 4.14 that

$$\frac{S}{2} = R \sin \frac{\theta_w^{tr}}{2} \tag{4.22}$$

Inserting S from equation (4.22) and x^{tr} from equation (4.19) into equation (4.15) results in

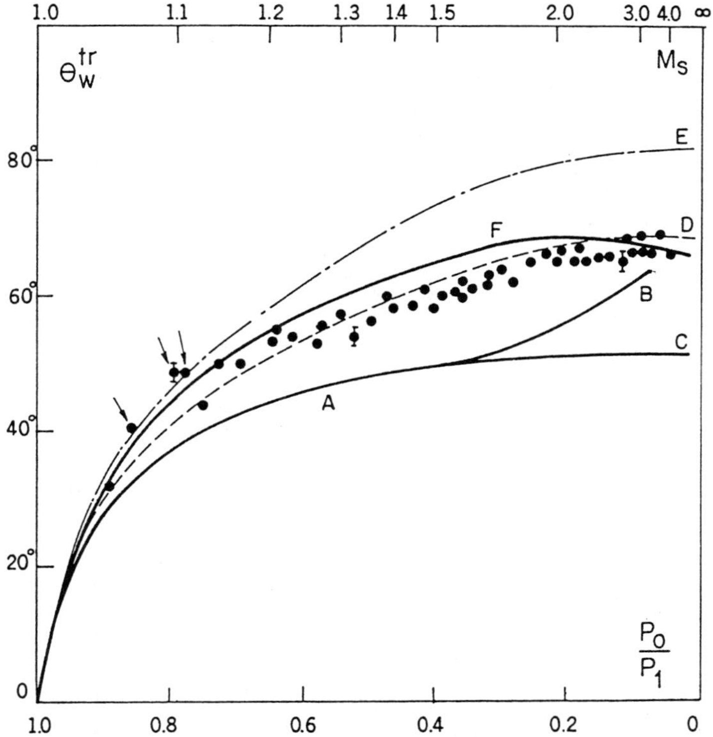

Figure 4.15 Experimental results of the transition wedge angle over a cylindrical concave surface with R = 50 mm and $\theta_w^{initial} = 0$ with different analytical transition lines.

$$\cos\frac{\theta_w^{tr}}{2} = \frac{V_s}{V_1 + a_1} \tag{4.23}$$

Again non-dimensionalizing the velocities on the right-hand side of equation (4.23) by the local speed of sound ahead of the incident shock wave results in

$$\cos\frac{\theta_w^{tr}}{2} = \frac{M_S}{V_{10} + A_{10}} \tag{4.24}$$

Experimental results of the transition wedge angle for R = 50 mm and $\theta_w^{initial} = 0$ are presented in figure 4.15 with the transition lines

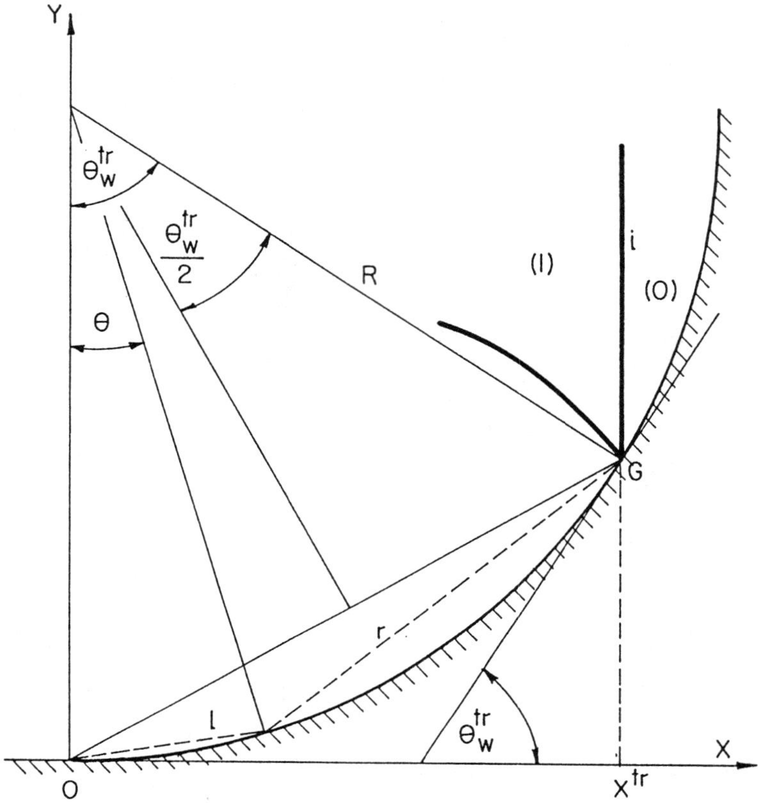

Figure 4.16 A Mach reflection over a cylindrical concave surface exactly at transition.

as obtained from the transition formulae given by equations (4.21) and (4.24).

Lines AB and AC are the RR⇔MR transition lines in steady and pseudo-steady flows, respectively. Lines D and E are the transition lines as predicted by equations (4.21) and (4.24), respectively. Line F will be referred to subsequently. As can be seen, the agreement between the experiments and the transition line predicted by equation (4.21), line D, is quite good in the range $1.125 < M_S < 4$. (The upper limit arises from the lack of experimental results for larger incident shock wave Mach numbers.) At the smaller Mach number range, $1 < M_S < 1.125$, the agreement between line D and the experiments is very poor (see the three experiments marked with an arrow which lie 4.5° to 7.5° above line D). The transition line predicted by equation (4.24), line E, shows excellent agreement with these experiments only, i.e., it is good in the range $1 < M_S < 1.125$. Outside this narrow range, the analytical predictions based on equation (4.24) are 10° to 15° larger than those obtained experimentally.

In order to improve the analytical prediction of the MR→TRR transition over cylindrical concave wedges the foregoing two analytical approaches of Takayama & Ben-Dor (1985) were reconsidered, and an additional analytical approach was put forward by Takayama & Ben-Dor (1989).

Consider figure 4.16 where an MR is illustrated exactly at its transition. Let us assume that a gas particle which propagated along the cylindrical wedge and reached an angular position, θ, generated a disturbance which propagated with the local speed of sound, a, and reached a circle having radius, r. Thus, the distance to which the corner-generated signals propagated can be obtained by a vector summation of the particle path, l, and the disturbance path, r, as shown by the dashed line in figure 4.16. Unlike the previous assumption that $V + a = V_1 + a_1$ let us assume that $V = V_1$ and $a = a_1$. Using these simplifying assumptions one can write

$$l = R\theta = V_1 \Delta t \tag{4.25}$$

$$r = a_1 \Delta t \tag{4.26}$$

From figure 4.16 it is clear that

$$(R \sin \theta_w^{tr} - R \sin \theta)^2 + (R \cos \theta - R \cos \theta_w^{tr})^2 = r^2 \tag{4.27}$$

Inserting equation (4.26) into equation (4.27) results in

$$\left(\sin \theta_w^{tr} - \sin \theta\right)^2 + \left(\cos \theta - \cos \theta_w^{tr}\right)^2 = \left(\frac{a_1 \Delta t}{R}\right)^2 \tag{4.28}$$

From equation (4.25) we have

$$\theta = \frac{V_1 \Delta t}{R} \tag{4.29}$$

Using the incident shock wave velocity, V_S, the following expression is self-explanatory

$$\sin \theta_w^{tr} = \frac{V_s \Delta t}{R} \tag{4.30}$$

Inserting equation (4.29) into equations (4.28) and (4.30), while eliminating the term $\Delta t/R$ results in

$$\theta = \frac{V_1}{V_s} \sin \theta_w^{tr} \tag{4.31}$$

$$\frac{2 - 2\cos(\theta - \theta_w^{tr})}{\sin^2 \theta_w^{tr}} = \left(\frac{a_1}{V_s}\right)^2 \tag{4.32}$$

Non-dimensionalizing the velocities on the right hand side of equations (4.31) and (4.32) by the local speed of sound ahead of the incident shock wave, and simplifying equation (4.32) by using well-known trigonometric relations, finally results in

$$\theta = \frac{V_{10}}{M_s} \sin\theta_w^{tr} \tag{4.33}$$

$$\frac{2 \sin \dfrac{\theta_w^{tr} - \theta}{2}}{\sin \theta_w^{tr}} = \frac{A_{10}}{M_s} \tag{4.34}$$

which for a given incident shock wave Mach number, M_s, can be solved iteratively to obtain θ_w^{tr} and θ. The transition line as obtained from equations (4.33) and (4.34) is shown in figure 4.15 as curve, F. Unlike the previous transition lines, D and E, which were good only at $M_s > 1.125$ and $M_s < 1.125$, respectively, and very poor beyond these ranges, line F shows, in general, fairly good agreement with the experiments in the entire range of M_s. In the range $1 < M_s < 1.125$ it lies slightly below line E and shows good agreement with the experiments marked with arrows. In the range $1.25 < M_s < 2$ line F is about 5° too high while for $2 < M_s < 4$ the agreement becomes good again. Note that in the range $1.125 < M_s < 2$ the predictions of line D are better than those of line F, but as mentioned earlier, in the lower Mach number regime the predictions of line D are very poor.

Unfortunately, the foregoing three models for predicting the analytical MR→TRR transition, which are all based on the length

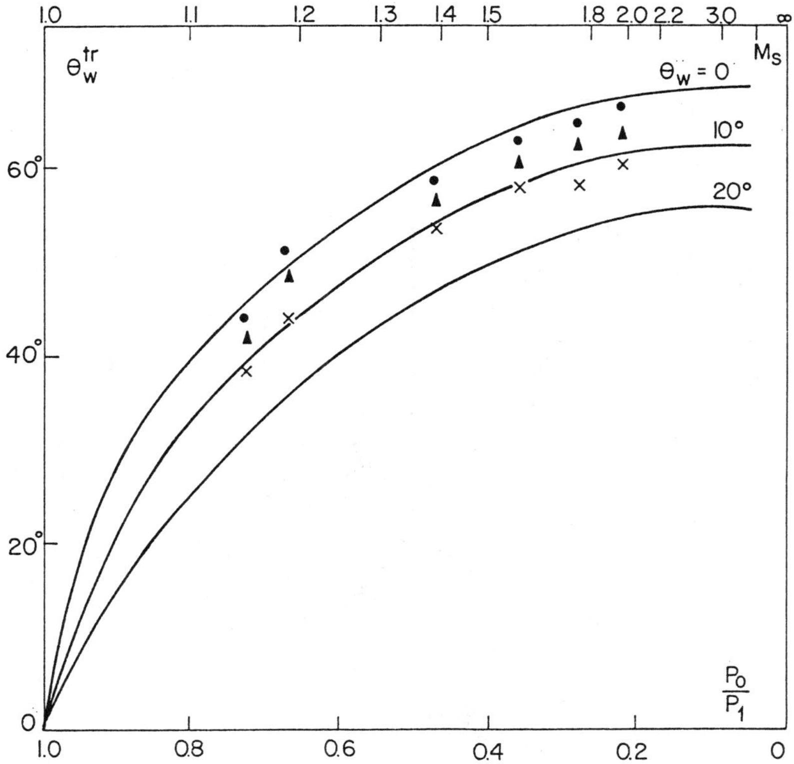

Figure 4.17 Comparison between the measured transition wedge angles over a cylindrical concave surface with diferent wedge angles and a simplified analytical transition criterion. (Data are for nitrogen.)

scale criterion, are independent of the radius of curvature of the reflecting wedge, R, and the initial angle of the cylindrical reflecting wedge, $\theta_w^{initial}$, which as shown in figure 4.3, have significant effects on the actual MR→TRR transition.

An approach partially overcoming these problems was suggested by Ben-Dor & Takayama (1985) who included the initial wedge angle in the transition criterion given by equations (4.21) to obtain

$$\frac{\sin \theta_w^{tr} - \sin \theta_w^{initial}}{\theta_w^{tr} - \theta_w^{initial}} = \frac{M_S}{V_{10} + A_{10}} \tag{4.35}$$

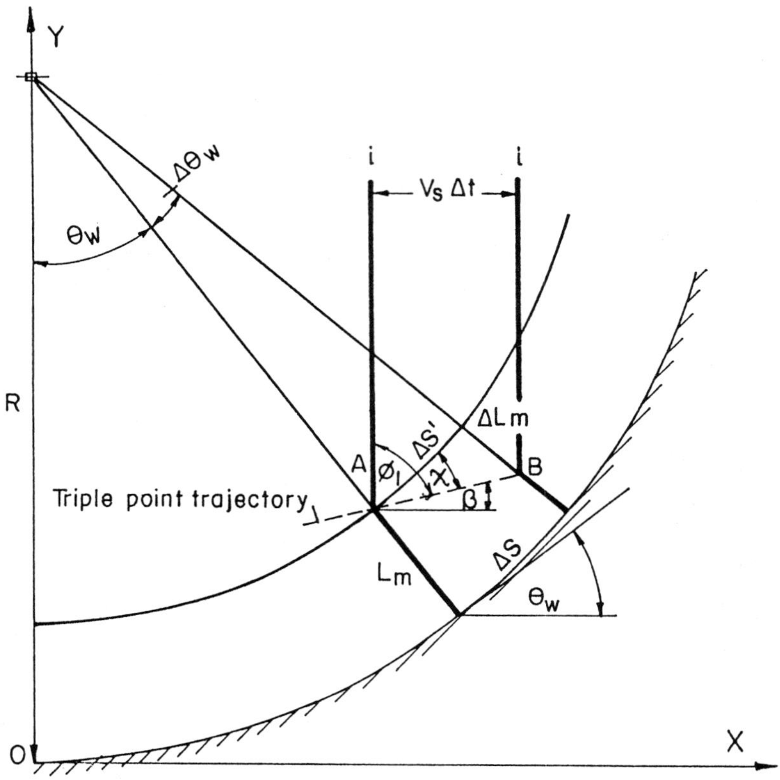

Figure 4.18 Schematic illustration of the wave configuration of an MR over a cylindrical concave surface at two different times.

The transition lines as predicted by equation (4.35) are compared with the experimental results of Itoh & Itaya (1980) in figure 4.17. The transition lines as predicted by equation (4.35) exhibit the same trend as the experiments. However, the agreement between the transition lines for a given value of $\theta_w^{initial}$ becomes poorer as $\theta_w^{initial}$ increases. This is probably because the higher is $\theta_w^{initial}$, the poorer is the assumption that $V + a = V_1 + a_1$. This indeed should be the case since when $\theta_w^{initial} > 0$ the reflected shock wave can no longer be assumed to be weak in the vicinity of the leading edge of the reflecting wedge.

When the initial wedge angle is included in equation (4.24) it becomes

$$\frac{\sin\theta_w^{tr} - \sin\theta_w^{initial}}{\sin\frac{1}{2}\left(\theta_w^{tr} - \theta_w^{initial}\right)} = 2\frac{M_s}{V_{10} + A_{10}} \tag{4.36}$$

Another analytical approach for calculating the transition wedge angle, θ_w^{tr}, which accounts for the radius of curvature of the reflecting cylindrical wedge, was suggested by Ben-Dor & Takayama (1986/7). Unfortunately, the suggested model is not complete, for it requires further information regarding the reflection which is as yet unobtainable.

Consider figure 4.18 where two wave configurations of an MR over a concave cylindrical wedge are drawn. For the sake of simplicity only the incident and the Mach stem shock waves are shown. When the triple point is at point A, the length of the Mach stem is L_m. During the time interval, Δt, the triple point moved to point B, and the length of the Mach stem decreases to $L_m - \Delta L_m$.

From simple geometry one can write

$$\overline{AB} = \frac{V_s\,\Delta t}{\sin\phi_1} \tag{4.37}$$

where ϕ_1 is the angle of incidence between the incident shock wave, i, and the oncoming flow, as measured from a frame of reference attached to the triple point when it is at point A.

In addition

$$\frac{\Delta S'}{\overline{AB}} \approx \cos\chi \tag{4.38}$$

where χ is the effective triple point trajectory angle. Combining equations (4.37) and (4.38) results in

$$\frac{\Delta S'\sin\phi_1}{V_s\,\Delta t} \approx \cos\chi \tag{4.39}$$

However, it is evident from figure 4.18 that

$$\theta_w - \chi = 90^\circ - \phi_1 \tag{4.40}$$

Consequently, equation (4.39) becomes

$$\frac{V_S\,\Delta t}{\Delta S'}\cos\chi \approx \cos(\theta_w - \chi) \tag{4.41}$$

From simple geometry $\Delta S' = (R - L_m)\Delta\theta_w$ and $\Delta S = R\Delta\theta_w$; thus

$$\frac{\Delta S'}{\Delta S} = \left(1 - \frac{L_m}{R}\right) \tag{4.42}$$

Combining equations (4.41) and (4.42) results in

$$\frac{V_S\,\Delta t}{\Delta S}\cos\chi \approx \left(1 - \frac{L_m}{R}\right)\cos(\theta_w - \chi)$$

or alternatively

$$\frac{V_S\,\Delta t}{R\Delta\theta_w}\cos\chi \approx \left(1 - \frac{L_m}{R}\right)\cos(\theta_w - \chi)$$

If $\Delta t \rightarrow 0$ then

$$\frac{d\theta_w}{dt} \approx \frac{V_S}{R}\,\frac{\cos\chi}{\left(1 - \frac{L_m}{R}\right)\cos(\theta_w - \chi)} \tag{4.43}$$

Equation (4.43) gives the rate of change of the wedge angle as the shock wave reflection sweeps it. Defining the transition wedge angle as θ_w^{tr} and recalling that at transition $L_m = 0$, results in

$$\frac{d\theta_w^{tr}}{dt} \approx \frac{V_S}{R}\,\frac{\cos\chi}{\cos\left(\theta_w^{tr} - \chi\right)} \tag{4.44}$$

If equation (4.44) could be integrated, i.e., if $\chi(t)$ were known, a relation of the form $\theta_w^{tr}(V_S, R, t)$ could be obtained. This relation could then be used to calculate the transition wedge angle, θ_w^{tr}, for a given incident shock wave Mach number, M_S, and a radius of curvature of the cylindrical reflecting surface, R. Unfortunately, however, $\chi(t)$ is as yet unknown, and hence equation (4.44) cannot be further used to analytically obtain θ_w^{tr}.

However, as is shown subsequently, equation (4.44) provides important information regarding the behavior of the triple point trajectory as the MR→TRR transition is approached. In order to deduce this information, let us first obtain another expression for $d\theta_w^{tr}/dt$, from the geometrical relation describing the cylindrical reflecting surface, i.e.,

$$x^2 + (y - R)^2 = R^2 \tag{4.45}$$

Differentiating this relation results in

$$\frac{dy}{dx} = \frac{x}{R - y} \tag{4.46}$$

Inserting y from equation (4.45) into equation (4.46) results in

$$\frac{dy}{dx} = \left[\left(\frac{R}{x}\right)^2 - 1\right]^{-\frac{1}{2}} \tag{4.47}$$

But $\theta_w = \tan^{-1}\frac{dy}{dx}$, and hence,

$$\theta_w = \tan^{-1}\left[\left(\frac{R}{x}\right)^2 - 1\right]^{-\frac{1}{2}} \tag{4.48}$$

or alternatively

$$\frac{d\theta_w}{dt} = \left[\left(\frac{R}{x}\right)^2 - 1\right]\frac{1}{x}\frac{dx}{dt} \tag{4.49}$$

However, at the transition point $V_s = dx/dt$ and $x/R = \sin\theta_w^{tr}$; therefore, at transition equation (4.49) assumes the following form

$$\frac{d\theta_w^{tr}}{dt} = \left[\left(\frac{1}{\sin\theta_w^{tr}}\right)^2 - 1\right]^{-\frac{1}{2}} \frac{1}{\sin\theta_w^{tr}} \frac{V_s}{R} \tag{4.50}$$

Comparing equations (4.44) and (4.50) results in

$$\frac{\cos\chi}{\cos\left(\theta_w^{tr} - \chi\right)} = \left[\left(\frac{1}{\sin\theta_w^{tr}}\right)^2 - 1\right]^{-\frac{1}{2}} \frac{1}{\sin\theta_w^{tr}}$$

This relation can be further simplified to obtain

$$\sin\,\theta_w^{tr}\,\sin\,\chi = 0 \tag{4.51}$$

Since at transition $\theta_w^{tr} \neq 0$, it is evident from equation (4.51) that at the moment of transition

$$\sin\,\chi = 0$$

This condition implies that at the MR→TRR transition

$$\chi = 0 \tag{4.52}$$

This result, which was obtained purely from analytical considerations, is fully supported by experimental results. Consider figure 4.6 where the experimentally measured triple point trajectory of an MR over a cylindrical concave surface is shown. It is evident that as the triple point approaches the point where the MR changes to a TRR, the triple point trajectory angle, χ, approaches zero.

4.1.1.3.2 Triple Point Trajectory for Weak Incident Shock Waves

Using the assumptions of the model by which equations (4.33) and (4.34) were developed, Ben-Dor, Takayama & Dewey (1987) were able to develop an analytical model for predicting the triple point trajectory for relatively weak incident shock waves.

If the location of the triple point is denoted by X_T and Y_T, measured from the leading edge of the reflecting cylindrical wedge, then

$$X_T = \frac{M_s}{V_{10}} R\theta \tag{4.53}$$

and

$$Y_T = R\theta \left[\frac{1}{M_1^2} - \left(\frac{M_s}{V_{10}} - \frac{\sin\theta}{\theta} \right)^2 \right]^{\frac{1}{2}} + R(1 - \cos\theta) \tag{4.54}$$

which for small values of θ adopts the following form

$$Y_T = R\theta \left[\frac{1}{M_1^2} - \left(\frac{M_s}{V_{10}} - 1 \right)^2 \right]^{\frac{1}{2}}$$

where $M_s = V_s / a_0$, $M_1 = V_1 / a_1$, $V_{10} = V_1 / a_0$, R is the radius of curvature, V_s is the incident shock wave velocity, V_1 the shock induced flow velocity, a_0 and a_1 are the local speeds of sound ahead of and behind the incident shock wave, and θ is the angular position of the flow particles which were initially at x = 0, as shown in figure 4.16. The triple point trajectory angle, χ, can thus be calculated from

$$\chi = \tan^{-1} \frac{dY_T}{dX_T} \tag{4.55}$$

Inserting equations (4.53) and (4.54) into equation (4.55) yields

$$\chi = \tan^{-1} \left\{ \frac{1}{M_s} \left[A_{10}^2 - (M_s - V_{10})^2 \right]^{\frac{1}{2}} + \frac{V_{10}}{M_s} \theta \right\} \tag{4.56}$$

The triple point trajectory angle at glancing incidence, χ_g, can be obtained from equation (4.56) by calculating $\chi_g = \lim_{\theta \to 0} \chi$. This results in

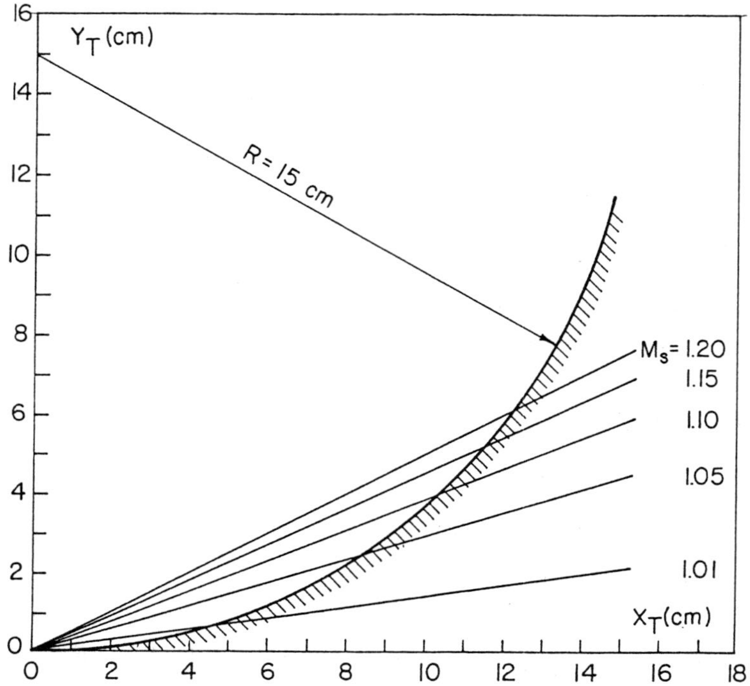

Figure 4.19 Analytical prediction of the triple point trajectories of MR with different incident shock wave Mach numbers over a cylindrical concave surface.

$$\chi_g = \tan^{-1}\left\{\frac{1}{M_S}\left[A_{10}^2 - (M_s - V_{10})^2\right]^{\frac{1}{2}}\right\} \tag{4.57}$$

The values of χ_g vs. M_S were shown earlier in figure 2.29. The angular position of the triple point, θ_T, can simply be obtained from

$$\theta_T = \tan^{-1}\left(\frac{X_T}{R - Y_T}\right) \tag{4.58}$$

The triple point trajectory angle as obtained from equation (4.56) is shown in figure 4.19 for M_S = 1.01, 1.05, 1.10, 1.15 and 1.20 over a cylindrical concave wedge with R = 15 cm. As can be seen, the predicted triple point trajectories are almost straight lines. This should indeed be the case, since for small values of M_S, the value of θ is less than 10° (see table 4 subsequently) and its influence in equation (4.56) is negligible.

Table 3: The triple point trajectory angles at glancing incidence, χ_g, and at transition, χ_{tr}, and values of θ_w^{tr} for incident shock wave Mach numbers in the range $1 < M_S \leq 1.15$.

M_S	χ_g	χ_{tr}	$\Delta\chi$	θ_w^{tr}
1.00	--	--	--	--
1.01	7.92	7.93	0.01	15.89
1.02	10.97	10.99	0.02	21.95
1.03	13.16	13.21	0.05	26.37
1.04	14.89	15.00	0.11	29.88
1.05	16.32	16.49	0.17	32.81
1.06	17.54	17.79	0.25	35.32
1.07	18.60	18.94	0.34	37.52
1.08	19.52	19.98	0.46	39.47
1.09	20.34	20.92	0.58	41.22
1.10	21.07	21.78	0.71	42.80
1.11	21.73	22.57	0.84	44.24
1.12	22.32	23.31	0.99	45.55
1.13	22.86	24.00	1.14	46.76
1.14	23.34	24.65	1.31	47.88
1.15	23.79	25.26	1.47	48.92

details of χ vs. θ_T for $M_S = 1.15$ are given in table 4.

The values of χ_g, χ_{tr}, and θ_w^{tr} for different values of M_S in the range $1 < M_S < 1.15$ are given in table 3. The values of θ_T and χ vs. θ for $M_S = 1.15$ (last line in table 3) are given in table 4.

Plotting these results shows that there is an almost perfect linear relationship between χ and θ_T so that one may write an empirical relation:

$$\chi = \frac{\chi_{tr} - \chi_g}{\theta_w^{tr}} \theta_T + \chi_g \tag{4.59}$$

Table 4: Comparison between the triple point trajectory angle as calculated from the simplified theory and the empirical relation for $M_S = 1.15$.

θ	θ_T	χ(theoretical) Eq. (4.56)	χ(empirical) Eq. (4.59)
0	0	23.788	23.790
1	5.101	23.957	23.943
2	10.535	24.127	24.107
3	16.231	24.296	24.279
4	22.096	24.465	24.454
5	28.016	24.633	24.631
6	33.871	24.800	24.808
7	39.548	24.968	24.978
8	44.953	25.134	25.141
8.777	48.920	25.264	25.260

The values of χ as calculated from the empirical relation (4.59) are also added to table 4 for comparison with the value of χ as obtained from the analytical equation (4.56). It is clear that the differences are insignificant.

A comparison of the analytical predictions of the triple point location [equations (4.53) and (4.54)] and the experimental results for two incident shock wave Mach numbers is shown in figure 4.20. The squares represent the triple point locations for $M_S = 1.103$ and 1.104 and the circles are for $M_S = 1.164$ and 1.166. The solid lines are the results of the analytical calculations for $M_S = 1.1035$ and 1.165. In the case of $M_S \approx 1.1035$ the recorded locations of the triple point lie almost on a straight line which excellently agrees with the analytical calculations for $M_S = 1.1035$. However, for the higher incident shock wave Mach number $M_S \approx 1.165$, the experimentally measured triple point trajectory is no longer a straight line and it is seen to approach the wedge at a glancing angle. Thus it can be concluded that the analytical model presented in this section is good for $M_S < 1.10$. In the range $M_S > 1.16$ the predictions of the model are not good enough. In order to accurately determine the upper limit of the incident shock wave Mach number for which the model presented in this section is valid, more experiments are needed in the range $1.10 < M_S < 1.16$.

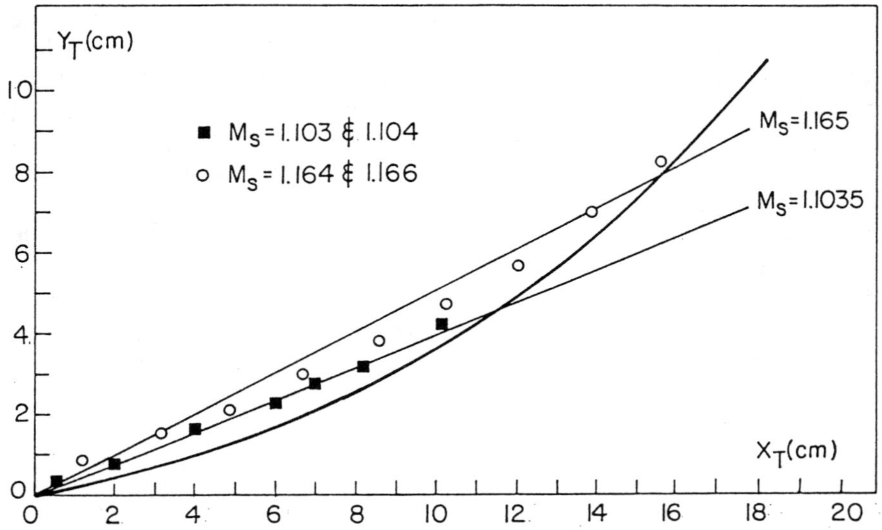

Figure 4.20 Comparison between the measured triple point locations and the analytical predictions for $M_s = 1.1035$ and 1.165. (Data are for air.)

4.1.1.3.3 Application of Steady Shock Polars to Unsteady Flows

In order to better understand the phenomenon of shock wave reflection in unsteady flows Ben-Dor, & Takayama (1986) suggested dividing the unsteady flow into a sequence of momentarily pseudo-steady states and thereby using the (p, θ)-shock polars. It is of interest to note that Marconi (1983) used shock polars in his study of the RR⇔MR transition in three dimensional steady flows. The third dimension in his study might be considered similar to the additional time dimension in the two dimensional unsteady flows.

A multi shock polar diagram is shown in figure 4.7. When the incident shock wave propagates along the reflecting wedge, its Mach number remains constant. However, the complementary wedge angle, θ_w^C, (equation 2.54) increases and hence the oncoming flow Mach number, M_0, with respect to the triple point increases. Thus, for a given location of the incident shock wave along the cylindrical concave surface, the momentary value of M_0 can be calculated, and the corresponding momentary pseudo-steady shock polar can be drawn.

Three combinations of the I-R shock polars are shown in figure 4.7. Since the incident shock wave moves with a constant velocity, the pressure jump across it remains constant. Therefore, all the R-polars emanate from their corresponding I-polars at the same pressure. Figure 4.7 shows a D_iMR at a point 'a' (the I_1- and R_1-polars intersect on the right branch of the I_1-polar), an S_tMR at point 'b' (the I_2- and R_2-polars intersect on the p-axis), and an I_nMR reflection at point 'c' (the I_3- and R_3-polars intersect on the left branch of the I_3-polar) and finally an RR at point 'd' (the intersection of the R_3-polar with the p-axis). Consequently, the multi shock polar suggests that the reflection goes through the following sequence of events; a D_iMR from point 'a' along the dashed line up to point 'b' where a momentary S_tMR is obtained. The momentary S_tMR changes immediately into an I_nMR. The I_nMR is maintained along the dashed line between points 'b' and 'c'. When the I_nMR reaches point 'c' a sudden transition from an I_nMR to an RR takes place at point 'd'.

Note that the multi-shock polar diagram suggests that when the I_nMR terminates at point 'c' and an RR forms at point 'd' the pressure drops from p_c to p_d. According to Henderson & Lozzi (1975): "If a pressure discontinuity occurs during transition then an unsteady wave of finite amplitude or a finite amplitude band of waves will be generated in the flow". Therefore, based on the shock polars shown in Fig. 4.7, one might expect that since the $I_nMR \rightarrow RR$ transition is associated with a sudden pressure drop, the RR, behind which the pressure is lower, should be followed by either a compression wave or a shock wave which should support the sudden drop in the pressure.

It was discussed in section 4.1.1.2.3 that the RR which is formed after an I_nMR terminates is followed by an additional shock wave. The overall wave configuration was termed transitional regular reflection - TRR. The additional shock wave of a TRR is the one hypothesized by Henderson & Lozzi (1975).

The fact that this additional shock wave does actually appear when an I_nMR terminates to form a TRR might serve as a justification for using steady shock polars for better understanding the reflection phenomena in unsteady flows as well as for explanatory purposes.

4.1.2 Shock Wave Reflections over Cylindrical Convex Surfaces

When a planar incident shock wave encounters a cylindrical convex surface it reflects over it either as an RR or as an MR depending upon the initial wedge angle and the incident shock wave Mach number. The initial type of reflection can be determined in a way similar to that presented in section 4.1.1.

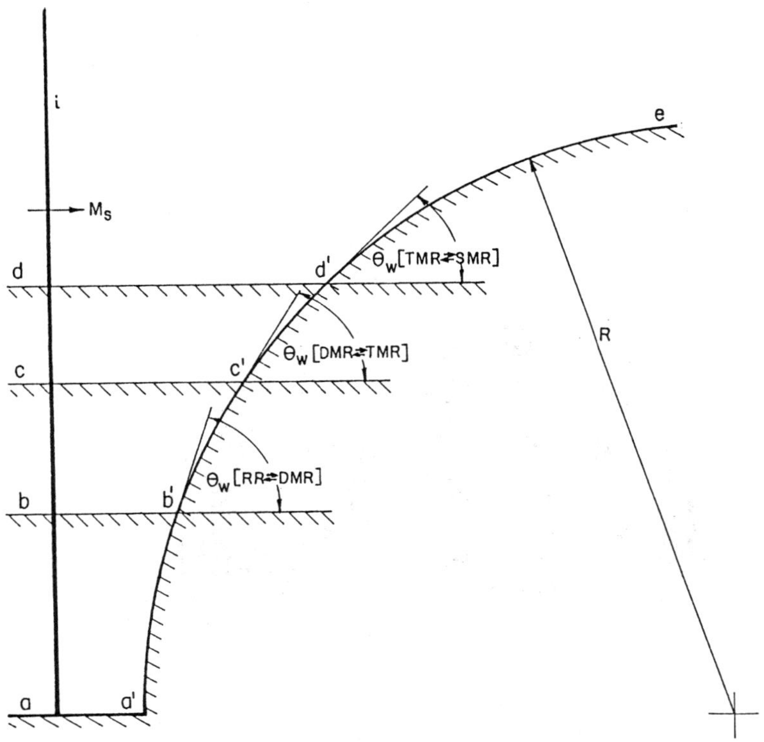

Figure 4.21 Schematic illustration of four different cylindrical convex surfaces over which different reflection processes are obtained.

Consider figure 4.21 where four different cylindrical convex wedges are drawn. While they all have the same radius of curvature, R, their initial wedge angles are different. The initial wedge angles are 90° for the wedge aa'e, θ_w[RR⇔DMR] for the wedge bb'e, θ_w[DMR⇔TMR] for the wedge cc'e and θ_w[TMR⇔SMR] for the wedge dd'e. Thus for all the cylindrical wedges with an initial angle in the range θ_w[RR⇔DMR] < $\theta_w^{\text{initial}} \leq 90°$ the initial reflection will be an RR, for all the wedges with an initial angle in the range θ_w[DMR⇔TMR] < $\theta_w^{\text{initial}}$ < θ_w[RR⇔DMR] the initial reflection will be a DMR, for all the wedges with an initial angle in the range θ_w[TMR⇔SMR] < $\theta_w^{\text{initial}}$ < θ_w[DMR⇔TMR] the initial reflection will be a TMR, and for all the wedges with an initial angle in the range $\theta_w^{\text{initial}}$ < θ_w[TMR⇔SMR] the initial reflection will be an SMR.

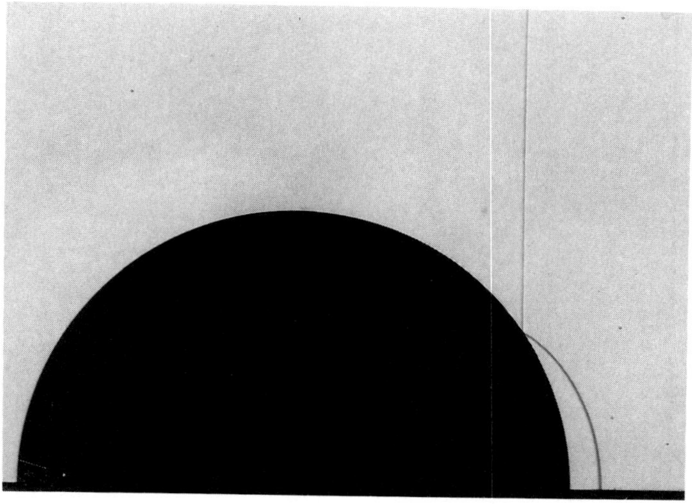

(a)

(b)

Figure 4.22 Shadowgraphs illustrating actual shock wave reflections over a cylindrical convex surface: a) regular reflection; b) double-Mach reflection (courtesy of Professor K. Takayama, Shock Wave Research Center, Institute of Fluid Science, Tohoku University, Sendai, Japan).

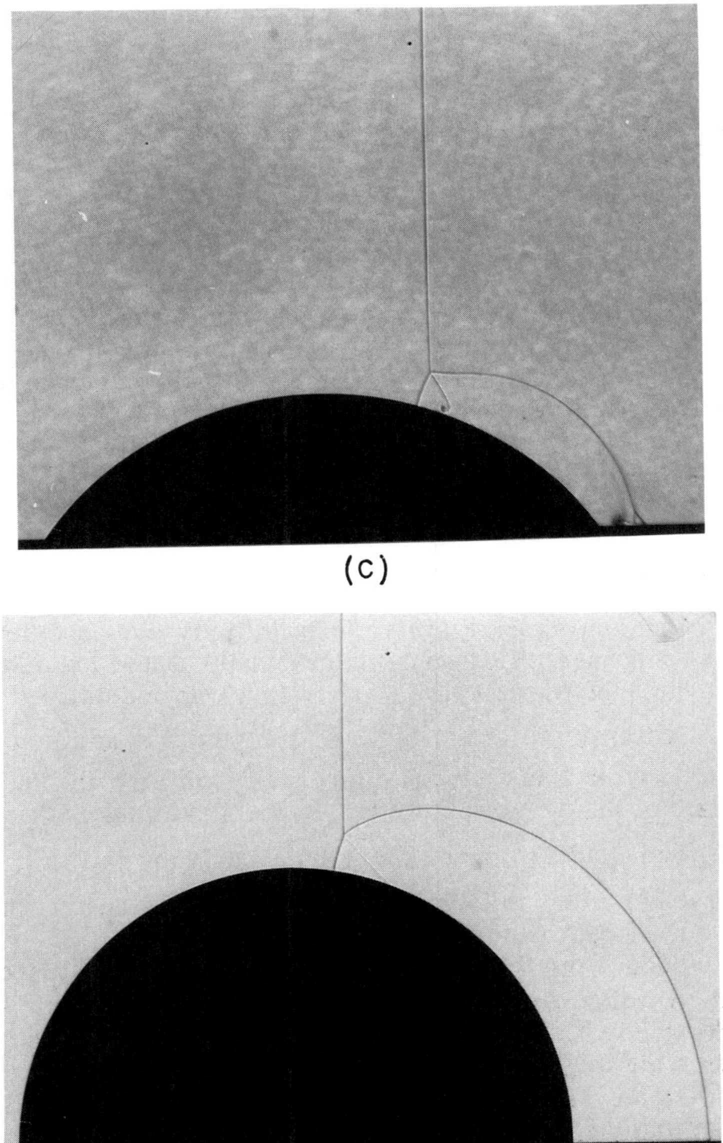

(c)

(d)

Figure 4.22 (cont.) Shadowgraphs illustrating actual shock wave reflections over a cylindrical convex surface: c) transitional-Mach reflection; d) single-Mach reflection (courtesy of Professor K. Takayama, Shock Wave Research Center, Institute of Fluid Science, Tohoku University, Sendai, Japan).

If the initial reflection is an RR (i.e., $90^o \geq \theta_w^{initial} > \theta_w[RR \Leftrightarrow DMR]$) then as the incident shock wave propagates along the cylindrical concave wedge it encounters an ever decreasing wedge angle, which will eventually force the RR to change into an MR. If, however, the initial reflection is an MR (i.e., $\theta_w^{initial} < \theta_w[RR \Leftrightarrow DMR]$) then the MR will persist.

The foregoing discussion implies that the initial reflection over a cylindrical convex wedge can be either an RR, a DMR, a TMR or an SMR depending upon the intitial wedge angle. If it is an RR then it will change first to a DMR, then to a TMR and finally to an SMR; if the initial reflection is a DMR then it will first change to a TMR and then to an SMR; if the initial reflection is a TMR it will change to an SMR; and if the initial reflection is an SMR it will remain an SMR.

Shadowgraphs of an RR, a DMR, a TMR and an SMR over cylindrical concave surfaces are shown in figures 4.22a to 4.22d, respectively.

4.1.2.1 RR→MR Transition

If an incident shock wave reflects initially over a cylindrical convex wedge as an RR then it will eventually change to an MR.

Takayama & Sasaki (1983) showed experimentally that the RR→MR transition wedge angle θ_w^{tr}, depends in addition to the incident shock wave Mach number, M_S, on both the radius of curvature of the cylindrical wedge, R, and the initial wedge angle, $\theta_w^{initial}$. The results of their experimental study are shown in figure 4.23 together with the RR⇔MR transition lines in steady (line AB) and pseudo-steady (line AC) flows.

It is evident from figure 4.23 that all the experimentally recorded transition wedge angles lie below the RR⇔MR transition line of pseudo-steady flows. Furthermore, as the radius of curvature increases the transition wedge angle increases and approaches the pseudo-steady RR⇔MR transition line. Similarly, the results indicate that the transition wedge angle decreases with decreasing initial wedge angles. The dependence of θ_w^{tr} on $\theta_w^{initial}$ is shown in a different way in figure 4.24 for $M_S = 1.6$. The results reveal a continuous non-linear decrease of θ_w^{tr} as $\theta_w^{initial}$ decreases. The solid hand-drawn line illustrates the dependence of θ_w^{tr} on $\theta_w^{initial}$. Since

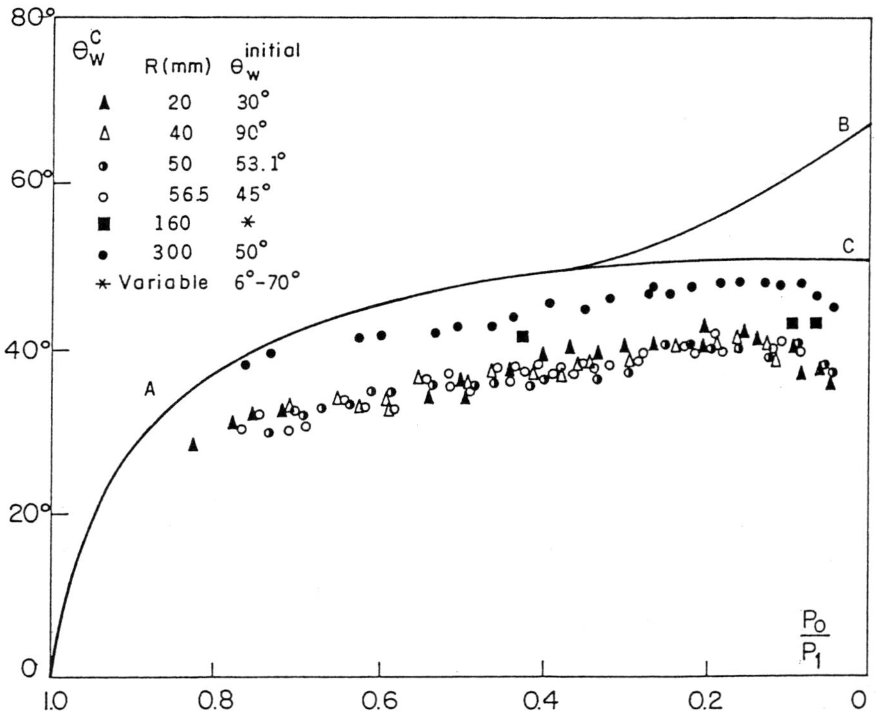

Figure 4.23 Experimental data with nitrogen illustrating the dependence of the transition wedge angle over a cylindrical convex surface on the radius of curvature and the initial wedge angle.

θ_w^{tr} cannot be greater than $\theta_w^{initial}$ this line must terminate at the point $\theta_w^{initial} = \theta_w^{tr}$. The initial wedge angle at this point will result in an MR for $M_S = 1.6$.

It should also be mentioned that the experimental results suggest that as the radius of curvature approaches infinity, i.e., $R \to \infty$, the actual transition wedge angles approach the transition line appropriate to pseudo-steady flows. Recall that in the case of a reflection of a planar shock wave over a cylindrical concave surface $R \to \infty$ resulted in an approach to the transition line appropriate to steady flows (see figure 4.3). Thus, one can conclude that the actual transition line in pseudo-steady flows, which is the case when $R \to \infty$ is approached, is not a unique one, but depends on the way the reflection was initiated and generated.

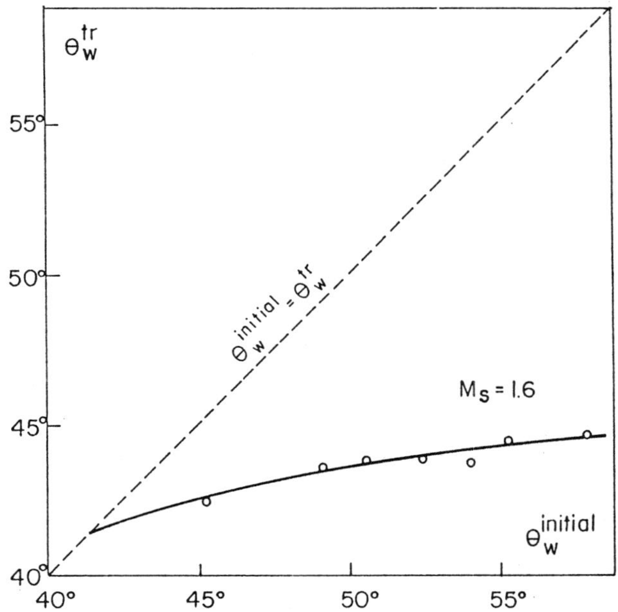

Figure 4.24 Experimental data with nitrogen illustrating the dependence of the transition wedge angle over a cylindrical convex surface on the initial wedge angle for $M_s = 1.6$.

4.1.2.1.1 Effect of Surface Roughness

The effect of surface roughness on the RR→MR transition was investigated experimentally by Takayama, Ben-Dor & Gotoh (1981). The roughness was imposed on the surface by pasting sand paper with different mesh sizes on the reflecting wedge surface. Their results for a cylindrical convex wedge with R = 50 mm, $\theta_w^{initial} = 90^\circ$ and three different roughnesses are shown in figure 4.25. Again, the RR⇔MR transition lines for steady (line AB) and pseudo-steady (line AC) flows are added.

It is evident from figure 4.25 that the transition wedge angle, θ_w^{tr}, decreases as the size of the roughness of the reflecting surface increases. In the case of the extremely rough surface (mesh No. 40 sand paper) θ_w^{tr} is almost independent of the incident shock wave Mach number since it assumes a value of about 28.7° throughout the entire range of the inverse pressure ratio across the incident shock wave - p_0/p_1. This behavior of reduction in the actual transition

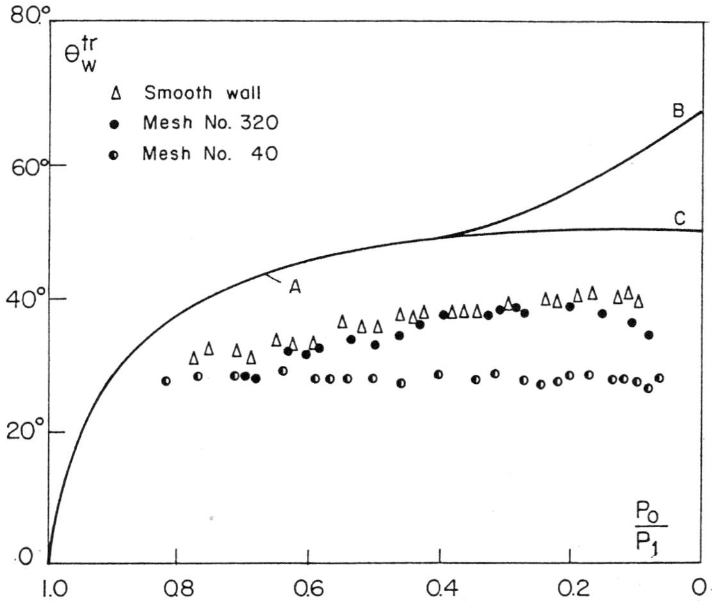

Figure 4.25 Experimental data with nitrogen illustrating the dependence of the transition wedge angle over a cylindrical convex surface on its roughness.

wedge angle with increasing roughness was also evident in the case of shock wave reflection over a cylindrical concave surface (see section 4.1.1.1.1).

Reichenbach (1985) experimentally investigated the RR→MR transition over rough cylindrical surfaces, using cylindrical wedges whose surfaces were machined to obtain step and cubically shaped roughnesses (see figure 2.62). Reichenbach (1985) defined the ratio between the transition wedge angle of rough reflecting surfaces and the transition wedge angle over smooth reflecting surfaces as - η, and presented the experimental results in the (η,ε)-plane shown in figure 4.26. Figure 4.26 clearly indicates that the roughness height, ε, rather than its shape, is the dominant factor in determining the transition wedge angle. For comparison purposes, the line from figure 2.63, which is appropriate to straight surfaces, is added as a dashed line to figure 4.26.

4.1.2.2 Analytical Considerations

In view of the fact that the length scale criterion of Hornung, Oertel & Sandeman (1979) was successful in predicting the transition lines in steady and pseudo-steady shock wave reflections

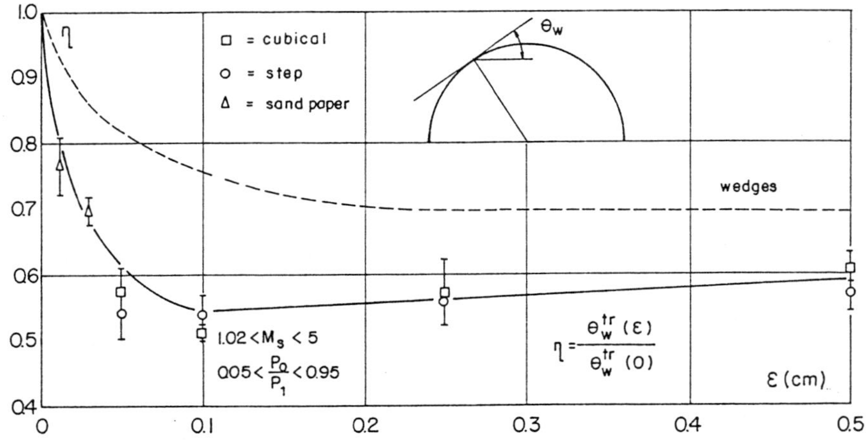

Figure 4.26 The dependence of the RR→MR transition angle ratio over cylindrical convex surfaces on the roughness height.

as well as in the case of unsteady shock wave reflections over cylindrical concave surfaces, it is the author's belief that the race between the corner-generated signals and the incident shock wave is the dominant factor in determining the RR→MR transition phenomenon over cylindrical convex surfaces as well. Thus, the RR→MR transition lines should be obtained, in principle, using models similar to those described in section 4.1.1.3.1. Unfortunately, however, similar analytical considerations for convex surfaces are much more difficult than for concave surfaces due to the fact that the reflected shock wave, at the leading edge of the reflecting wedge, is much stronger when the incident shock wave reflects over a convex surface. Therefore, the simplifying assumptions used to analyze shock wave reflections over concave surfaces, namely; $V + a = V_1 + a_1$ or $V = V_1$ and $a = a_1$ cannot be applied in the case of shock wave reflection over convex surfaces. Thus if one wishes to calculate analytically the catch-up condition then V and a must be known in the flow field in order to carry out the integration $\int(V + a)dt$, which appears on the right hand side of equation (4.13)

4.1.2.2.1 Numerical Predictions of the RR→MR Transition

Although there are no available analytical predictions of the RR→MR transition over cylindrical convex surfaces, there are fortunately, relatively simple numerical codes capable of predicting this transition.

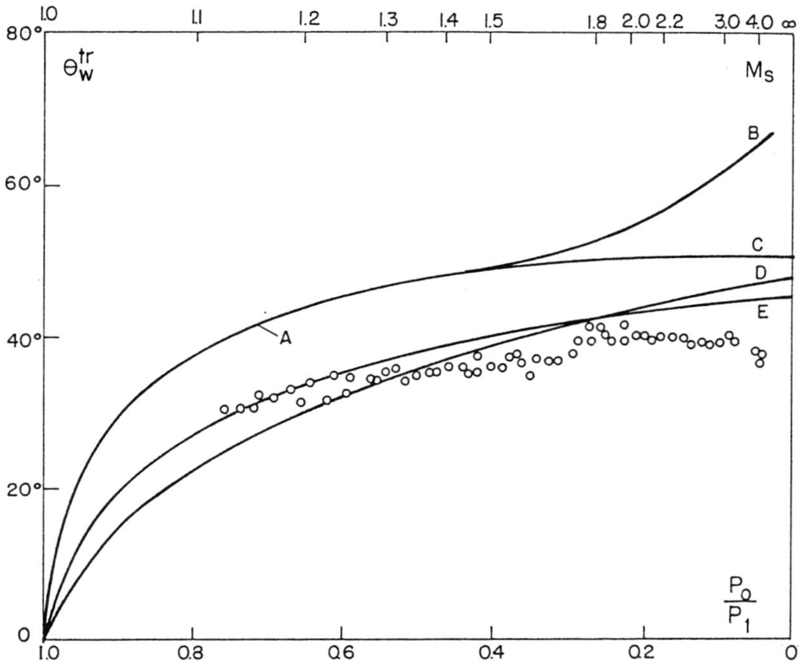

Figure 4.27 Comparison between the measured transition wedge angles over a cylindrical convex surface and various numerical criteria. (Data are for nitrogen.)

Consider figure 4.27 where the recorded transition wedge angles, θ_w^{tr}, over a convex cylindrical wedge with R = 50 mm and $\theta_w^{initial} = 90^\circ$ are plotted with some analytical and numerical transition lines. Line AB is the RR⇔MR transition in steady flows and line AC is the RR⇔MR transition in pseudo-steady flows. Line D was numerically calculated by Heilig (1969) who applied Whitham's (1957) classical ray-shock theory. Line E is a similar calculation of Itoh, Okazaki & Itaya (1981) who used Milton's (1975) modification of Whitham's (1957) theory. A comparison between curves D and E and the experimental results indicates that Milton's modification of Whitham's theory improves the agreement between the numerical predictions and the experimental results. The numerical predictions of Heilig (1969) and Itoh, Okazaki & Itaya (1981) become poorer as M_S increase beyond $M_S \approx 2$, because the transition lines gradually increase with increasing M_S while the experimental results level out and even exhibit a slight decrease in

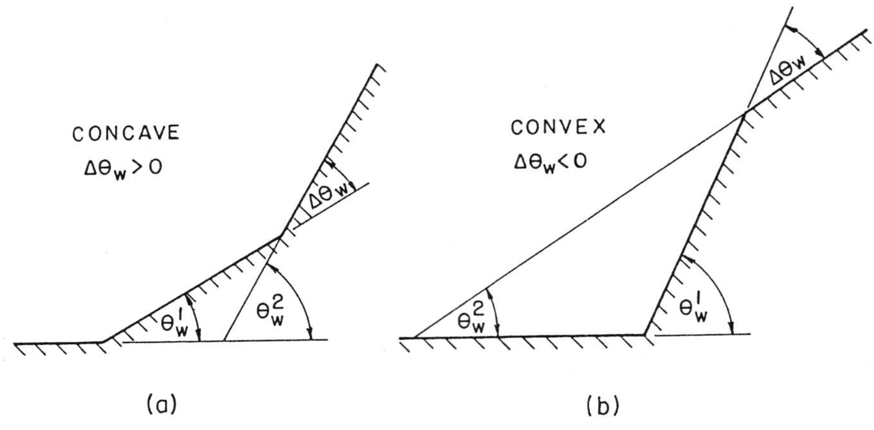

Figure 4.28 Schematic illustrations of a double wedge: a) concave; b) convex.

the actual transition wedge angle. This could be due to the fact that for $M_S > 2$ the shock induced flow becomes supersonic.

4.1.2.3 Additional Remark

Heilig (1969) has also experimentally investigated the shock wave reflection over elliptical convex surfaces. The experimental results were, in general, similar to those obtained over a cylindrical convex surfaces with an appropriate radius of curvature.

4.1.3 Shock Wave Reflections over Double Wedges

Ginzburg & Markov (1975) initiated an experimental investigation of the reflection of a planar shock wave over a two-facet concave wedge. Srivastava & Deschambault (1984) experimentally investigated a two-facet convex wedge. A detailed experimental and analytical study of this reflection phenomenon was reinitiated by Ben-Dor, Dewey & Takayama (1987) who termed it shock wave reflection over double concave or convex wedges. Schematic drawings of a concave and a convex double wedge are shown in figures 4.28a and 4.28b, respectively. For both wedges, θ_w^1, θ_w^2 are the angles of the first and second surfaces and $\Delta\theta_w$ is the inclination of the second surface with respect to the first surface, i.e., $\Delta\theta_w = \theta_w^2 - \theta_w^1$. For a concave double wedge $\Delta\theta_w > 0$ while for a convex double wedge $\Delta\theta_w < 0$. $\Delta\theta_w = 0$ corresponds to a single face straight wedge.

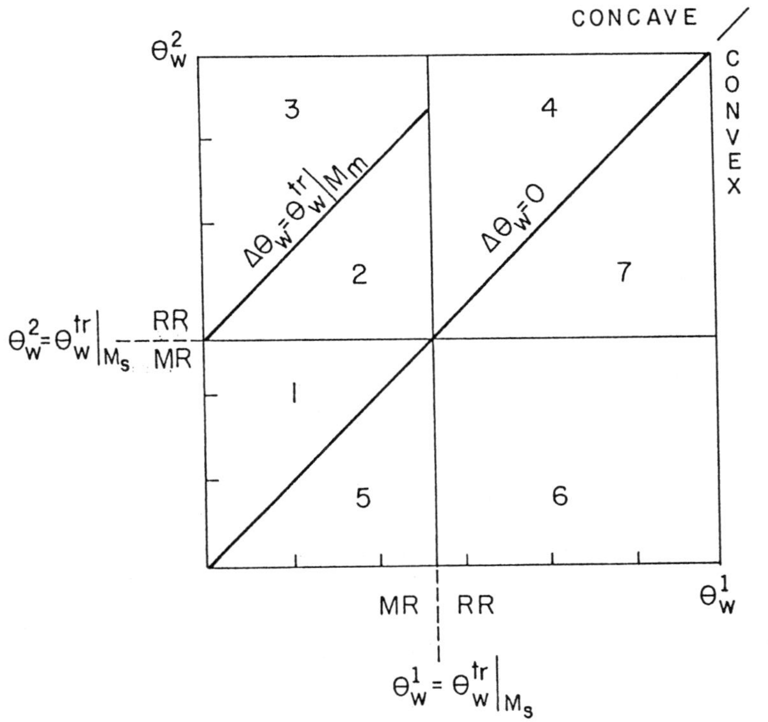

Figure 4.29 Domains and analytical transition boundaries of different types of reflection processes over a double wedge in the (θ_w^1, θ_w^2)-plane for a given M_S.

4.1.3.1 Domains of Different Types of Reflection Processes

The analytical results of Ben-Dor, Dewey & Takayama (1987) are shown in figure 4.29 in the (θ_w^1, θ_w^2)-plane. The third parameter, M_S, is kept constant. Thus for any given value of incident shock wave Mach number, M_S, there is a specific plot such as the one shown in figure 4.29.

There are four transition lines in figure 4.29 which divide the (θ_w^1, θ_w^2)-plane into seven domains of different types of reflection processes over any double wedge.

The line $\Delta\theta_w = 0$ divides the (θ_w^1, θ_w^2)-plane into two domains of different geometries, i.e., concave and convex double wedges. The line $\theta_w^1 = \theta_w^{tr}|_{M_S}$ determines the type of the reflection of the incident shock wave over the first reflecting surface. If $\theta_w^1 < \theta_w^{tr}|_{M_s}$ then the reflection of the incident shock wave over the first reflecting surface is an MR, and if $\theta_w^1 > \theta_w^{tr}|_{M_s}$ then the incident shock wave reflection over the first reflecting surface is an RR. The line $\theta_w^2 = \theta_w^{tr}|_{M_s}$ determines the final type of reflection of the incident shock wave over the second reflecting surface. If $\theta_w^2 < \theta_w^{tr}|_{M_s}$ then the final reflection of the incident shock wave over the second reflecting surface is an MR and if $\theta_w^2 > \theta_w^{tr}|_{M_s}$ then the final incident shock wave reflection over the second reflecting surface is an RR.

When $\theta_w^1 < \theta_w^{tr}|_{M_s}$, then as mentioned earlier, the incident shock wave reflects over the first reflecting surface as an MR; if in addition, the wedge is concave, i.e., $\Delta\theta_w > 0$, then the Mach stem of this MR encounters a compressive corner from which it can reflect either as an MR or as an RR depending on whether $\Delta\theta_w$ is smaller or greater than $\theta_w^{tr}|_{M_m}$. If $\Delta\theta_w < \theta_w^{tr}|_{M_m}$ then the Mach stem reflects over the second reflecting surface as an MR and if $\Delta\theta_w > \theta_w^{tr}|_{M_m}$ then the Mach stem reflects over the second reflecting surface as an RR. Note that if the Mach stem is assumed to be straight and perpendicular to the reflecting surface then

$$M_m = M_s \frac{\cos\chi}{\cos\left(\theta_w^1 + \chi\right)}$$

where χ is the triple point trajectory angle and M_s and M_m are the Mach numbers of the incident and Mach stem shock waves, respectively.

It is clear from this relation that $M_m > M_s$. However, from figure 2.39 it is evident that the RR⇔MR transition wedge angle does not depend very strongly on the shock wave Mach number for $M_s > 2$. Hence, it can be assumed that

$$\theta_w^{tr}|_{M_m} \approx \theta_w^{tr}|_{M_s}$$

Table 5: The reflection processes in each of the seven domains shown in figure 4.29

Region	Type of Double Wedge	Initial Reflection over the 1st surface	Initial Reflection over the 2nd surface	Final Reflection over the 2nd surface
1	Concave	MR	MR	MR
2	Concave	MR	MR	TRR
3	Concave	MR	RR	RR
4	Concave	RR	RR	RR
5	Convex	MR	MR	MR
6	Convex	RR	MR	MR
7	Convex	RR	RR	RR

For example, for $M_S = 2.5$ and $\theta_w^1 = 20°$ the MR solution results in $\chi = 12.88°$. This results in $M_m = 2.902$. The corresponding transition wedge angles for these values of M_S and M_m are 50.77° and 50.72°, respectively. This example indicates that although the difference in the shock wave Mach numbers is about 15%, the difference in the corresponding RR⇔MR transition wedge angles is only 0.05°.

The seven domains of different types of reflection processes, numbered in figure 4.29 from 1 to 7, are described in table 5.

4.1.3.2 RR⇔MR Transition over Double Wedges

The experimental study of Ben-Dor, Dewey & Takayama (1987) indicated that the RR⇔MR transition wedge angle, θ_w^{tr}, over a double wedge depends on $\Delta\theta_w$.

Typical results for a concave double wedge with $M_S = 2.45$ are shown in figure 4.30. When $\Delta\theta_w \doteq 0$, i.e., a straight single surface

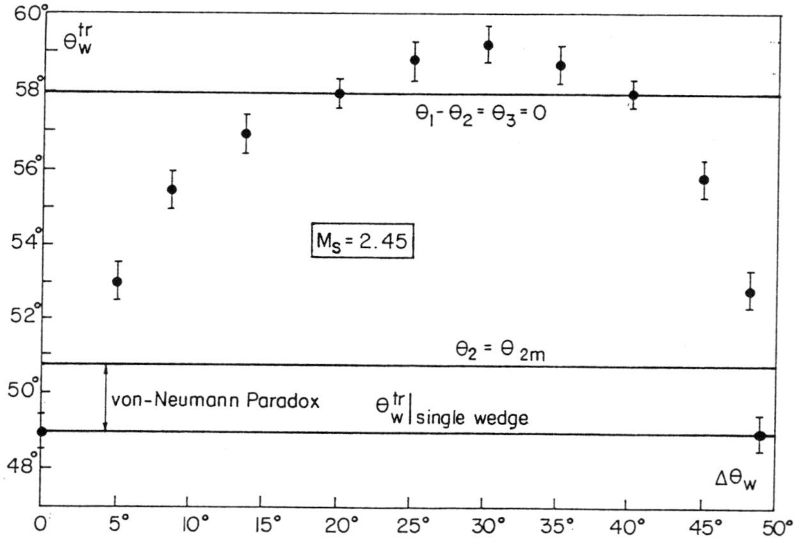

Figure 4.30 The MR⇔RR transition wedge angle over a concave double wedge in nitrogen with $M_s = 2.45$.

wedge, the value of θ_w^{tr} is about 49°. This value is about 1.75° below the RR⇔MR transition wedge angle as predicted by the sonic criterion. As $\Delta\theta_w$ increases, θ_w^{tr} increases until a maximum of $\theta_w^{tr} \approx 59.5°$ at $\Delta\theta_w = 30°$. From there on θ_w^{tr} decreases with increasing $\Delta\theta_w$ until it again reaches the value of 49° at about $\Delta\theta_w = 48.8°$. Note that in this plot the value of θ_w^{tr} is actually the value of θ_w^2 at which transition from MR to RR was recorded. Thus for each experimental point $\theta_w^2 = \theta_w^{tr}$ and $\theta_w^1 = \theta_w^{tr} - \Delta\theta_w$.

The measured transition wedge angles for a convex double wedge with $M_S = 1.3$ are shown in figure 4.31. At $\Delta\theta_w = 0$ the value of θ_w^{tr} is about 44.5°. This value is about 1.65° below the RR⇔MR transition wedge angle as predicted by the sonic criterion. As $\Delta\theta_w$ decreases (note $\Delta\theta_w$ is negative), θ_w^{tr} decreases until a minimum of $\theta_w^{tr} \approx 43°$ at $\Delta\theta_w = -25°$. From there on θ_w^{tr} increases with decreasing $\Delta\theta_w$ until it reaches again the value of 44.5° at about

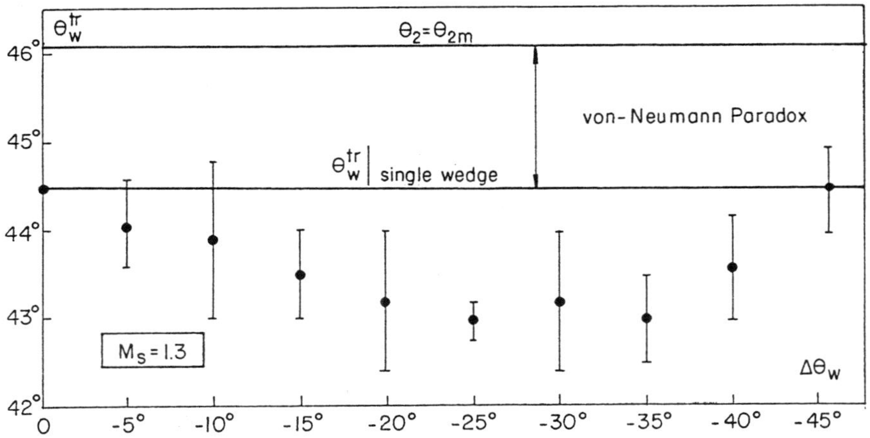

Figure 4.31 The RR⇔MR transition wedge angle over a convex double wedge in nitrogen with $M_s = 1.3$.

$\Delta\theta_W = -45.5°$. Similar to the previous case of a concave double wedge, here again: $\theta_w^2 = \theta_w^{tr}$ and $\theta_w^1 = \theta_w^{tr} - \Delta\theta_W$, where $\Delta\theta_W$ is negative and hence $\theta_w^1 > \theta_w^2$.

It is evident from figures 4.30 and 4.31 that for a concave double wedge θ_w^{tr} is greater than the one predicted by the sonic criterion, while for a convex double wedge θ_w^{tr} is smaller than that predicted by the sonic criterion. This behavior is similar to that shown earlier for cylindrical concave and convex wedges (see figures 4.3 and 4.23). However, while for the case of shock wave reflections over cylindrical concave wedges the transition wedge angles were above those predicted by the steady flow RR⇔MR transition line (see line B in figure 4.3), for the case of shock wave reflections over concave double wedges it is possible to obtain transition wedge angles below this transition line as shown in figure 4.30.

Figure 4.32 illustrates the transition wedge angle, θ_w^{tr}, for four cases: reflection over a cylindrical concave wedge with R = 300 mm, over a cylindrical convex wedge with R = 300 mm, over a concave double wedge with various values of $\Delta\theta_W$ and a concave double wedge with $\Delta\theta_W = 20°$. Note that for R = 300 mm the transition wedge angles over a cylindrical concave wedge approach those appropriate to pseudo steady flows, while the transition wedge

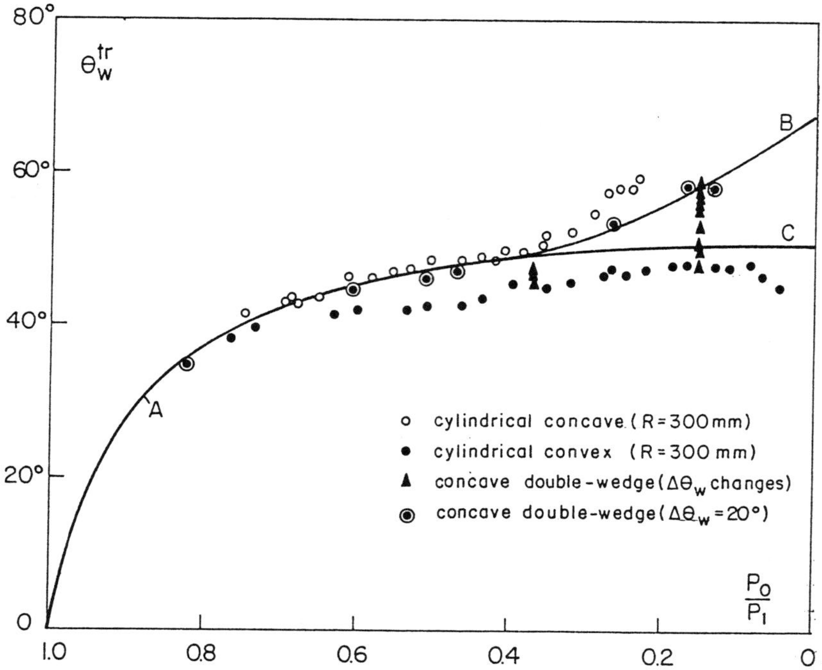

Figure 4.32 Actual transition wedge angles over various cylindrical concave, cylindrical convex and double wedges.

angles over a cylindrical convex wedge approach those appropriate to steady flows. The transition wedge angles are expected to reach the appropriate transition line at $R \to \infty$. In view of the earlier discussion the other results shown in figure 4.32 are self-explanatory.

4.1.3.3 Additional Remarks

As mentioned earlier the shock wave reflection phenomena over double wedges is a relatively new line of research which still requires further analytical and experimental studies. The experimental studies concerning the RR⇔MR transition over double wedges as well as the experimentally recorded transition wedge angles over single wedges are sufficient for replotting two of the transition lines shown in figure 4.29, namely; the lines $\theta_w^1 = \theta_w^{tr}|_{M_s}$ and $\theta_w^2 = \theta_w^{tr}|_{M_s}$. The $\Delta\theta_w = 0$ line is a geometrical condition, and hence is absolutely correct. Thus, more data is needed to obtain the actual boundary separating domains 2 and 3. Figure 4.33 is a

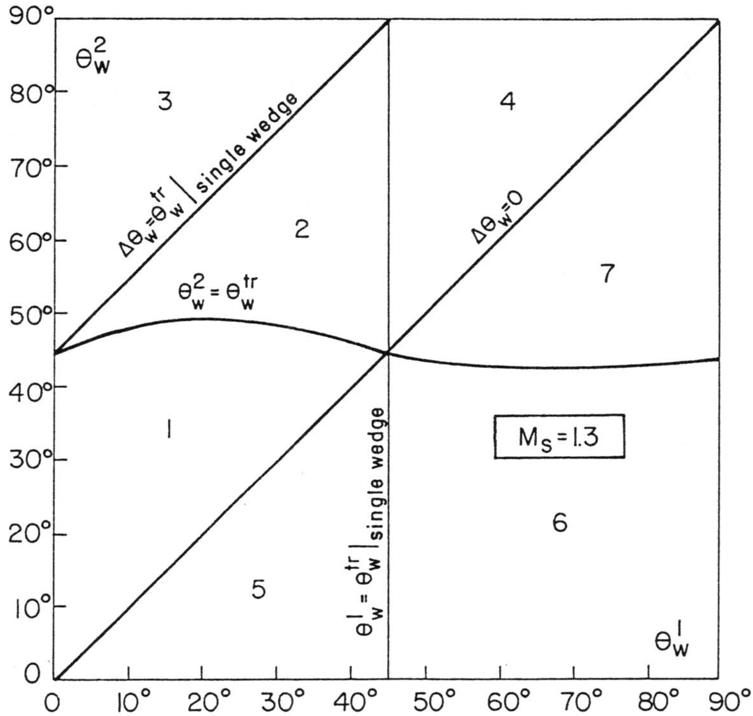

Figure 4.33 Domains and experimental transition boundaries of the different types of reflection processes over a double wedge in the (θ_w^1, θ_w^2)-plane for a given M_S.

reproduction of figure 4.29 with the actual transition lines replacing the corresponding analytical transition lines for $M_S = 1.3$.

It is of interest to note, that although the reflection of a shock wave over a double wedge has not yet been completely explained from a theoretical point of view, the fact that the reflection over concave or convex double wedges shows similar behavior to the reflection over concave or convex cylindrical surfaces should be encouraging, since the analytical study of the former might be easier. Once the reflection over a double wedge is understood, a natural step would be to go over to a triple wedge or a multiple facetted wedge.

As a final remark it is should be mentioned that Ben-Dor, Dewey, McMillin & Takayama (1988) showed experimentally that if the second surface of a double wedge is long enough then the final reflection which develops over it (i.e., an RR on an MR) asymptotically approaches the wave configuration which would

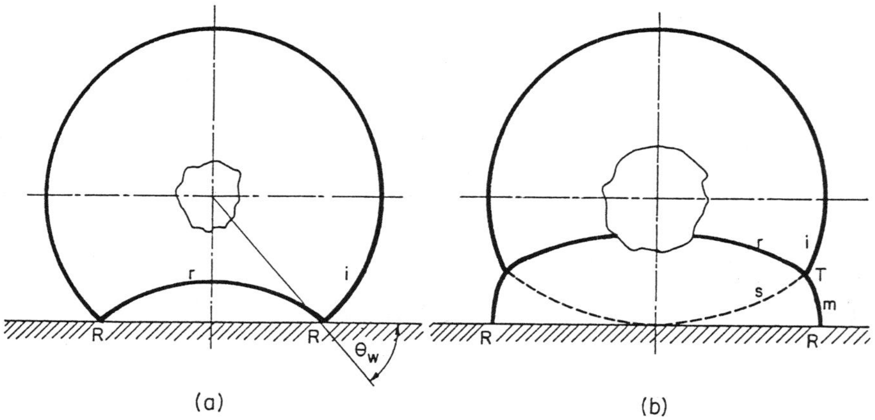

Figure 4.34 Schematic illustrations of spherical shock wave reflections over a straight surface: a) regular reflection; b) Mach reflection.

have been obtained, had the incident shock wave reflected over a straight wedge having an angle equal to θ_w^2.

4.2 Non-Constant Velocity Shock Wave Reflections over Straight Surfaces

The reflection of a non-constant velocity shock wave over straight surfaces is another example of a truly unsteady process. Although the analytical treatment of a reflection of this type might be important to better understand the more complex reflection of a spherical shock wave (non-constant velocity shock wave) reflection over the ground surface (straight surface), which is dealt with in the next section, unfortunately no work has been done on this reflection process. This is in spite of the fact that experimental investigations of this type of shock wave reflection processes can be performed relatively easily in a shock tube. The only requirement is to generate an accelerating or decelerating (decaying) planar shock wave and reflect it over a wedge with a straight surface. Techniques for generating decaying planar shock waves in shock tubes can be found in the literature related to shock tubes.

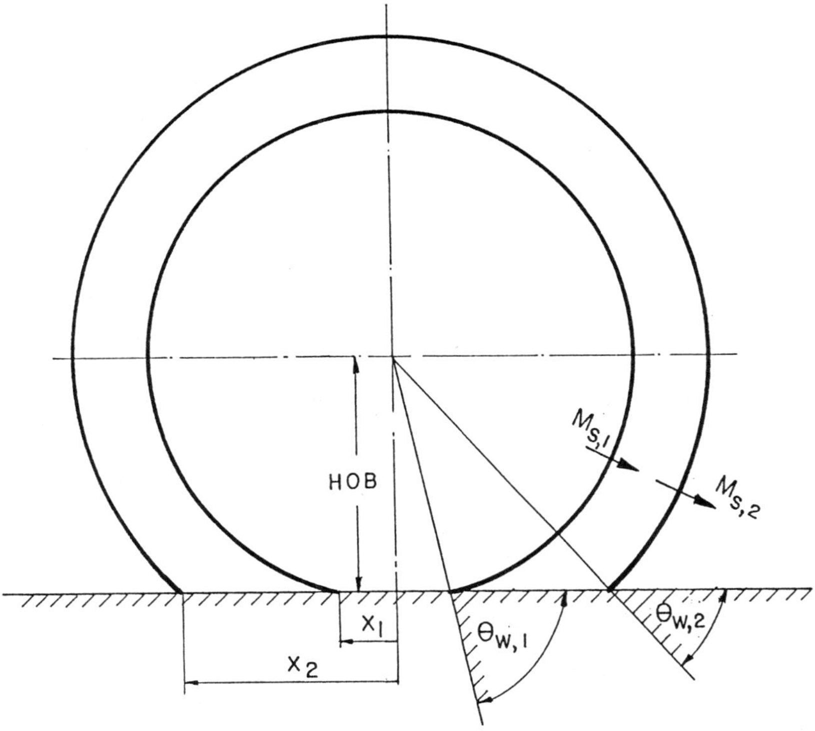

Figure 4.35 Schematic illustration of the instantaneous incident shock wave Mach number and the instantaneous reflecting wedge angle of a spherical shock wave propagating over a straight surface.

4.3 Spherical Shock Wave Reflections over Straight and Non-Straight Surfaces

The reflection of a spherical shock wave, e.g., a blast wave which is generated when energy is released in short times, over straight and non-straight surfaces has been experimentally investigated quite extensively in the past few decades, e.g., Dewey, McMillin & Classen (1977), Dewey & McMillin (1981), Hu & Glass (1986), etc.

When a spherical shock wave is generated from an explosion detonated above the ground it reflects over the surface initially as an

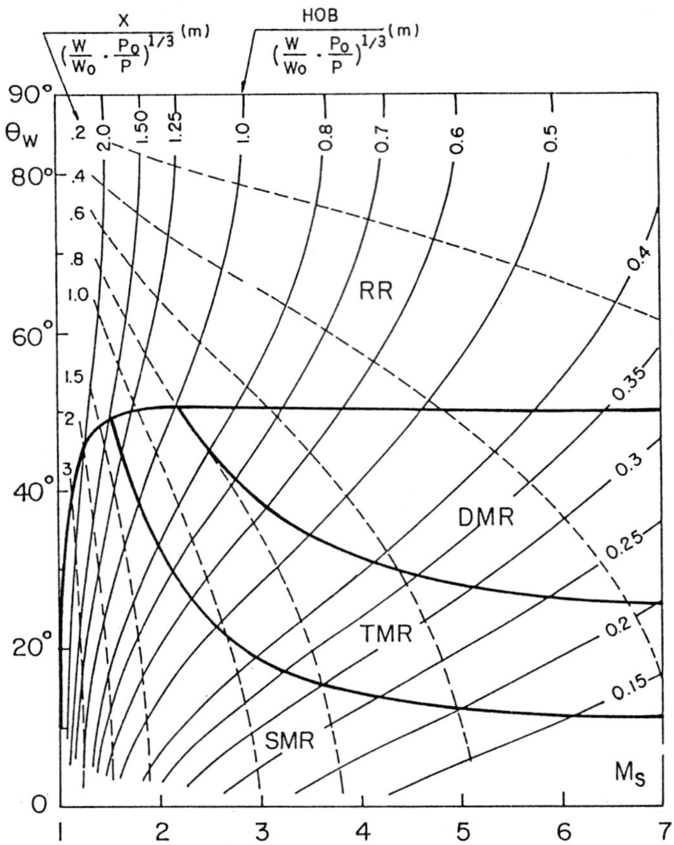

Figure 4.36 Domains of different types of reflections in pseudo-steady flows and spherical shock wave trajectories (courtesy of Professor I. I. Glass).

RR, which then, depending upon the strength of the spherical shock wave, changes into a DMR, a TMR and finally to an SMR as it propagates outwards.

Figures 4.34a and 4.34b illustrate an RR and an MR of a spherical shock wave over a straight surface. Consider figure 4.35 where the spherical shock wave at two different times is drawn, and note that as the spherical shock wave propagates outwards the point where it touches the reflecting surface encounters an ever decreasing effective reflecting wedge angle, i.e., $\theta_{w,2} < \theta_{w,1}$. Thus beside the fact that the instantaneous Mach number M_S decreases with time ($M_{S,2} < M_{S,1}$), the effective reflecting wedge angle also decreases with time.

Figure 4.36 [originally plotted by Hu & Glass (1986)] shows the domain of different types of reflections (in pseudo-steady flows) in the (M_S, θ_W)-plane for air. Spherical shock wave trajectories, i.e.,

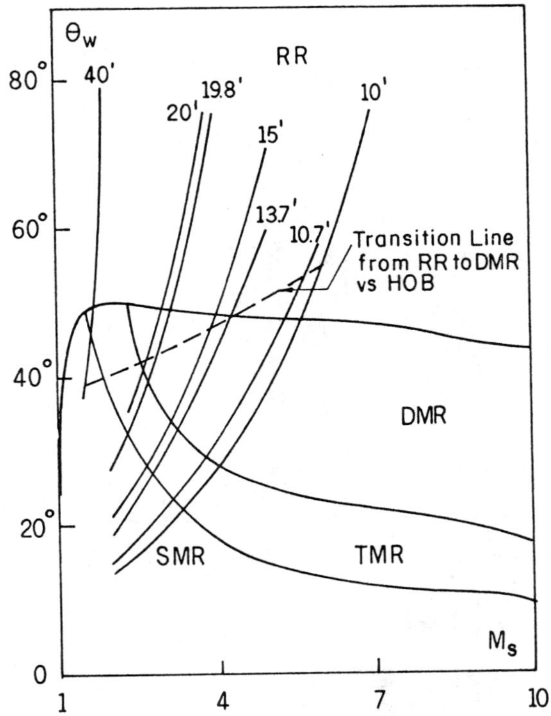

Figure 4.37 Comparison of average transition lines of actual blast wave experiments as compared to the transition boundaries appropriate to pseudo-steady reflections in inperfect nitrogen (courtesy of Mr. J. Wisotski, Denver Research Institute, Denver, Colorado).

M_S vs. θ_W, are added to figure 4.36. The trajectories differ from each other by the non-dimensionalized height of burst, HOB (see figure 4.35). The non-dimensionalizing factor is $(Wp_0/W_0p)^{1/3}$ where W and p are the weight (equivalent to TNT) of the explosive charge, used to generate the blast (spherical shock) wave, and the atmospheric pressure, respectively, and W_0 = 1Kg TNT and p_0 = 1 atm. The non-dimensionalized distance X denotes the radial location (measured from the center of explosion on the ground, see figure 4.35). The X = 0 line coincides with the $\theta_W = 90°$ line. It is seen that for an explosive charge of 1 Kg TNT and a height of burst of 0.8 m, at 1 atm, the blast wave has a strength equivalent to $M_S \approx 3.7$ upon colliding with the reflecting surface. The initial reflection for this case is an RR that changes later to a DMR, then to a TMR and finally to an SMR until it degenerates to a sound wave, $M_S \rightarrow 1$. On

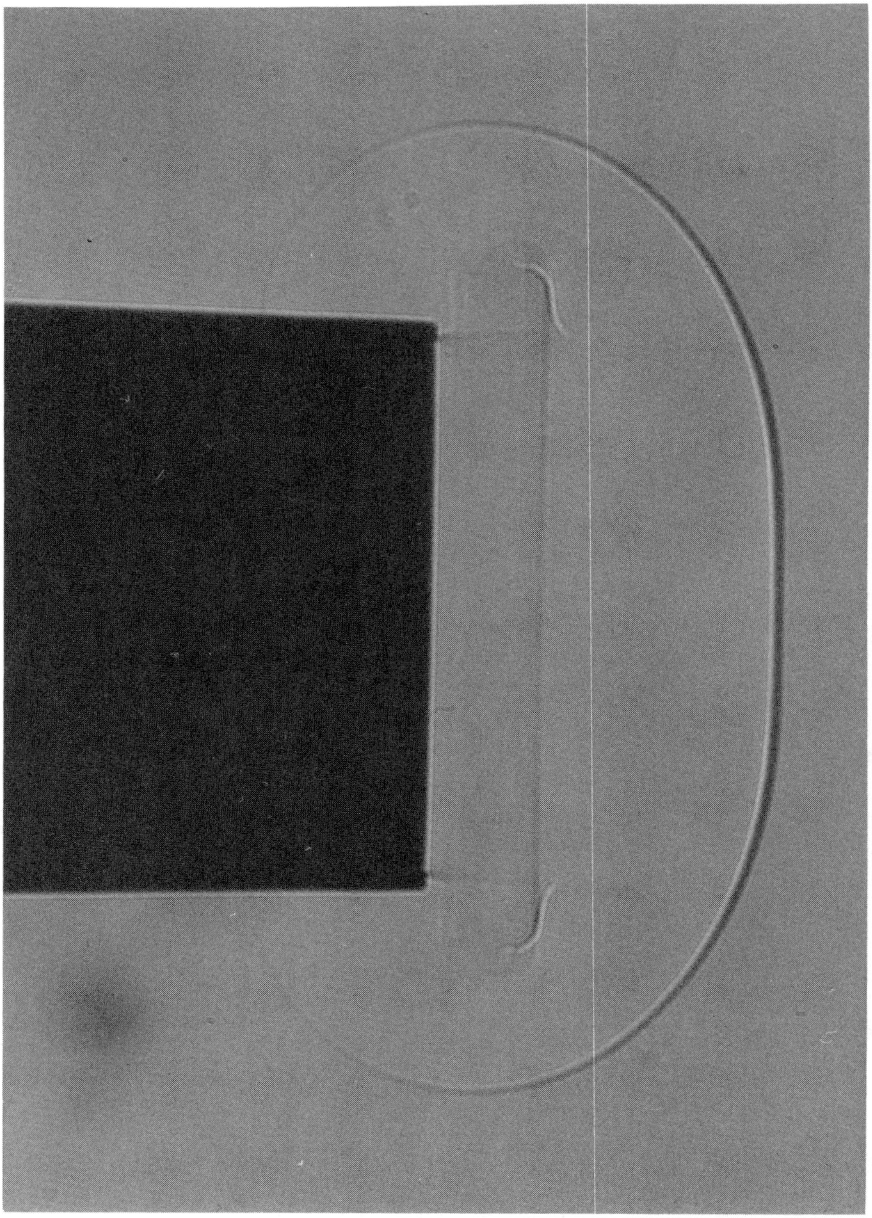

a)

Figure 4.38 The generation of a spherical shock wave by means of a shock tube: a) The shape of the incdient shock wave a short time after it emerges out of the open ended circular shock tube (courtesy of Professor K. Takayama, Shock Wave Research Center, Institute of Fluid Science, Tohoku University, Sendai, Japan).

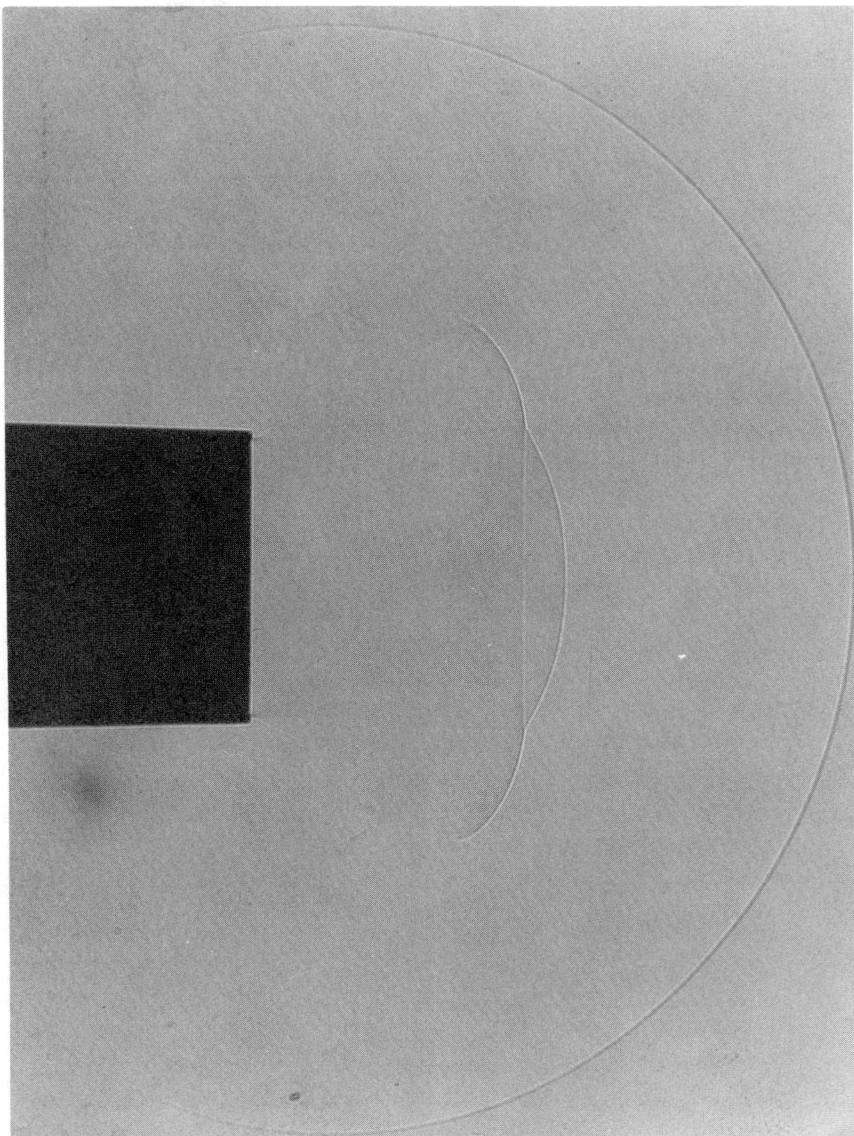

b)

Figure 4.38(cont.) b) After a relatively long time a perfect spherical shape is approached (courtesy of Professor K. Takayama, Shock Wave Research Center, Institute of Fluid Science, Tohoku University, Sendai, Japan).

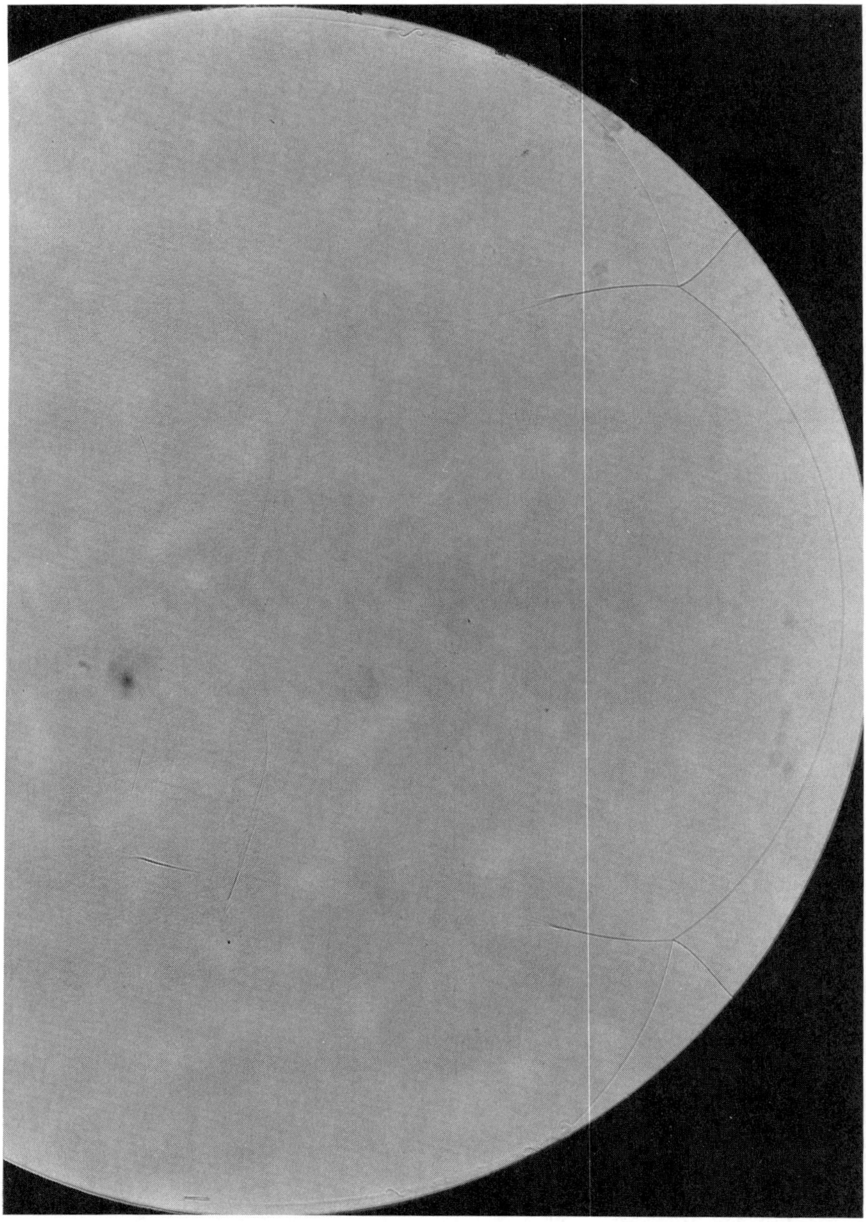

Figure 4.39 The reflection of a spherical shock wave as generated by the experimental set-up shown in figure 4.38, from a cylindrical reflecting surface (courtesy of Professor K. Takayama, Shock Wave Research Center, Institute of Fluid Science, Tohoku University, Sendai, Japan).

the other hand if the HOB for the same charge is 2m, the instantaneous Mach number when the blast wave collides with the ground surface is $M_S \approx 1.5$ and the RR changes directly to an SMR before it degenerates to a sound wave. The trajectories shown in figure 4.36 are fits based on the best experimental data. However, the foregoing discussion should be regarded only as explanatory from a phenomenological point of view, since the pseudo-steady transition lines shown in figure 4.36 are not applicable for unsteady flows.

An experimental test of the applicability of the transition lines between the various types of reflections in pseudo-steady flows, for the case of blast wave reflections is shown in figure 4.37, where the actual RR→MR transition angles for seven different heights of burst experiments are connected by a dashed line. The experimental RR→MR transition wedge angles are up to 10° smaller or greater than the pseudo-steady RR⇔MR transition line. Furthermore, while the pseudo-steady RR⇔MR transition line slightly decreases as M_S increases, the unsteady experimental RR→MR transition line shows a gradual increase with increasing M_S. Thus, beside the fact that the (M_S, θ_W)-plot is helpful in determining the sequence of events during the RR→MR transition, it is quite obvious that it cannot predict the actual transition angles.

Spherical shock waves can be generated in shock tubes by producing a planar shock wave inside a tube and allowing it to emerge from it. The formation of such a blast wave as it emerges from the tube is shown in figures 4.38a and 4.38b. While at early stages the shape of the shock wave emerging from the end of the tube is not spherical, at later times (figure 4.38b) it attains a perfectly spherical shape. Takayama & Sekiguchi (1981a and 1981b) have used this technique of producing spherical shock waves in a shock tube to investigate the reflection of spherical shock waves from various surfaces, in general, and cones in particular. The reflection of the spherical shock wave shown in figure 4.38b from a cylindrical reflecting surface is shown in figure 4.39.

4.4 References

Bazhenova, T. V., Fokeev, V. P. & Gvozdeva, L. G., "Regions of Various Forms of Mach Reflection and Its Transition to Regular Reflection", Acta Astro., Vol. 3, pp. 131-140, 1976.

Ben-Dor, G., "Analytical Solution of a Double-Mach Reflection", AIAA J., Vol. 18, pp. 1036-1043, 1980.

Ben-Dor, G., Dewey, J.M., McMillin, D.J. & Takayama, K., "Experimental Investigation of the Asymptotically Approached Mach Reflection over the Second Surface in the Reflection over a Double Wedge", Experiments in Fluids, Vol. 6, pp. 429-434, 1988.

Ben-Dor, G., Dewey, J.M. & Takayama, K., "The Reflection of a Planar Shock Wave over a Double Wedge", J. Fluid Mech., Vol. 176, pp. 483-520, 1987.

Ben-Dor, G. & Takayama, K., "Analytical Prediction of the Transition from Mach to Regular Reflection over Cylindrical Concave Wedges", J. Fluid Mech., Vol. 158, pp. 365-380, 1985.

Ben-Dor, G. & Takayama, K., "Application of Steady Shock Polars to Unsteady Shock Wave Reflections", AIAA J., Vol. 24, pp. 682-684, 1986.

Ben-Dor, G. & Takayama, K., "The Dynamics of the Transition from Mach to Regular Reflection over Concave Cylinders", Israel J. Tech., Vol. 23, pp. 71-74, 1986/7.

Ben-Dor, G., Takayama, K. & Dewey, J.M., "Further Analytical Considerations of the Reflection of Weak Shock Waves over a Concave Wedge", Fluid Dyn. Res., Vol. 2, pp. 77-85, 1987.

Courant, R. & Friedrichs, K.O., *Hypersonic Flow and Shock Waves*, Wiley Interscience, New York, U.S.A., 1948.

Dewey, J.M. & McMillin, D.J., "An Analysis of the Particle Trajectories in Spherical Blast Waves Reflected from Real and Ideal Surfaces", Canadian J. Phys., Vol. 59, pp. 1380-1390, 1981.

Dewey, J.M., McMillin, D.J. & Classen, D.F., "Photogrammetry of Spherical Shocks Reflected from Real and Ideal Surfaces", J. Fluid Mech., Vol. 81, pp. 701-717, 1977.

Ginzburg, I.P. & Markov, Y.S., "Experimental Investigation of the Reflection of a Shock Wave from a Two-Facet Wedge", Fluid Mech.-Soviet Res., Vol. 4, pp. 167-172, 1975.

Heilig, W.H., "Diffraction of Shock Wave by a Cylinder", Phys. Fluids Supll. I., Vol. 12, pp. 154-157, 1969.

Henderson, L.F. & Lozzi, A., "Experiments on Transition of Mach Reflection", J. Fluid Mech., Vol. 68, pp. 139-155, 1975.

Hornung, H.G., Oertel, H. Jr. & Sandeman, R.J., "Transition to Mach Reflection of Shock Waves in Steady and Pseudo-Steady Flow with and without Relaxation", J. Fluid Mech., Vol. 90, pp. 541-560, 1979.

Hu, T.C.J. & Glass, I.I., "Blast Wave Reflection Trajectories from a Height of Burst", AIAA J., Vol. 24, pp. 607-610, 1986.

Itoh, S. & Itaya, M., "On the Transition Between Regular and Mach Reflection" in "Shock Tubes and Waves", Ed. A. Lifshitz and J. Rom, Magnes Press, Jerusalem, pp. 314-323, 1980.

Itoh, S., Okazaki, N. & Itaya, M., "On the Transition Between Regular and Mach Reflection in Truly Non-Stationary Flows", J. Fluid Mech., Vol. 108, pp. 383-400, 1981.

Law, C.K. & Glass, I.I., "Diffraction of Strong Shock Waves by a Sharp Compressive Corner", CASI Trans., Vol. 4, pp. 2-12, 1971.

Marconi, F., "Shock Reflection Transition in Three-Dimensional Flow About Interfering Bodies", AIAA J., Vol. 21, pp. 107-113, 1983.

Milton, B.E., "Mach Reflection Using Ray-Shock Theory", AIAA J., Vol. 13, pp. 1531-1533, 1975.

Reichenbach, H., "Roughness and Heated Layer Effects on Shock Wave Propagation and Reflection - Experimental Results", Ernst Mach Institute Rep. E 24/85, Freiburg, West Germany, 1985.

Srivastava, R.S. & Deschambault R.L., "Pressure Distribution Behind a Nonstationary Reflected-Diffracted Oblique Shock Wave", AIAA J., Vol. 22, pp. 305-306, 1984.

Takayama, K. & Ben-Dor, G., "A Reconsideration of the Transition Criterion from Mach to Regular Reflection over Cylindrical Concave Surfaces", KSME J., Vol. 3, pp. 6-9, 1989.

Takayama, K., Ben-Dor, G. & Gotoh, J., "Regular to Mach Reflection Transition in Truly Nonstationary Flows - Influence of Surface Roughness", AIAA J., Vol. 19, pp. 1238-1240, 1981.

Takayama, K. & Sasaki, M., "Effects of Radius of Curvature and Initial Angle on the Shock Transition over Concave and Convex Walls", Rep. Inst. High Speed Mech., Tohoku Univ., Sendai, Japan, Vol. 46, pp. 1-30, 1983.

Takayama, K. & Sekiguchi, H., "Triple-Point Trajectory of Strong Spherical Shock Wave", AIAA J., Vol. 19, pp. 815-817, 1981a.

Takayama, K. & Sekiguchi, H., "Formation and Diffraction of Spherical Shock Waves in Shock Tube", Rep. Inst. High Speed Mech., Tohoku Univ., Sendai, Japan, Vol. 43, pp. 89-119, 1981b.

Whitham, G.B., "A New Approach to Problems of Shock Dynamics. Part 1. Two Dimensional Problems", J. Fluid Mech., Vol. 2, pp. 145-171, 1957.

5

Source List

List of References

In the following a comprehensive list of papers and reports dealing with the various aspects of shock wave reflection phenomenon is given.

The list is divided into two parts. The first contains papers which were published in refereed scientific journals and which are, in principle, available to the scientific community. The second contains departmental reports from institutions and universities which have been involved in the study of shock wave reflection.

An effort was made to try and include most of the papers and the reports which have been published so far in the area of shock wave reflection. In addition, an effort was made to provide full reference to each of the listed references.

Finally, since only papers published in refereed scientific journals and departmental reports are listed in the following, the interested reader is advised to look for more papers on the subject of shock-wave reflections in the proceedings of the following conferences:

1) The International Symposium on Shock Tubes and Waves (ISTS).
2) The International Symposium on Military Applications of Blast Simulations (MABS).
3) The International Colloquium on Dynamics of Explosions and Reactive Systems (ICDERS).
4) The Mach Reflection Symposium (MRS).

5.1 Scientific Journals

Adachi, T., Kobayashi, S. & Suzuki, T., "An Analysis of the Reflected Shock Wave Structure over a Wedge with Surface Roughness", Dyn. Image Anal., Vol. 1, pp. 30-37, 1988.

Adachi, T., Kobayashi, S. & Suzuki, T., "Reflection of Plane Shock Waves over a Dust Layer - The Effects of Permeability", Theor. & Appl. Mech., Vol. 37, pp. 23-30, 1989.

Adachi, T., Kobayashi, S. & Suzuki, T., "An Experimental Study of Reflected Shock Wave Structure over a Wedge with Surface Roughness", Trans. Japan Soc. Mech. Eng., Vol. B56, pp. 1244-1249, 1990.

Adachi, T., Kobayashi, S. & Suzuki, T., "Unsteady Behavior of Mach Reflection of a Plane Shock Wave over a Wedge with Surface Roughness", Theor. & Appl. Mech., Vol. 39, pp. 285-291, 1990.

Adachi, T., Suzuki, T. & Kobayashi, S., "Unsteady Oblique Reflection of a Plane Shock Wave over a Wedge with Surface Roughness", Dyn. Image Anal., Vol. 2, pp. 31-36, 1990.

Arutyunyan, G. M., "On Interaction of Shock Waves with a Wedge", Dokl. Akad. Nauk Arm. SSSR, Vol. 46, pp. 160-167, 1968.

Arutyunyan, G. M., "On Diffraction of Shock Waves", Prikl. Mat. & Mekh., Vol. 34, pp. 693-699, 1970.

Arutyunyan, G. M., "Theory of Anomalous Regimes in the Regular Reflection of Shock Waves", Izv. Akad. Nauk SSSR, Mekh. Zh. i Gaza, Vol. 5, pp. 117-125, 1973.

Arutyunyan, G. M., Belokon, V. A. & Karchevsky, L. V., "Effect of Adiabatic Index on Reflection of Shocks", Zh. Prikl. Mekh. i Teor. Fiz., Vol. 4, 62-66, 1970.

Arutyunyan, G. M. & Karchevsky, L. F., *Reflected Shock Waves*, Mashinostroyeniye Press, Moscow, SSSR, 1975.

Auld, D. J. & Bird, G. A., "Monte Carlo Simulation of Regular and Mach Reflection", AIAA J., Vol. 15, pp. 638-641, 1977.

Bazhanov, K. A., "The Regular Case of Shock Wave Diffraction on a Wedge Partially Submerged in Fluid", Appl. Math. & Mech., Vol. 46, pp. 628-634, 1982.

Bazhenova, T. V., Fokeev, V. P. & Gvozdeva, L. G., "Regions of Various Forms of Mach Reflection and Its Transition to Regular Reflection", Acta Astro., Vol. 3, pp. 131-140, 1976.

Bazhenova, T. V., Fokeev, V.P. & Gvozdeva, L.G., "Reply to the Comments of C. K. Law & I. I. Glass", Acta Astro., Vol. 4, pp. 943-944, 1977.

Bazhenova, T. V. & Gvozdeva, L. G., *Unsteady Interactions of Shock Waves*, Nauka, Moscow, SSSR, 1977.

Bazhenova, T. V. & Gvozdeva, L. G., "The Reflection and Diffraction of Shock Waves", Fluid Dyn. Trans., Vol. 11, pp. 7-16, 1983.

Bazhenova, T. V., Gvozdeva, L. G., Komarov, V. S. & Sukhov, B. G., "Pressure and Temperature Change on the Wall Surface in Strong Shock Wave Diffraction", Astro. Acta, Vol. 17, pp. 559-566, 1972.

Bazhenova, T. V., Gvozdeva, L. G., Komarov, V. S. & Sukhov, B. G., "Diffraction of Strong Shock Waves by Convex Corners", Izv. Akad. Nuak SSSR, Mekh. Zh. Gaza, Vol. 4, pp. 122-134, 1973.

Bazhenova, T. V., Gvozdeva, L. G. & Nettleton, M. A., "Unsteady Interactions of Shock Waves", Prog. Aero. Sci., Vol. 21, pp. 249-331, 1984.

Bazhenova, T. V., Gvozdeva, L. G. & Zhilin, Yu. V., "Change in the Shape of a Diffracting Shock Wave at a Convex Corner", Acta Astro., Vol. 6, pp. 401-412, 1979.

Ben-Dor, G., "A Reconsideration of the Shock Polar Solution of a Pseudo-Steady Single-Mach Reflection", CASI Trans., Vol. 26, pp. 98-104, 1980.

Ben-Dor, G., "Analytical Solution of Double-Mach Reflection", AIAA J., Vol. 18, pp. 1036-1043, 1980.

Ben-Dor, G., "Steady Oblique Shock Wave Reflections in Perfect and Imperfect Monatomic and Diatomic Gases", AIAA J., Vol. 18, pp. 1143-1145, 1980.

Ben-Dor, G., "Relation Between First and Second Triple Point Trajectory Angles in Double-Mach Reflection", AIAA J., Vol. 19, pp. 531-533, 1981.

Ben-Dor, G., "A Reconsideration of the Three-Shock Theory of a Pseudo-Steady Mach Reflection", J. Fluid Mech., Vol. 181, pp. 467-484, 1987.

Ben-Dor, G., "Steady, Pseudo-Steady and Unsteady Shock Wave Reflections", Prog. Aero. Sci., Vol. 25, pp. 329-412, 1988.

Ben-Dor, G., "Structure of the Contact Discontinuity of Nonstationary Mach Reflections", AIAA J., Vol. 28, pp. 1314-1316, 1990.

Ben-Dor, G., "Interaction of a Planar Shock Wave with a Double Wedge-Like Structure", AIAA J., Vol. unavailable, pp. unavailable, 1991.

Ben-Dor, G. & Dewey, J.M., "The Mach Reflection Phenomenon - A Suggestion for an International Nomenclature," AIAA J., Vol. 23, pp. 1650-1652, 1985.

Ben-Dor, G., Dewey, J. M., McMillin, D. J. & Takayama, K., "Experimental Investigation of the Asymptotically Approached Mach Reflection over the Second Surface in a Reflection over a Double Wedge", Exp. in Fluids, Vol. 6, pp. 429-434, 1988.

Ben-Dor, G., Dewey, J. M. & Takayama, K., "The Reflection of a Planar Shock Wave over a Double Wedge", J. Fluid Mech., Vol. 176, pp. 483-520, 1987.

Ben-Dor, G. & Glass, I. I., "Nonstationary Oblique Shock Wave Reflections: Actual Isopycnics and Numerical Experiments", AIAA J., Vol. 16, pp. 1146-1153, 1978.

Ben-Dor, G. & Glass, I. I., "Domains and Boundaries of Nonstationary Oblique Shock Wave Reflections:1. Diatomic Gas", J. Fluid Mech., Vol. 92, pp. 459-496, 1979.

Ben-Dor, G. & Glass, I. I., "Domains and Boundaries of Nonstationary Oblique Shock Wave Reflections:2. Monatomic Gas", J. Fluid Mech. Vol. 96, pp. 735-756, 1980.

Ben-Dor, G., Mazor, G., Takayama, K. & Igra, O., "The Influence of Surface Roughness on the Transition from Regular to Mach Reflection in a Pseudo Steady Flow", J. Fluid Mech., Vol. 176, pp. 333-356, 1987.

Ben-Dor, G. & Takayama, K., "Analytical Prediction of the Transition from Mach to Regular Reflection over Cylindrical Concave Wedges", J. Fluid Mech., Vol. 158, pp. 365-380, 1985.

Ben-Dor, G. & Takayama, K., "Application of Steady Shock Polars to Unsteady Shock Wave Reflections," AIAA J., Vol. 24, pp. 682-684, 1986.

Ben-Dor, G. & Takayama, K., "The Dynamics of the Transition from Mach to Regular Reflection over Concave Cylinders", Israel J. Tech., Vol. 23, pp. 71-74, 1986/7.

Ben-Dor, G. & Takayama, K., "The Reflection of a Planar Shock Wave over a Water Wedge", Israel J. Tech., Vol. 23, pp. 169-173, 1986/7.

Ben-Dor, G. & Takayama, K., "Streak Camera Photography with Curved Slits for the Precise Determination of Shock Wave Transition Phenomena", CASI J., Vol. 27, pp. 128-134, 1981.

Ben-Dor, G., Takayama, K. & Dewey, J. M., "Further Analytical Considerations of the Reflection of a Weak Shock Wave over a Concave Wedge", Fluid Dyn. Research, Vol. 2, pp. 77-85, 1987.

Ben-Dor, G., Takayama, K. & Kawauchi, T., "The Transition from Regular to Mach Reflection and from Mach to Regular Reflection in Truly Nonstationary Flows", J. Fluid Mech. Vol. 100, pp. 147-160, 1980.

Ben-Dor, G., Takayama, K. & Needham, C. E., "The Thermal Nature of the Triple Point of a Mach Reflection", Phys. Fluids, Vol. 30, pp. 1287-1293, 1987.

Berezkina, M. K., Syshchikova, M. P. & Semenov, A. N., "Interaction of Two Consecutive Shock Waves with a Wedge", Sov. Phys. Tech. Phys., Vol. 27, pp. 835-841, 1982.

Beryozkina, M. K., Syshchikova, M. P., Semenov, A. N. & Krassovskaya, I. V., "Some Properties of the Nonstationary Interaction of Two Shock Waves with a Wedge", Arch. Mech., Vol. 32, pp. 621-631, 1980.

Beryozkina, M. K., Syshchikova, M. P. & Semenov, A. N., "Interaction of Two Successive Shock Waves with a Wedge", Zh. Tekh. Fiz., Vol. 52, pp. 1375-1385, 1982.

Bleakney, W., Fletcher, C.H. & Weimer, D.K., "The Density Field in Mach Reflection of Shock Waves", Phys. Rev. 76, 323-324, 1949.

Bleakney, W. & Taub, A. H. "Interaction of Shock Waves", Rev. Mod. Phys., Vol. 21, pp.584-605, 1949.

Bleakney, W., White, D. R. & Griffith, W. C., "Measurement of Diffraction of Shock Waves and Resulting Loading of Structure", J. Appl. Mech., Vol. 17, pp. 439-445, 1950.

Bogatko, V. I. & Kolton, G. A., "On Regular Reflection of Strong Shock Waves by a Thin Wedge", Izv. Akad. Nauk SSSR, Mekh. Zh. i Gaza, Vol. 5, pp. 55-61, 1974.

Bogatko, V. I. & Kolton, G. A., "Regular Reflection of a Strong Shock Wave by the Surface of a Wedge", Izv. Akad. Nauk SSSR, Mekh. Zh. i Gaza, Vol. 2, pp. 92-96, 1983.

Bryson, A. E. & Gross, R. F. W., "Diffraction of Strong Shocks by Cones, Cylinders and Spheres", J. Fluid Mech., Vol. 10, pp. 1-16, 1961.

Buggisch, H., "The Steady Two-Dimensional Reflection of an Oblique Partly Dispersed Shock Wave from a Plane Wall", J. Fluid Mech., Vol. 61, pp. 159-172, 1973.

Chang, K. S. & Kim, S. W., "Reflection of Shock Wave from a Compression Corner in the Particulate Laden Gas Region", Shock Waves, Vol. 1, pp. 65-74, 1991.

Chang, T.S. & Laporte, O., "Reflection of Strong Blast Waves", Phys. Fluids, Vol. 7, pp. 1225-1232, 1964.

Chester, W., "The Diffraction and Reflection of Shock Waves", Quart. J. Mech. & Appl. Math., Vol. 7, pp. 57-82, 1954.

Chopra, M. G., "Diffraction and Reflection of Shock from Corners", AIAA J., Vol. 11, pp. 1452-1453, 1973.

Clarke, J. F., "Regular Reflection of a Weak Shock Wave from a Rigid Porous Wall", Quar. J. Mech. & Appl. Math., Vol. 37, pp. 87-111, 1984.

Clarke, J. F., "Reflection of a Weak Shock Wave From a Perforated Plug", J. Eng. Math., Vol. 18, pp. 335-350, 1984.

Clarke, J. F., "The Reflection of Weak Shock Waves from Absorbent Surfaces", Proc. Roy. Soc. Lond., Ser. A, Vol. 396, pp. 365-382, 1984.

Colella, P. & Henderson, L.F., "The von-Neumann Paradox for the Diffraction of Weak Shock Waves", J. Fluid Mech., Vol. 213, pp. 71-94, 1990.

Courant, R. & Friedrichs, K. O., *Supersonic Flow and Shock Waves* , Wiley Interscience, New York, U.S.A., 1948.

Deschambault, R. L. & Glass I. I., "An Update on Non-Stationary Oblique Shock Wave Reflections. Actual Isopycnics and Numerical Experiments", J. Fluid Mech., Vol. 131, pp. 27-57, 1983.

Davis, J. L., "On the Nonexistence of Four Confluent Shock Waves", J. Aero. Sci., Vol. 20, pp. 501-502, 1953.

Dewey, J. M., "The Mach Reflection of Spherical Blast Waves", Nucl. Eng. & Des., Vol. 55, pp. 363-373, 1979.

Dewey, J. M. & McMillin, D. J., "An Analysis of the Particle Trajectories in Spherical Blast Waves Reflected from Real and Ideal Surfaces", Canadian J. Phys., Vol. 59, pp. 1380-1390, 1981.

Dewey, J. M. & McMillin D. J., "Observation and Analysis of the Mach Reflection of Weak Uniform Plane Shock Waves. Part 1. Observations", J. Fluid Mech., Vol. 152, pp. 49-66, 1985.

Dewey, J. M. & McMillin D. J., "Observation and Analysis of the Mach Reflection of Weak Uniform Plane Shock Waves. Part 2. Analysis", J. Fluid Mech., Vol. 152, pp. 67-81, 1985.

Dewey, J. M., McMillin, D. J. & Classen, D. F., "Photogrammetry of Spherical Shocks Reflected from Real and Ideal Surfaces", J. Fluid Mech., Vol. 81, pp. 701-717, 1977.

Dewey, J. M. & Walker, D. K., "A Multiply Pulsed Double-Pass Laser Schlieren System for Recording the Movement of Shocks and Particle Tracers within a Shock Tube", J. Appl. Phys., Vol. 46, pp. 3454-3458, 1975.

Dunayev, Yu. A., Syschikova, M. P., Berezkina, M. K. & Semenov, A. N., "The Interaction of a Shock Wave with a Body in the Presence of Ionization Relaxations", Acta Astro., Vol. 14, 491-495, 1969.

Dunne, B. B., "Mach Reflection of 700-kbar Shock Waves in Gases", Physics Fluids, Vol. 4, pp. 1565-1566, 1961.

Dunne, B. B., "Mach Reflection of Detonation Waves in Condensed High Explosives", Phys. Fluids, Vol. 4, pp. 918-924, 1961.

Dunne, B. B., "Mach Reflection of Detonation Waves in Condensed High Explosives. II", Phys. Fluids, Vol. 7, pp. 1707-1712, 1964.

Duong, D. Q. & Milton, B. E., "The Mach Reflection of Shock Waves in Converging Cylindrical Channels", Exp. in Fluids, Vol. 3, pp. 161-168, 1985.

Fletcher, C. H., Taub, A. H. & Bleakney, W., "The Mach Reflection of Shock Waves at Nearly Glancing Incidence", Rev. Mod. Phys., Vol. 23, pp. 271-286, 1951.

Fletcher, C. H., Weimer, D. K. & Bleakney, W., "Pressure Behind a Shock Wave Diffracted Through a Small Angle", Phys. Rev., Vol. 78, pp. 634-635, 1950.

Ginzberg, I. P. & Markov, Yu. S., "Experimental Investigation of the Reflection of a Shock Wave from a Two-Facet Wedge", Fluid Mech.-Sov. Res., Vol. 4, pp. 167-172, 1975.

Glass, I. I., "Present Status of Oblique-Shock-Wave Reflections", Physicochemical Hydrodynamics, Vol. 6, pp. 863-864, 1985.

Glass, I. I., "Some Aspects of Shock-Wave Reseaech", AIAA J., Vol. 25, pp. 214-229, 1987.

Glass, I. I., "Over Forty Years of Continuous Research at UTIAS on Nonstationary Flows and Shock Waves", Shock Waves, Vol. 1, pp. 75-86, 1991.

Glass, I. I. & Heuckroth, L. G., "Head-On Collision of Spherical Shock Waves", Phys. Fluids, Vol. 2, pp. 542-546, 1959.

Glaz, H. M., Colella, P., Collins, J. P. & Ferguson, R. E., "Nonequilibrium Effects in Oblique Shock Wave Reflection", AIAA J., Vol. 26, pp. 698-705, 1988.

Glaz, H. M., Colella, P., Glass, I. I. & Deschambault, R. L., "A Numerical Study of Oblique Shock-Wave Reflections with Experimental Comparisoms", Proc. Roy. Soc. Lond., Ser. A, Vol. 398, pp. 117-140, 1985.

Griffith, W. C., "Shock Waves", J. Fluid Mech., Vol. 106, pp. 81-101, 1981.

Griffith, W. C. & Bleakney, W., "Shock Waves in Gases", Amer. J. Phys., Vol. 22, pp. 597-612, 1954.

Griffith, W. C. & Brickle, D. E., "The Diffraction of Strong Shock Waves", Phys. Rev., Vol. 89, pp. 451-453, 1953.

Grudnitskii, V. G. & Prokhorchuk, Yu. A., "Analysis of the Diffraction of a Shock Wave at a Curvilinear Surface", J. Num. Anal. & Math. Phys., Vol. 15, pp. 1525-1534, 1975.

Gvozdeva, L. G., Bazhenova, T. V., Lagutov, Yu. P. & Fokeev, V. P., "Shock Wave Interaction with Cylindrical Surfaces", Arch. Mech., Vol. 32, pp. 693-702, 1980.

Gvozdeva, L. G., Bazhenova, T. V., Predvoditeleva, O. A. & Fokeev, V. P., "Mach Reflection of Shock Waves in Real Gases", Astro. Acta, Vol. 14, pp. 503-508, 1969.

Gvozdeva, L. G. Bazhenova, T. V., Predvoditeleva, O. A. & Fokeev, V. P., "Pressure and Temperature at the Wedge Surface for Mach Reflection of Strong Shock Waves", Astro. Acta, Vol. 15, pp. 503-510, 1970.

Gvozdeva, L. G. & Fokeev, V. P., "Experimental Study of Irregular Shock Wave Reflection from the Surface of a Wedge", Fiz. Gorenia i Vzryva, Vol. 12, pp. 260-269, 1976.

Gvozdeva, L. G. & Fokeev, V. P., "Transition from Mach to Regular Reflection and Domains of Various Mach Reflection Configurations", Comb. Exp. & Shock Waves, Vol. 13, pp. 86-93, 1977.

Gvozdeva, L. G., Lagutov, Yu. P. & Fokeev, V. P., "Transition from Regular Reflection to Mach Reflection in the Interaction of Shock Waves with a Cylindrical Surface", Sov. Tech. Phys. Lett., Vol. 5, pp. 334-336, 1979.

Gvozdeva, L. G., Lagutov, Yu. P. & Fokeev, V. P., "Transition from Mach Reflection to Regular Reflection when Strong Shock Waves Interact with Cylindrical Surfaces", Izv. Akad. Nauk SSSR, Mekh. Zh. i Gaza, Vol. 2, pp. 132-138, 1982.

Gvozdeva, L. G. & Predvoditeleva, O. A., "Experimental Investigation of Mach Reflection of Shock Waves with Velocities of 1000-3000 m/sec in Carbon Dioxide Gas, Nitrogen and Air", Sov. Phys. Dokl., Vol. 10, pp. 694-697, 1965.

Gvozdeva, L. G., Predvoditeleva, O. A. & Fokeev, V. P., "Double Mach Reflection of Strong Shock Waves", Izv. Akad. Nauk SSSR, Mekh. Zh. i Gaza, Vol. 14, pp. 112-119, 1968.

Gvozdeva, L. G., Predvoditeleva, O. A., "A Study of Triple Configuration of Detonation Waves in Gases", Comb. Exp. & Shock Waves, Vol. 4, pp. 451-461, 1969.

Handke, E. & Obermeier, F., "Theoretical Results from the Mach Reflection of Weak Shock Waves", Z. Angew. Math. & Mech., Vol. 65, pp. 202-204, 1985.

Heilig, W. H., "Diffraction of Shock Wave by a Cylinder", Phys. Fluids Supll. I, Vol. 12, pp. 154-157, 1969.

Henderson, L. F., "On the Confluence of Three Shock Waves in a Perfect Gas", Aero. Quart., Vol. 15, pp. 181-197, 1964.

Henderson, L. F., The Three Shock Confluence on a Simple Wedge Intake", Aero. Quart., Vol. 16, pp. 42-54, 1965.

Henderson, L. F., "The Reflection of a Shock Wave at a Rigid Wall in the Presence of a Boundary Layer", J. Fluid Mech., Vol. 30, pp. 699-723, 1967.

Henderson, L. F., "On the Whitham Theory of Shock Wave Diffraction at Concave Corners", J. Fluid Mech., Vol. 99, pp. 801-811, 1980.

Henderson, L. F., "Exact Expressions for Shock Reflection Transition Criteria in a Perfect Gas", Z. Ang. Math. & Mech., Vol. 62, pp. 258-261, 1982.

Henderson, L. F., "Regions and Boundaries for Diffracting Shock Wave Systems", Z. Ang. Math. & Mech., Vol. 67, pp. 1-14, 1987.

Henderson, L. F. & Gray, P. M., "Experiments on the Diffraction of Strong Blast Waves", Proc. Roy. Soc. Lond., Ser. A, Vol. 377, pp. 363-378, 1981.

Henderson, L. F. & Lozzi, A., "Experiments on Transition of Mach Reflection", J. Fluid Mech., Vol. 68, pp. 139-155, 1975.

Henderson, L. F. & Lozzi, A., "Further Experiments on Transition to Mach Reflection", J. Fluid Mech., Vol. 94, pp. 541-559, 1979.

Henderson, L.F., Ma, J-H, Sakurai, A. & Takayama, K., "Refraction of a Shock Wave at an Air-Water Interface", Fluid Dyn. Res., Vol. 5, pp. 337-350, 1990.

Henderson, L. F. & Siegenthaler, A., "Experiments on the Diffraction of Weak Blast Waves: The von Neumann Paradox", Proc. Roy. Soc. Lond., Ser, A, Vol. 369, pp. 537-555, 1980.

Higashino, F., Henderson, L. F. & Shimizu, F., "Experiments on the Interaction of a Pair of Cylindrical Weak Blast Waves in Air", Shock Waves, Vol. unavailable, pp. unavailable, 1991.

Honma, H. & Henderson, L. F., "Irregular Reflections of Weak Shock Waves in Polyatomic Gases", Phy. Fluids A, Vol. 1, pp. 597-599, 1989.

Honma, H., Maekawa, H. & Usui, T., "Numerical Analysis of Non Stationary Oblique Reflection of Weak Shock Waves", Computer & Fluids, Vol. unavailable, pp. unavailable, 1991.

Hornung, H. G., "The Effect of Viscosity on the Mach Stem Length in Unsteady Shock Reflection", Flow of Real Liquids, Vol. 235, pp. 82-91, 1985.

Hornung, H. G., "Regular and Mach Reflection of Shock Waves", Ann. Rev. Fluid Mech., Vol. 18, pp. 33-58, 1986.

Hornung, H. G., Oertel, H. Jr. & Sandeman, R. J., "Transition to Mach Reflection of Shock Waves in Steady and Pseudosteady Flow with and without Relaxation", J. Fluid Mech., Vol. 90, pp. 541-560, 1979.

Hornung, H. G. & Robinson, M. L., "Transition from Regular to Mach Reflction of Shock Waves. Part 2. The Steady Flow Criterion", J. Fluid Mech., Vol. 123, pp. 155-164, 1982.

Hornung, H. G. & Taylor, J. R., "Transition from Regular to Mach Reflection of Shock Waves. Part 1. The Effect of Viscosity in the Pseudosteady Case", J. Fluid Mech., Vol. 123, pp. 143-153, 1982.

Hu, T. C. J. & Glass, I. I., "Blast Wave Trajectories from Height of Burst", AIAA J., Vol. 24, pp. 607-610, 1986.

Hu, T. C. J. & Glass, I. I., "Pseudostationary Oblique Shock Wave Reflections in Sulphur Hexafluoride (SF_6): Interferometric and Numerical Results", Proc. Roy. Soc. London, Ser. A, Vol. 408, pp. 321-344, 1986.

Hunt, B. L., "Calculation of Two-Dimensional and Three-Dimensional Regular Shock Interactions", Aero. J., Paper No. 828, pp. 285-289, 1980.

Hunt, B. L. & Lamont, P. J., "The Confluence of Three Shock Waves in a Three-Dimensional Flow", Aero. Quart., Vol. 29, pp. 18-27, 1978.

Ikui, T., Matsuo, K., Aoki, T. & Kondoh, N., "Investigations of Mach Reflection of a Shock Wave. Part 1. Configurations and Domains of Shock Reflection", Bull. Japan Soc. Mech. Eng., Vol. 25, pp. 1513-1520, 1982.

Itoh, S., Okazaki, N. & Itaya, M., "On the Transition Between Regular and Mach Reflection in Truly Non-Stationary Flows", J. Fluid Mech., Vol. 108, pp. 383-400, 1981.

Itoh, K., Takayama, K. & Ben-Dor, G., "Numerical Simulation of the Reflection of a Planar Shock Wave over a Double Wedge", Int. J. Numerical Methods Fluids, Vol. 13, pp. unavailable, 1991.

Jahn, R. G., "Transition Process in Shock Wave Interactions", J. Fluid Mech., Vol. 2, pp. 33-48, 1956.

Jones, D. M., Martin, P. M. E. & Thornhill, C. K., "A Note on the Pseudo-Stationary Flow Behind a Strong Shock Diffracted or Reflected at a Corner", Proc. Roy. Soc. Lond., Ser. A, Vol. 209, pp. 238-248, 1951.

Kaliski, S. & Wlodarczyk, E., "The Influence of the Parameters of State on the Characteristics of Regular Reflection of Oblique Shock Waves", Proc. Vibr. Problems, Vol. 14, pp. 4-5, 1973.

Kaliski, S. & Wlodarczyk, E., "Regular Reflection of Intense Oblique Shock Waves from a Rigid Wall in a Solid Body", Proc. Vib. Problems, Vol. 15, pp. 271-282, 1974.

Kawamura, R. & Saito, H., "Reflection of Shock Waves-1. Pseudo-Stationary Case", J. Phys. Soc. Japan, Vol. 11, pp. 584-592, 1956.

Kireev, V. T., "On the Reflection of a Strong Shock Wave by a Sphere or a Cylinder", Izv. Akad. Nauk SSSR, Mekh. Zh. i Gaza, Vol. 3, pp. 31-40, 1969.

Klein, E. J., "Interaction of a Shock Wave and a Wedge: An Application of the Hydraulic Analogy", AIAA J., Vol. 3, pp. 801-808, 1965.

Kobayashi, S., Adachi, T. & Suzuki, T., "An Investigation on the Transition Criterion for Reflection of a Shock over a Dusty Surface", Dyn. Image Anal., Vol. 1, pp. 67-73, 1988.

Kolgan, V. P. & Fonarev, A. C., "Development of the Flow in Shock Incidence on a Cylinder or Sphere", Izv. Akad. Nauk SSSR, Mekh. Zh. i Gaza, Vol. 5, pp. 97-103, 1972.

Krehl, P. & van der Geest, M., "The Discovery of the Mach Reflection Effect and Its Demonstration in an Auditorium", Shock Wave, Vol. 1 , pp. 3-15, 1991.

Krishnan, S., Brochet, C. & Cheret, R., "Mach Reflection in Condensed Explosives", Propellants & Explosives, Vol. 6, pp. 170-172, 1981.

Kutler, P. & Sakell, L., "Three-Dimensional Shock on Shock Interaction Problem", AIAA J., Vol. 13, pp. 1360-1367, 1975.

Kutler, P., Sakell, L. & Aiello, G., "Two-Dimensional Shock on Shock Interaction Problem", AIAA J., Vol. 13, pp. 361-367, 1975.

Kutler, P. & Shankar, V., "Diffraction of a Shock Wave by a Compression Corner: I. Regular Reflection", AIAA J., Vol. 15, pp. 197-203, 1977.

Law, C. K. & Glass, I. I., "Diffraction of Strong Shock Waves by a Sharp Compressive Corner", CASI Trans., Vol. 4, pp. 2-12, 1971.

Law, C. K. & Glass, I. I., "Coments on Regions of Various Forms of Mach Reflection and Its Transition to Regular Reflection", Acta Astro., Vol. 4, pp. 939-941, 1977.

Le Kuok Khyu, "Geometrical Investigation of the Regular Reflection of Shock Waves from a Wall", Izv. Akad. Nauk SSSR, Mekh. Zh. i Gaza, Vol. 5, pp. 182-185, 1980.

Lee, J. -H. & Glass, I. I., "Pseudo-Stationary Oblique Shock Wave Reflections in Frozen and Equilibrium Air", Prog. Aero. Sci., Vol. 21, pp. 33-80, 1984.

Li, Y., Zhang, D. & Cao, Y., "The Computational Numerical Experimental Research of Mach Reflection", Acta Mech. Sin., Vol. 17, pp. 162-167, 1985.

Lighthill, M. J., "The Diffraction of Blast. I", Proc. Roy. Soc. Lond., Ser. A, Vol. 198, pp. 454-470, 1949.

Lighthill, M. J., "The Diffraction of Blast. II", Proc. Roy. Soc. Lond., Ser. A, Vol. 200, pp. 554-565, 1950.

Lipnitskii, Yu. M. & Lyakhov, V. N., "Numerical Solution of the Problem of Shock Wave Diffraction at a Wedge", Izv. Akad. Nauk SSSR, Mekh. Zh. i Gaza, Vol. 6, pp. 88-93, 1974.

Lipnitskii, Yu. M. & Panasenko, A. V., "Investigation into the Interaction of a Shock Wave with an Acute Case", Izv. Akad. Nauk SSSR, Mech. Zh. i Gaza, Vol. 3, pp. 98-104, 1980.

Lock, G. D. & Dewey, J. M., "An Experimental Investigation of the Sonic Criterion for Transition from Regular to Mach Reflection of Weak Shock Waves", Exp. in Fluids, Vol. 7, pp. 289-292, 1989.

Logvenov, A. Yu., Misonochnikov, A. L. & Rumyanstev, B. V., "Mach Reflection of Shock Waves in Condensed Media", Sov. Tech. Phys. Lett., Vol. 13, pp. 131-132, 1987.

Ludloff, H. F. & Friedman, M. B., "Mach Reflection of Shocks at an Arbitrary Incidence", J. Appl. Phys., Vol. 24, pp. 1247-1248, 1953.

Ludloff, H. F. & Friedman, M. B., "Aerodynamics of Blasts-Diffraction of Blast Around Finite Corners", J. Aero. Sci., Vol. 22, pp. 27-34, 1955.

Ludloff, H. F. & Friedman, M. B., "Difference Solution of Shock Diffraction Problem", J. Aero. Sci., Vol. 22, pp. 139-140, 1955.

Lyakhov, V. N., "Unsteady Loads in Shock Wave Diffraction", Izv. Akad. Nauk SSSR, Mekh. Zh. i Gaza, Vol. 4, 123-129, 1975.

Lyakhov, V. N., "Mathematical Simulation of Mach Reflection of Shock Waves in Media with Different Adiabatic Indicies", Izv. Akad. Nauk SSSR, Mekh. Zh. i Gaza, Vol. 3, 90-94, 1976.

Lyakhov, V. N., "Concerning the Evaluation of Pressure in Unsteady Reflection of Shock Waves", Izv. Akad. Nauk SSSR, Mekh. Zh. i Gaza, Vol. 2, 100-106, 1977.

Lyakhov, V. N., "Interaction of Shock Waves of Moderate Intensity with Cylinders", Izv. Akad. Nauk SSSR, Mekh. Zh. i Gaza, Vol. 2, 113-119, 1979.

Lyakhov, V. N. & Ryzhov, A. S., "On the Law of Similarity in Nonlinear Reflection of Shock Waves by a Rigid Wall", Izv. Akad. Nauk SSSR, Mekh. Zh. i Gaza, Vol. 3, 116-123, 1977.

Mach, E., "Über den Verlauf von Funkenwellen in der Ebene und im Raume", Sitzungsber. Akad. Wiss. Wien, Vol. 78, pp. 819-838, 1878.

Makarevich, G. A., Lisenkova, G. S., Tikhomirov, N. A. & Khodstev, A. V., "Experimental Study of Mach Reflection of Weak Shock Waves", Sov. Phys.-Tech. Phys., Vol. 29, pp. 370-372, 1984.

Makarevich, G. A., Predvoditeleva, O. A. & Lisenkova, G. S., "Reflection of Shock Waves from a Wedge", Tepl. Vys. Temp., Vol. 12, pp. 1318-1321, 1974.

Makomaski, A. H., "Some Effects of Surface Roughness on Two Dimensional Mach Reflection of Moving Plane Shock Waves in Air", Can. Aeron. Space J., Vol. 12, pp. 109-111, 1966.

Marconi, F., "Shock Reflection Transition in Three-Dimensional Steady Flow About Interfering Bodies", AIAA J., Vol. 21, pp. 707-713, 1983.

Matsuo, K., Aoki, T. & Hirahara, H., "Visual Studies of Characteristics of Slipstream in Mach Reflection of a Shock Wave", Flow Visualization, Vol. 4, pp. 543-548, 1987.

Matsuo, K., Aoki, T., Hirahara, H., Kondoh, N. & Kuriwaki, K., "Investigations of Mach Reflection of a Shock Wave. Part 3: Strength of a Reflected Shock Wave", Bull. Japan Soc. Mech. Eng., Vol. 29, pp. 428-433, 1986.

Matsuo, K., Aoki, T., Hirahara, H., Kondoh, N. & Kishigami, T., "Investigations of Mach Reflection of a Shock Wave. Part 4: The Transition Between Regular and Mach Reflection", Bull. Japan Soc. Mech. Eng., Vol. 29, pp. 434-438, 1986.

Matsuo, K., Aoki, T., Hirahara, H. & Kondoh, N., "Nonlinear Problem on Interaction Between a Shock Wave and Rigid Wall", in Theor. & Appl. Mech., Vol. 33, pp. 51-57, 1985.

Matsuo, K., Aoki, T., Kondoh, N. & Hirahara, H., "Investigations of Mach Reflection of a Shock Wave. Part 2: Shape of a Reflected Shock Wave", Bull. Japan Soc. Mech. Eng., Vol. 29, pp. 422-427, 1986.

Matsuo, K., Ikui, T. Aoki, T.,. & Kondoh, N., "Interaction of a Propagating Shock Wave with an Inclined Wall", Theor. & Appl. Mech., Vol. 31, pp. 429-437, 1983.

Merritt, D. L., "Mach Reflection on a Cone", AIAA J., Vol. 6, pp. 1208-1209, 1968.

Milton, B. E. & Archer, R. D., "Generation of Implosions by Aera Change in a Shock Tube", AIAA J., Vol. 7, pp. 779-780, 1969.

Milton, B. E., "Mach Reflection Using Ray-Shock Theory", AIAA J., Vol. 13, pp. 1531-1533, 1975.

Mirels, H., "Mach Reflection Flowfields Associated with Strong Shocks", AIAA J., Vol. 23, pp. 522-529, 1985.

Mogilevich, L. I. & Shindyapin, G. P., "On Nonlinear Diffraction of Weak Shock Waves", Prikl. Mat. & Mekh., Vol. 35, pp. 492-498, 1971.

Mölder, S., "Reflection of Curved Shock Waves in Steady Supersonic Flow", CASI Trans., Vol. 4, pp. 73-80, 1971.

Mölder, S., "Polar Streamline Directions at the Triple Point of a Mach Interaction of Shock Waves", CASI Trans., Vol. 5, pp. 88-89, 1972.

Mölder, S., "Particular Conditions for the Termination of Regular Reflection of Shock Waves", CASI Trans., Vol. 25, pp. 44-49, 1979.

Morgan, K., "The Diffraction of a Shock Wave by a Slender Body", J. Appl. Math. & Phys., Vol. 26, pp. 13-29, 1975.

Morgan, K., "The Diffraction of a Shock Wave by an Inclined Body of Revolution", J. Appl. Math. & Phys., Vol. 26, pp. 299-306, 1975.

Moran, T. P. & Moorhem, W. K., " Diffraction of a Plane Shock by an Analytic Blunt Body", J. Fluid Mech., Vol. 38, pp. 127-136, 1969.

Morro, A., "Oblique Interaction of Waves with Shocks", Acta Mech., Vol. 38, pp. 241-248, 1981.

Murdoch, J. W., "Shock Wave Interaction with Two Dimensional Bodies", AIAA J., Vol. 13, pp. 15-17, 1975.

Neumann, J. von, *Collected Works*, Vol. 6, pp. 238-308, Pergamon press, 1963.

Obermeier, F. & Handke, E., "Theoretische Ergebnisse zur Mach Reflexion Schwacher Stoβwellen", Z. Ang. Math. & Mech., Vol. 65, T202-T204, 1985.

Oertel, H. Jr., "Oxygen Vibrational and Dissociation Relaxation Behind Regular Reflected Shocks", J. Fluid Mech. Vol. 74, pp. 477-495, 1976.

Oertel, H. Jr., "Oxygen Dissociation Relaxation Behind Oblique Reflected and Stationary Oblique Shocks", Arch. Mech., Vol. 30, pp. 123-133, 1978.

Pack, D. C., "The Reflection and Diffraction of Shock Waves", J. Fluid Mech., Vol. 18, pp. 549-576, 1964.

Pant, J. C., "Regular Reflection of a Shock Wave in the Presence of a Transverse Magnetic Field", Z. Ang. Math. & Mech., Vol. 47, pp. 73-85, 1968.

Pant, J. C., "Reflection of a Curved Shock from a Straight Rigid Boundary", Phys. Fluids, Vol. 14, pp. 534-538, 1971.

Pekurovskii, L. E., "Shock Wave Diffraction on a Thin Wedge Moving with a Slip Relative to the Wave Front with Irregular Shock Interaction", Prikl. Mat. & Mekh., Vol. 44, pp. 183-190, 1981.

Piechór, K., "Reflection of a Weak Shock Wave from an Isothermal Wall", Arch. Mech., Vol. 32, pp. 233-249, 1980.

Piechór, K., "Regular Interaction of Two Weak Shock Waves", Arch. Mech., Vol. 33, pp. 829-843, 1981.

Piechór, K., "Regular Reflection of a Weak Shock Wave from an Inclined Plane Isothermal Wall", Arch. Mech., Vol. 33, pp. 337-346, 1981.

Podlubnyi, V. V. & Fonarev, A. S., "Reflection of a Spherical Blast Wave from Plane Surface", Izv. Akad. Nauk SSSR, Mekh. Zh. i Gaza, Vol. 6, pp. 166-172, 1974.

Poluboyarinov, A. K. & Tsirkunov, V. E., "Concerning the Reflection of a Shock Wave by a Cylinder or a Sphere", Fluid Mech.-Sov. Res., Vol. 10, pp. (unavailble), 1981.

Ram, R. & Sharma, V. D., "Regular Reflection of a Shock Wave from a Rigid Wall in a Steady Plane Flow of an Ideal Dissociating Gas", Tensor, Vol. 26, pp. 185-190, 1972.

Rawling, G. & Polachek, H., "On the Three Shock Configuration", Phys. Fluids, Vol. 1, pp. 572-577, 1950.

Reichenbach, H., "Contribution of Ernst Mach to Fluid Mechanics", Ann. Rev. Fluid Mech., Vol. 15, pp. 1-28, 1983.

Rusanov, V. V., "Calculation of the Interaction of Non-Stationary Shock Waves and Obstructions", J. Num. Anal. & Math. Phys., Vol. 1, pp. 267-279, 1961.

Ryzhov, O. S. & Khristianovich, S. A., "Nonlinear Reflection of Weak Shock Waves", Prikl. Mat. & Mekh., Vol. 22, pp. 586-599 1958.

Sakurai, A., "On The Problem of Weak Mach Reflection", J. Phys. Soc. Japan, Vol. 19, pp. 1440-1450, 1964.

Sakurai, A., Henderson, L. F., Takayama, K., Walenta, Z. & Colella, P., "On the von Neumann Paradox of Weak Mach Reflection", Fluid Dyn. Res., Vol. 4, pp. 333-345, 1989.

Schmidt, B., "Shock-Wave Transition from Regular to Mach Reflection on a Wedge-Shaped Edge", Z. Ang. Math. & Mech., Vol. 65, pp. 234-236, 1985.

Schneyer, G. P., "Numerical Simulation of Regular and Mach Reflections", Phys. Fluids, Vol. 18, pp. 1119-1124, 1975.

Semenov, A. N. & Syshchikova, M. P., "Properties of Mach Reflection in the Interaction of Shock Waves with a Stationary Wedge", Comb. Expl. & Shock Waves, Vol. 11, pp. 506-515, 1975.

Semenov, A. N., Syshchikova, M. P. & Berezkina, M. K., "Experimental Investigation of Mach Reflection in a Shock Tube", Sov. Phys.- Tech. Phys., Vol. 15, pp. 795-803, 1970.

Shankar, V., Kutler, P. & Anderson D. A., "Diffraction of a Shock Wave by a Compression Corner: Part II: Single Mach Reflection", AIAA J., Vol. 16, pp. 4-5, 1978.

Sharma, V. D. & Shyman, R., "Regular Reflection of a Shock Wave from a Rigid Wall in a Steady Plane Flow of a Vibrationally Relaxing Gas", Indian J. Pure & Appl. Math., Vol. 15, pp. 15-22, 1984.

Shindyapin, G. P., "Irregular Reflection of Weak Shock Waves from a Rigid Wall", Zh. Prikl. Mekh. i. Tekh. Fiz., Vol. 2, pp. 22-28, 1964.

Shindyapin, G. P., "Regular Reflection of Weak Shock Waves from a Wall", Prikl. Mat. Mekh., Vol. 29, pp. 114-121, 1965.

Shindyapin, G. P., "Irregular Interaction of Weak Shock Waves of Different Intensity", Prikl. Mat. Mech., Vol. 38, pp. 105-114, 1974.

Shindyapin, G. P., "Numerical Solution of the Problem of Non-Regular Reflection of a Weak Shock Wave by a Rigid Wall in a Perfect Gas", Zh. Vyehisl. Mat. Mat. Fiz., Vol. 20, pp. 249-254, 1980.

Shirouzu, M. & Glass, I. I., "Evaluation of Assumptions and Criteria in Pseudostationary Oblique Shock Wave Reflections", Proc. Roy. Soc. Lond., Ser. A, Vol. 406, pp. 75-92, 1986.

Skews, B. W., "The Shape of a Diffracting Shock Wave", J. Fluid Mech., Vol. 29, pp. 297-304, 1967.

Skews, B. W., "Shock-Shock Reflection", CASI Trans., Vol. 4, pp. 16-19, 1971.

Skews, B. W., "The Flow in the Vicinity of the Three-Shock Intersection", CASI Trans., Vol. 4, pp. 99-107, 1972.

Skews, B. W., "The Shape of a Shock in Regular Reflection from a Wedge", CASI Trans., Vol. 5, pp. 28-32, 1972.

Skews, B. W., "Shock Wave Shaping", AIAA J., Vol. 10, pp. 839-841, 1972.

Smith, W. R., "Mutual Reflection of Two Shock Waves of Arbitrary Strengths", Phys. Fluids, Vol. 2, pp. 533-541, 1959.

Sokolov, V. B., "Reflection of a Strong Shock Wave by an Ellipsoid of Revolution and Elliptical Cylinder", Izv. Akad. Nauk SSSR Mekh. Zh. i Gaza, Vol. 4, pp. 173-174, 1974.

Srivastava, R. & Chopra, M. G., "Diffraction of Blast Waves for the Oblique Case", J. Fluid Mech., Vol. 40, pp. 821-831, 1970.

Srivastava, R. S. & Deschambault, R. L., "Pressure Distribution Behind a Nonstationary Reflected-Diffracted Oblique Shock Wave", AIAA J., Vol. 22, pp. 305-306, 1984.

Sternberg, J., "Triple-Shock-Wave Interactions", Phys. Fluids, Vol. 2, pp. 179-206, 1959.

Suhindyapin, G. P., "On Irregular Reflection of Weak Shock Waves by a Rigid Wall", Zh. Prikl. Mat. Tekh. Fiz., Vol. 2, pp. 22-28, 1964.

Suhindyapin, G. P., "On the Regular Reflection of Weak Shock Waves by a Rigid Wall", Prikl. Mat. & Mekh., Vol. 29, pp. 114-121, 1965.

Suhindyapin, G. P., "Irregular Interaction of Weak Shock Waves of Different Intensity", Prikl. Mat. & Mekh., Vol. 38, pp. 105-114, 1974.

Suhindyapin, G. P., "Numerical Solution of the Problem of Non Regular Reflection of a Weak Shock Wave by a Rigid Wall in a Perfect Gas", J. Num. Anal. & Math. Phys., Vol. 20, pp. 249-254, 1980.

Suzuki, T. & Adachi, T., "The Reflection of a Shock Wave over a Wedge with Dusty Surface", Trans. Japan Soc. Aero. & Astro., Vol. 28, pp. 132-139, 1985.

Suzuki, T. & Adachi, T., "Comparison of Shock Reflection from a Dust Layer with Those from a Smooth Surface", Theo. & Appl. Mech., Vol. 35, pp. 345-352, 1987.

Suzuki, T. & Adachi, T., "An Experimental Study of Oblique Shock Wave Reflection over a Dusty Surface", Trans. Japan Soc. Aero. & Astro., Vol. 31, pp. 104-110, 1988.

Suzuki, T., Adachi, T. & Kobayashi, S., "Image Analysis of Oblique Shock Reflection over a Model Wedge", J. Visualization Soc. Japan, Vol. 10, Suppl. 2, pp. 7-10, 1990.

Suzuki, T., Adachi, T. & Tanabe, K., "Structure of the Mach-Type Reflection on a Dust Layer", Theor. & Appl. Mech., Vol. 34, pp. 73-80, 1986.

Syshchikova, M. P. & Krassovskaya, I. V., "Some Properties of Regular and Irregular Interaction of Shock Waves", Arch. Mech., Vol. 31, pp. 135-145, 1979.

Syshchikova, M. P., Semenov, A. N. & Berezkina, M. K., "Shock Wave Reflection by a Curved Concave Surface", Sov. Phys. Tech. Phys., Vol. 2, pp. 61-66, 1976.

Sysoev, N. N., "Unsteady Reflection of Shock Waves by a Sphere or Cylinder", Ser. Phys. & Astron., Vol. 20, pp. 90-91, 1979.

Sysoev, N. N. & Shugaev, F. V., "Transient Reflection of a Shock Wave from a Sphere and a Cylinder", Vestnik Moskovskogo Universiteta. Fiz., Vol. 34, pp. 90-91, 1979.

Takayama, K. & Ben-Dor, G., "A Reconsideration of the Hysteresis Phenomenon in the Regular⇔Mach Reflection Transition in Truly Nonstationary Flows," Israel J. Tech., Vol. 21, pp. 197-204, 1983.

Takayama, K. & Ben-Dor, G., "The Inverse-Mach Reflection", AIAA J., Vol. 23, pp. 1853-1859, 1985.

Takayama, K. & Ben-Dor, G., "Application of Streak Photgraphy for the Study of Shock Wave Reflections over a Double Wedge", Exp. in Fluids, Vol. 6, pp. 11-15, 1987.

Takayama, K. & Ben-Dor, G., "A Reconsideration of the Transition Criterion from Mach to Regular Reflection over Cylindrical Concave Surfaces", Korean Soc. Mech. Eng. J., Vol. 3, pp. 6-9, 1989.

Takayama, K. & Ben-Dor, G., "Pseudo-Steady Oblique Shock Wave Reflections over Water Wedges", Exp. in Fluids, Vol. 8, pp. 129-136, 1989.

Takayama, K., Ben-Dor, G. & Gotoh, J., "Regular to Mach Reflection Transition in Truly Nonstationary Flows - Influence of Surface Roughness", AIAA J., Vol. 19, pp. 1238-1240, 1981.

Takayama, K. & Sekiguchi, H., "Triple-Point Trajectory of a Strong Spherical Shock Wave", AIAA J., Vol. 19, pp. 815-817, 1981.

Tan, H. S., "Strengh of Reflected Shock in Mach Reflection", J. Aero. Sci., Vol. 18, pp. 768-770, 1951.

Taub, A. H., "Refraction of Plane Shock Waves", Phys. Rev., Vol. 72, pp. 51-60, 1947.

Ter-Minassiants, S. M., "The Diffraction Accompanying the Regular Reflection of a Plane Obliquely Impinging Shock Wave from the Walls of an Obtuse Wedge", J. Fluid Mech., Vol. 35, pp. 391-410, 1969.

Teshukov, V. M., "On Regular Reflection of a Shock Wave from a Rigid Wall", Prikl. Mat. & Mekh., Vol. 46, pp. 225-234, 1982.

Ting, L., & Ludloff, H. F., "Difference Solution of Shock Diffraction Problem", J. Aero. Sci., Vol. 19, pp. 317-328, 1952.

Tsu-Sien-Shao, "Numerical Solution of Plane Viscous Shock Reflections", J. Comp. Phys., Vol. 1, pp. 367-381, 1977.

Voinovich, P. A., Popov, F. D. & Fursenko, A. A., "Numerical Simulation of Shock Wave Reflection from a Concave Surface", Sov. Tech. Phys. Lett., Vol. 4, pp.127-128, 1978.

Voloshinov, A. V., Kovalev, A. D. & Shindyapin, G. P., "Transition from Regular Reflection to Mach Reflection when a Shock Wave Interacts with a Wall in Two-Phase Gas-Liquid Medium", Izv. Akad. Nauk SSSR, Mekh. Zh. i Gaza, Vol. 5, pp. 190- 192, 1983.

Voloshinov, A. V., Kovalev, A. D. & Shindyapin, G. P., "Transition from Regular Reflection to Mach Reflection When a

Shock Wave Interacts with a Wall in a Two- Phase Gas-Liquid Medium", Fluid Dyn., Vol. 18, pp. 827-829, 1983.

Walenta, Z. A., "Regular Reflection of the Plane Shock Wave from an Inclined Wall", Arch. Mech., Vol. 26, pp. 825-832, 1974.

Walenta, Z. A., "Microscopic Structure of the Mach-Type Reflection of the Shock Wave", Arch. Mech., Vol. 32, pp. 819-825, 1980.

Walenta, Z. A., "Formation of the Mach-Type Reflection of Shock Waves", Arch. Mech., Vol, 35, pp. 187-196, 1983.

Walker, D. K., Dewey, J. M. & Scotten, L. N., "Observation of Density Discontinuities Behind Reflected Shocks Close to the Transition from Regular to Mach Reflection", J. Appl. Phys., Vol. 53, pp. 1398-1400, 1982.

White, D. R., "Reflection of Strong Shock at Nearly Glancing Incidence", J. Aero. Sci., Vol. 18, pp. 633-634, 1951.

Whitham, G. B., "A New Approach to Problems of Shock Dynamics. Part 1. Two Dimensional Problems", J. Fluid Mech. Vol. 2, pp. 145-171, 1957.

Woodward, P. R. & Colella, P., "Numerical Solution of Two-Dimensional Fluid Flow with Strong Shocks", J. Compt. Phys., Vol. 54, pp. 115-173, 1984.

Xu, D. Q. & Honma, H., "Numerical Simulation for Nonstationary Mach Reflection of a Shock Wave: A Kinetic Model Approach", Shock Waves, Vol. 1, pp. 43-50, 1991.

Yang, J.Y., Liu, Y. & Lomax, H., "Computation of Shock Wave Reflection by Circular Cyliners", AIAA J., Vol. 25, pp. 683-689, 1987.

Yu, Sh. & Gronig, H., "A Simple Approximation for Axially Symmetric Diffraction of Plane Shocks by Cones", Z. Naturforsch, Vol. 39a, pp. 320-324, 1984.

Zaslavskii, B. I. & Safarov, R. A., "Mach Reflection of Weak Shock Waves from a Rigid Wall", Zh. Prikl. Mekh. i Tekh. Fiz., Vol. 5, pp. 26-33, 1973.

Zaslavskii, B. I. & Safarov, R. A., "On the Similarity of Flows Occurring in Weak Shock Wave Reflections from a Rigid Wall or Free Surface", Fiz. Goreniya i Vzryva, Vol. 4, pp. 579-585, 1973.

Zhigalko, Y. F., "Concerning the Problem of Shock Wave Diffraction and Reflection", Vestnik Moskovskogo Universiteta. Fiz., Vol. 1, pp. 78-82, 1967.

Zhigalko, Y. F., "A Linear Approximation of Shock Wave Diffraction and Reflection", Vestnik Moskovskogo Universiteta. Fiz., Vol. 3, pp. 94-104, 1969.

Zhigalko, Y. F., "Approximate Locally-Nonlinear Solutions of the Problem on Interaction Between a Shock Wave and a Rigid Wall", Fluid Mech. Sov. Res., Vol. 4, pp. 81-91, 1975.

Zhigalko, Y. F., "Linear Approximation of the Reflection of Shock Waves from a Concave Wall", Fluid Mech. Sov. Res., Vol. 10, pp. 60-71, 1981.

Zhigalko, Y. F., Kolyshkina, L. L. & Shevtsov, V. D., "Vortical Singularity of Self-Similar Gas Flow in the Reflection of a Shock Wave", Zh. Prikl. Mekh. i Tekh. Fiz., Vol. 1, pp. 107-111, 1982.

Znamenskaya, I. A., Ryazin, A. P. & Shugaev, F. V., "On Some Features of Gas Parameter Distribution in the Early Stages of Shock Reflection by a Sphere or Cylinder", Izv. Akad. Nauk SSSR Mekh. Zh. i Gaza, Vol. 3, pp. 103-110, 1979.

Zumwalt, G. W., Weak Shock Reflections at Near 90° Angle of Incidence", J. Appl. Mech., Vol. 41, pp. 1142-1143, 1974.

5.2 Reports

Adachi, T., Kobayashi, S. & Suzuki, T., "Unsteady Behavior of Mach Reflection over a Wedge with Surface Roughness", Ann. Rep. Japan Soc. Heat Fluid Eng., Vol. 4, pp. 53-57, 1989.

Ando, S., "Pseudo-Stationary Oblique Shock Wave Reflection in Carbon Dioxide -Domains and Boundaries", UTIAS Tech. Note 231, Institute for Aerospace Studies, Univ. Toronto, Toronto, Ont., Canada, 1981.

Bargmann, V., "On Nearly Glancing Reflection of Shocks", Sci. Res. Dev. OSRD Rep. 5171, Washington, DC, U.S.A. or Nat. Defence Res. Committee, NDRC, 1945.

Bargmann, V. & Montgomery, D., "Prandtl-Meyer Zones in Mach Reflection", Off. Sci. Res. Dev., OSRD Rep. 5011, Washington, DC, U.S.A., 1945.

Ben-Dor, G., "Regions and Transitions of Nonstationary Oblique Shock Wave Diffractions in Perfect and Imperfect Gases", UTIAS Rep. 232, Institute for Aerospace Studies, Univ. Toronto, Toronto, Ont., Canada, 1978.

Ben-Dor, G., "Nonstationary Oblique Shock Wave Reflection in Nitrogen and Argon: Experimental Results", UTIAS Rep. 237, Institute for Aerospace Studies, Univ. Toronto, Toronto, Ont., Canada, 1978.

Bertrand, B. P., "Measurement of Pressure in Mach Reflection of Strong Shock Waves in a Shock Tube", Mem. Rep. 2196, Ballistic Research Laboratory, U.S.A., 1972.

Bleakney, W., "The Effect of Reynolds Number on the Diffraction of Shock Waves", Princeton Univ., Dept. Phys. Tech. Rep. II-8, Princeton, N.J., U.S.A., 1951.

Clarke, J. F., "Regular Reflection of a Weak Shock Wave From a Rigid Porous Wall. Some Additional Results", CoA Memo 8225, Cranfield Inst. Tech., Cranfield, England, 1982.

Clarke J. F., "Inertia Effects on the Regular Reflection of a Weak Reflection of a Weak Shock Wave from a Rigid Porous Wall", CoA Rep. NFP84/2, Cranfield Inst. Tech. Cranfield, England, 1984.

Deschambault, R. L., "Nonstationary Oblique-Shock-Wave Reflections in Air", UTIAS Rep. 270, Institute for Aerospace Studies, Univ. Toronto, Toronto, Ont., Canada, 1984.

Fletcher, C. H., "The Mach Reflection of Weak Shock Waves", Princeton Univ., Dept. Phys. Tech. Rep. II-4, Princeton, N.J., U.S.A., 1950.

Friend, W. H., "The Interaction of a Plane Shock Wave with an Inclined Perforated Plate", UTIAS Tech. Note 25, Institute for Aerospace Studies, Univ. Toronto, Toronto, Ont., Canada, 1958.

Fry, M., Picone, J. M., Boris, J. P. & Book, D. L., "Transition to Double-Mach Stem for Nuclear Explosion at 104-ft Height of Burst", NRL MR 4630, 1981.

Fuchs, J. & Schmidt, B., "Enstehungsort des Tripelpunktes und Seine Anschließende Spur Beim Ubergang von Regularer Reflektion zur Mach Reflektion", Stromangsmechanik und Stromeungsmaschinen Rep. 40/89, Universitat Karlsruhe, Karlsruhe, Germany, 1989.

Glass, I. I., "Beyond Three Decades of Continuous Research at UTIAS on Shock Tubes and Waves", UTIAS Rev. 45, Institute for Aerospace Studies, Univ. Toronto, Toronto, Ontario, Canada, 1981.

Glass, I. I., "Some Aspects of Shock Wave Research", UTIAS Rev. 48, Institute for Aerospace Studies, Univ. Toronto, Toronto, Ont., Canada, 1986.

Glaz, H. M., Colella, P., Glass, I. I. & Deschambault, R. L., "A Detailed Numerical, Graphical and Experimental Study of Oblique Shock Wave Reflections", LBL-20033, Lawrence Berkely Lab., Berkely, CA, U.S.A., 1985.

Glaz, H. M., Colella, P., Glass, I. I. & Deschambault, R. L., " A Numerical and Experimental Study of Pseudostationary Oblique-Shock-Wave Reflections in Argon and Air", UTIAS Rep. 285, Institute for Aerospace Studies, Univ. Toronto, Toronto, Ont., Canada, 1986.

Guo Chang-Ming, "Numerical Calculation of Mach Reflection of Spherical Shock Wave on Rigid Wall", Rep. 9/85, Dept. Modern Mech., Univ. Sci. & Tech. China, Hefei, China, 1985.

Handke, E. & Obemeier, F., "Some Theoretical Results on Mach Reflection of Moderately Strong Shock Waves", Max Planck Inst. Rep. 14, 1983.

Harrison, F. B. & Bleakney, W., "Remeasurement of Reflection Angles in Regular and Mach Reflection of Shock Waves", Princeton Univ., Dept. Phys. Tech. Rep. II-0, Princeton, N.J., U.S.A., 1947.

Hisley, D. M., "BLAST2D Computations of the Reflection of Planar Shocks from Wedge Surfaces with Comparison to SHARC and SLEATH Results", Tech. Rep. BRL-TR-3147, Ballistic Research Laboratory, Aberdeen Proving Ground, Maryland, U.S.A., 1990.

Hu, T. C. J., "Pseudo-Stationary Oblique-Shock-Wave-Reflections in a Polyatomic Gas-Sulfur Hexafluoride", UTIAS Tech. Note 253, Institute for Aerospace Studies, Univ. Toronto, Toronto, Ont., Canada, 1985.

Hu, T. C. J. & Shirouzu, M., "Tabular and Graphical Solutions of Regular and Mach Reflections in Pseudo-Stationary Frozen and Vibrational Equilibrium Flows", UTIAS Rep. 283, Parts 1 & 2, Institute for Aerospace Studies, Univ. Toronto, Toronto, Ont., Canada, 1985.

Ikui, T., Matsuo, K. Aoki, T. & Kondoh, N., "Mach Reflection of a Shock Wave from an Inclined Wall", Memoirs Faculty Eng., Kyushu Univ., Vol. 41, pp. 361-380, Fukuoka, Japan, 1981.

Jahn, R. G., "The Refraction of Shock Waves at a Gaseous Interface; Regular Reflection of Weak Shocks", Princeton Univ., Dept. Phys. Tech. Rep. II-16, Princeton, N.J., U.S.A., 1954.

Jahn, R. G., "The Refraction of Shock Waves at a Gaseous Interface; Regular Reflection of Strong Shocks", Princeton Univ., Dept. Phys. Tech. Rep. II-18, Princeton N.J., U.S.A., 1955.

Jahn, R. G., "The Refraction of Shock Waves at a Gaseous Interface; Irregular Refraction", Princeton Univ., Dept. Phys. Tech. Rep. II-19, Princeton, N.J., U.S.A., 1955.

Kaca, J., "An Interferometric Investigation of the Diffraction of a Planar Shock Wave over a Semicircular Cylinder", UTIAS Tech. Note 269, Institute for Aerospace Studies, Univ. Toronto, Toronto, Ont., Canada, 1988.

Law, C. K. "Diffraction of Strong Shock Waves by a Sharp Compressive Corner", UTIAS Tech. Note 150, Institute for Aerospace Studies, Univ. Toronto, Toronto, Ont., Canada, 1970.

Lean, G. H., "Report on Further Experiments on the Reflection of Inclined Shock Waves", Nat. Phys. Lab. London, England, 1946.

Lee, J. H. & Glass, I. I. "Domains and Boundaries of Pseudo Stationary Oblique Shock Wave Reflections in Air", UTIAS Rep. 262, Institute for Aerospace Studies, Univ. Toronto, Toronto, Ont., Canada, 1982.

Li, J. -C. & Glass, I. I., "Collision of Mach Reflections with a 90-Degree Ramp in Air and CO_2", UTIAS Rep. 290, Institute for Aerospace Studies, Univ. Toronto, Toronto, Ont., Canada, 1985.

Lottero, R. E. & Wortman, J. D., "Evaluation of the HULL and SHARC Hydrocodes in Simulating the Reflection of a Mach 2.12 Non-Decaying Shock on Wedges of Various Angles", Tech. Rep. BRL-TR-****, Ballistic Research Laboratory, Aberdeen Proving Ground, Maryland, U.S.A., 1991.

Lozzi, A., "Double Mach Reflection of Shock Waves", M.Sc. Thesis, The Univ. Sydney, Sydney, New South Wales, Australia, 1971.

Meiburg, E., Oertel, H. Jr., Walenta, Z. & Fiszdon, W.,"Quasistationary Mach Reflection of Shock Waves, Preliminary Results of Experimental and Monte Carlo Investigations", DFVLR-AVA Rep. IB-221-83A, 1983.

Mirels, H., "Mach Reflection Flow Fields Associated with Strong Shocks", Aerospace Corp., TR-0083(3785)-1, El Segundo, CA, U.S.A.

Mölder, S., "Head on Interaction of Oblique Shock Waves", UTIAS Tech. Note 38, Institute for Aerospace Studies, Univ. Toronto, Toronto, Ont., Canada, 1960.

Neumann, J. von, "Oblique Reflection of Shocks", Explos. Res. Rep. 12, Navy Dep., Bureau of Ordnance, Washington, DC., U.S.A, 1943.

Neumann, J. von, "Refraction, Intersection and Reflection of Shock Waves", NAVORD Rep. 203-45, Navy Dep., Bureau of Ordnance, Washington, DC, U.S.A., 1945.

Olim, M., "A Study of the Reflection of Weak ($1.1 < M_i < 1.5$) Shocks and the Development of a Revised Three Shock Solution", Ph.D Thesis, Dept. Phys., Univ. Victoria, Victoria, B.C., Canada, 1990.

Onodera, H. & Takayama, K., "Shock Wave Propagation over Slitted Wedges", Rep. Inst. Fluid Sci., Tohoku Univ., Sendai, Japan, Vol. 1, pp. 45-66, 1990.

Polachek, H. & Seeger, R. J., "Regular Reflection of Shocks in Water-Like Substances", Exples. Res. Rep. 14, Navy Dept., Bureau of Ordnance, Washington, DC., U.S.A., 1943.

Prasse, H. G., "An Analytic Description of the Expansion of the Reflected Shock Wave at the Mach Reflection on Wedges", Rep. 4/72, Ernst Mach Institute, Freiburg, Germany, 1972.

Reichenbach, H., "Roughness and Heated Layer Effects on Shock Wave Propagation and Reflection - Experimental Results", Rep. 24/85, Ernst Mach Institute, Freiburg, Germany, 1985.

Richtmyer, R. D., "Progress Report on the Mach Reflection Calculation", Rep. NYU 9764, Courant Inst., New York Univ., N.Y., U.S.A., 1961.

Schultz-Grunow, F., "Diffuse Reflexion einer Stoßwelle", Rep. 4/72, Ernst Mach Institute, Freiburg, Germany, 1972.

Seeger, R. J. & Polachek, H., "Regular Reflection of Shocks in Ideal Gases", Explos. Res. Rep. 13, Navy Dep., Bureau of Ordnance, Washington, DC., U.S.A., 1943.

Shankar, V., Kutler, P. & Anderson, D., "Diffraction of a Shock Wave by a Compression Corner; Regular and Single Mach Reflection", NASA TM X-73,178, 1976.

Shirouzu, M. & Glass, I.I. "An Assessment of Recent Results on Pseudo Statioinary Oblique-Shock-Wave Reflections", UTIAS Rep. 264, Institute for Aerospace Studies, Univ. Toronto, Toronto, Ont., Canada, 1982.

Skews, B. W., "Profiles of Diffracting Shock Waves (an Analysis Based on Whitham's Theory)", Univ. Witwatersrand, Dept. Mech. Eng., Rep. 35, Johannesburg, South Africa, 1966.

Skews, B. W., "Shock Wave Diffraction a Review", Univ. Witwatersrand, Dept. Mech. Eng. Rep. 32, Johannesburg, South Africa, 1966.

Skews, B. W., "The Perturbed Region Behind a Diffracting Shock Wave", Univ. Witwatersrand, Dept. Mech. Eng. Rep. 44, Johannesburg, South Africa, 1967.

Skews, B. W., "The Deflection Boundary Condition in the Regular Reflection of Shock Waves", Departmental Report, McMaster Univ., Hamilton, Ontario, Canada, 1972.

Skews, B. W., "The Effect of Angular Slipstream on Mach Reflection", Departmental Report, McMaster Univ., Hamilton, Ontario, Canada, 1972.

Smith, L. G., "Photographic Investigation of the Reflection of Plane Shocks in Air", Off. Sci. Res. Dev. OSRD Rep. 6271, Washington, DC, U.S.A. or Nat. Defence Res. Committee NDRC Rep. A-350, 1945.

Srivastava, R.S., "Diffraction of Blast Waves for the Oblique Case", British Aero. Res. Council Rep. No. 1008, 1968.

Suzuki, T. & Adachi, K., "Oblique Reflection of a Plane Shock on a Rough and Porous Wedge", Ann. Rep. Japan Soc. Heat Fluid Eng., Vol. 2, pp. 23-26, 1987.

Takayama, K. & Ben-Dor, G., "Reflection and Diffraction of Shock Waves over a Circular Concave Wall", Rep. Inst. High Speed Mech., Tohoku Univ., Sendai, Japan, Vol. 51, pp. 44-87, 1986.

Takayama, K., Miyoshi, H. & Abe, A., "Shock Wave Reflection over Gas/Liquid Interface", Rep. Inst. High Speed Mech., Tohoku Univ., Sendai, Japan, Vol. 57, pp. 1-25, 1989.

Takayama, K., Onodera, O. & Gotoh, J., "Shock Wave Reflections over a Rough Surface Wedge and a Curved Rough Surface", Rep. Inst. High Speed Mech., Tohoku Univ., Sendai, Japan, Vol. 48, pp. 1-21, 1982.

Takayama, K. & Sasaki, M., "Effects of Radius of Curvature and Initial Angle on the Shock Transition over Concave and Convex Walls", Rep. Inst. High Speed Mech., Tohoku Univ., Sendai, Japan, Vol. 46, pp. 1-30, 1983.

Takayama, K. & Sekiguchi H., "An Experiment on Shock Diffraction by Cones", Rep. Inst. High Speed Mech., Tohoku Univ., Sendai, Japan, Vol. 36, pp. 53-74, 1977.

Takayama, K. & Sekiguchi, H., "Formation and Diffraction of Spherical Shock Waves in a Shock Tube", Rep. Inst. High Speed Mech., Tohoku Univ., Sendai, Japan, Vol. 43, pp. 89-119, 1981.

Tepe, F. R. Jr. & Tabakoff, W., "The Interaction of Unequal Oblique Shock Waves in a Hypersonic Flow", Aero. Res. Lab., ARL Rep. 63-146, U.S.A.

Urbanowicz, J. T., "Pseudo Stationary Oblique Shock Wave Reflections in Low Gamma Gases - Isobutane and Sulphur Hexafluoride", UTIAS Tech. Note 267, Institute for Aerospace Studies, Univ. Toronto, Toronto, Ont., Canada, 1988.

Weynants, R. R., "An Experimental Investigation of Shock-Wave Diffraction over Compression and Expansion Corners", UTIAS Tech. Note 126, Institute for Aerospace Studies, Univ. Toronto, Toronto, Ont., Canada, 1968.

Wheeler, J., "An Interferometric Investigation of the Regular to Mach Reflection Transition Boundary in Pseudostationary Flow in Air", UTIAS Tech. Note 256, Institute for Aerospace Studies, Univ. Toronto, Toronto, Ont., Canada, 1986.

White, D. R., "An Experimental Survey of the Mach Reflection of Shock Waves", Princeton Univ., Dept. Phys. Tech. Rep. II-10, Princeton, N.J., U.S.A., 1951.

White, D. R., "An Experimental Study of Shock Waves", Proc. 2nd Midwest Conf. on Fluid Mech., Vol. 21, No. 3, pp. 252-263, 1952.

Yagla, J. J., "Machine Calculation of Unsteady Mach Reflections and Prandtl-Meyer Supersonic Flows", NWL Tech. Rep. TN-2897, 1973.

Zhang, D. L. & Glass, I. I., "An Interferometric Investigation of the Diffraction of Planar Shock Waves over a Half Diamond Cylinder in Air", UTIAS Rep. 322, Institute for Aerospace Studies, Univ. Toronto, Toronto, Ont., Canada, 1988.

Index

About the Author

Professor Gabi Ben-Dor was born on June 10, 1950 in Beer Sheva, Israel to Shoshana and Paul Sandor, who survived the Holocaust and emigrated to Israel from Hungary in 1949.

Professor Ben-Dor received his B.Sc. degree (cum laude) in Mechanical Engineering from the Ben-Gurion University of the Negev in 1975. He obtained his M.Sc. degree (cum laude) in Mechanical Engineering from the Ben-Gurion University of the Negev in 1976. In 1978 he obtained his Ph.D. from the Institute of Aerospace Studies of the University of Toronto.

Professor Ben-Dor joined the Department of Mechanical Engineering of the Ben-Gurion University of the Negev in 1979 as a Lecturer. In 1981 he was promoted to the rank of Senior Lecturer, in 1985 to the rank of Associate Professor, and since 1990 he has been a Full Professor.

Since 1978 Professor Ben-Dor has published about 70 papers in scientific journals, and presented about 100 papers in scientific conferences and symposia. Of his journal papers, about 30 are directly related to the reflection phenomenon of shock waves.

Professor Ben-Dor's current areas of interest are: shock wave reflection phenomena, shock wave propagation in dusty gases, shock wave propagation in porous media, and coal combustion.